Comprehensive Experiments for Materials Science and Engineering

Comprehensive Experiments for Materials Science and Engineering

Fei Ye, Chengzhu Liao,
Hua Cheng, Jianbo Zhang,
Haiou Wang, Yanyan Li
Huili Li, Jing Ming

Southern University of Science and Technology, China

Published by

Higher Education Press Limited Company
4 Dewai Dajie, Xicheng District
Beijing 100120, P. R. China
and
World Scientific Publishing Co. Pte. Ltd.
5 Toh Tuck Link, Singapore 596224
USA office: 27 Warren Street, Suite 401-402, Hackensack, NJ 07601
UK office: 57 Shelton Street, Covent Garden, London WC2H 9HE

British Library Cataloguing-in-Publication Data
A catalogue record for this book is available from the British Library.

COMPREHENSIVE EXPERIMENTS FOR MATERIALS SCIENCE AND ENGINEERING

Copyright © 2023 by Higher Education Press Limited Company and
World Scientific Publishing Co. Pte. Ltd.

All rights reserved. This book, or parts thereof, may not be reproduced in any form or by any means, electronic or mechanical, including photocopying, recording or any information storage and retrieval system now known or to be invented, without written permission from the publishers.

For photocopying of material in this volume, please pay a copying fee through the Copyright Clearance Center, Inc., 222 Rosewood Drive, Danvers, MA 01923, USA. In this case permission to photocopy is not required from the publishers.

ISBN 978-981-127-404-6 (hardcover)
ISBN 978-981-127-405-3 (ebook for institutions)
ISBN 978-981-127-406-0 (ebook for individuals)

For any available supplementary material, please visit
https://www.worldscientific.com/worldscibooks/10.1142/13345#t=suppl

Desk Editor: Rhaimie B Wahap

Preface

Materials science and engineering (MSE) combines physics and chemistry principles to solve practical problems associated with major engineering disciplines, such as information technology, energy, manufacturing, nanotechnology, and biotechnology. The experimental teaching of this specialty is particularly prominent because it is an important way to train students to apply multidisciplinary knowledge comprehensively. In order to improve students' ability to solve practical problems in scientific research and engineering development, it is necessary to eliminate obsolete experiments and increase the number of comprehensive experiments that involve multiple principles, skills, and methods.

Besides the essential basic experiments in the first several chapters, most experiments designed in this book are comprehensive. This is why this book is titled *Comprehensive Experiments*. In this book, a comprehensive experiment can be one experiment or divided into several experiments to facilitate teaching implementation. The experiments involve the forefront of scientific research and the materials industry with appropriate modification.It is worth noting that this book does not specifically design experiments on material characterization, although it is an essential basis of MSE. Instead, the principles, methods, and apparatuses for material characterization are integrated into relevant experiments.

This book covers the main contents of experimental courses of MSE. It intends to serve as a textbook for undergraduate students and aims to help teachers find a wide enough variety of experiments to construct an experimental course.

Organization of the book

In recent years, many universities and colleges have put forward modularized teaching mode. The experimental knowledge and skills can be modularized to form a systematic, independent curriculum system so that students can systematically master the complete experimental knowledge and skills of MSE. Therefore, this book establishes a modularized structure. The experiments in this book are divided into five parts.

Part I is Preliminary Exploration of Materials Science and Engineering, including preliminary experiments of metallographic analysis, material processing, electronic materials, energy materials, and biomaterials.

Part II is Fundamentals of Chemistry and Crystallography, including basic experiments of physical chemistry, organic chemistry, and crystallography. Chemical synthesis is an important way to obtain materials, and chemical reaction is closely related to a variety of material treatment and processing methods. In contrast, the students majoring in MSE generally lack the knowledge and experiment skills in chemistry, especially in organic chemistry. Therefore, the chemical experiments are strengthened in this part to support the study of subsequent experiments.

Part III is Material Properties, including experiments of mechanical and physical properties of materials.

Part IV is Material Preparation and Treatment, including experiments of forming and heat treatment of metal materials, synthesis of organic and polymeric materials,and synthesis of nanomaterials.

Part V is Material Applications, including experiments of energy materials, electronic materials and devices, bio materials, and materials and environment.

Besides the experiments in the text, the appendices describe the most relevant aspects of experimental safety, error, and data presentation in a general way. The contents and requirements of the experimental report are suggested.

At the end of each chapter, a list of books, journal articles, and websites is provided for extended reading on topics covered in the chapter.

Organization of the experiments

The main contents of each experiment consist of the following sections:

Type of the experiment (cognitive, comprehensive, or designing) and *Recommended credit hours* are first described. These may facilitate teaching implementation. For the experiment that is further divided into several experiments, the type and credit hours for each separated experiment are given.

Brief introduction compendiously introduces the main contents of the experiment, the materials used, and the principles involved.

Objectives describe the main contents to be completed and the knowledge and experimental skills to be mastered in the experiment.

Principles introduce the materials, methods, apparatuses involved in the experiment. It is noted that these principles are in the form of a brief introduction or a summary since we pay more attention to applying the relevant knowledge and the cultivation of experiment skills.

Experimental includes two parts. They are the "preparation", which lists materials, chemicals, apparatuses, tools, and others for the experiment, and the "procedure", which describes the main steps of the experiment. The manufacturers and models of the main apparatuses are given to facilitate the introduction of the experimental steps.

Requirements for experimental report put forward the specific requirements in the report. These include description and explanation of experimental phenomena, analysis of photographs, and data processing.

Questions and further thoughts guide students to expand their thinking or knowledge based on the experiment through several questions.

Cautions list key operating points or critical safety matters that should be paid attention to during the experiment.

Division of the authors

The authors all come from the teaching laboratory of the Department of Materials Science and Engineering at Southern University of Science and Technology (SUSTech). Most experiments were designed or reformed from the classic experiments by the authors during teaching practice.

Professor Fei Ye is the chief editor of the book and is responsible for the compilation and review of the book.He studied at Tsinghua University from 1994 to 2004 and successively received a bachelor's degree, a master's degree, and a Ph.D. degree in materials science and engineering. Then, he worked at National Institute for Materials Science, Japan, and became a regular researcher in 2008. Professor Fei Ye worked at Dalian University of Technology as a full professor from 2010 to 2018. After that, he joined Southern University of Science and Technology in July 2018. Professor Fei Ye has taught the course "Fundamentals of Materials Science and Engineering" for years and has rich teaching experience.

The basic information and division of the other authors are as follows:

Chengzhu Liao received a Ph.D. degree in Materials Science from City University of Hong Kong in 2011. She wrote Experiments 5.1, 14–19, 21.1, 21.3, 22, 23, 27, 34, 37.1, 38, 39, 57. She also participated in the writing of Chapter 12, Biomaterials.

Hua Cheng received a Ph.D. degree in materials science from City University of Hong Kong in 2011. She wrote Experiments 13, 29, 30, 41, 47, 50, 52, 59.

Jianbo Zhang received a master's degree in material physics and chemistry from Shenzhen University in 2011. He wrote Experiments 2, 6–12, 37.2, 37.3, 58.

Haiou Wang received a master'sdegree in materials science and engineering from Tsinghua University in 2009. She wrote Experiments 20, 24, 25, 31, 48, 49.

Yanyan Li received a master'sdegree in physics and chemistry from Xiamen University in 2013. She wrote Experiments 4, 21.2, 26, 40, 42–46.

Huili Li received a master'sdegree in biological materials from Shenzhen University in 2013. She wrote Experiments 5.2, 53–56.

Jing Ming received a master's degree in materials processing engineering from Harbin Institute of Technology in 2011. He wrote Experiments 1, 3, 28, 32, 33, 35, 36, 51.

Acknowledgments

We would like to express our appreciation to the College of Engineering and the Department of Materials Science and Engineering at SUSTech. I am especially grateful to Professor Bi Zhang for his help in preparing the book.

We are also indebted to the sponsoring editor, Mr. Zhanwei Liu, from Higher Education Press, for his assistance and guidance in developing and producing this work.

Fei Ye
In Shenzhen, April 2021

Contents

Preface v

Part I Preliminary Exploration of Materials Science and Engineering 1

Chapter 1 Practice for Materials Science and Engineering 3

Experiment 1 Practice for Optical Metallography — Preparation and Observation of Metallographic Sample 3
Experiment 2 Practice for Material Processing — 3D Printing 8
Experiment 3 Practice for Energy Materials and Devices — Preparation of Proton Exchange Membrane Fuel Cell 12
Experiment 4 Practice for Electronic Materials and Devices — Preparation and Characterization of Organic Photodetector 15
Experiment 5 Practice for Biomaterials — Observation of Biological Samples 19
   Experiment 5.1 Preparation and Observation of Biological Samples by Scanning Electron Microscope 19
   Experiment 5.2 Observation of Biological Samples by Fluorescence Microscope 24

Part II Fundamentals of Chemistry and Crystallography 29

Chapter 2 Fundamentals of Physical Chemistry 31

Experiment 6 Determination of Combustion Heat 31
Experiment 7 Determination of Saturated Vapor Pressure of Pure Liquid 35
Experiment 8 Construction of Binary Gas-Liquid Phase Diagram 38

Experiment 9   Determination of Rate Constant for Hydrolysis Conversion of
                Sucrose   44
Experiment 10 Determination of Electromotive Force of Galvanic Cells   48
Experiment 11 Determination of Surface Tension of
                Ethanol-Water Solutions   53
Experiment 12 Determination of Relative Molecular Weight of
                Polymer by Viscosity Method   59
Experiment 13 Determination of Crosslink Density of Rubbers by Swelling
                Method   64

Chapter 3 Fundamentals of Organic Chemistry   71

Experiment 14 Isolation of Alcohol from Red Wine by Simple Distillation   71
Experiment 15 Purification of Naphthalene by Recrystallization   77
Experiment 16 Separation of Liquid Mixture by Liquid-Liquid Extraction   82
Experiment 17 Extraction of Oil from Sunflower Seeds by
                Solid-Liquid Extraction   87
Experiment 18 Separation of Pigments from Spinach Leaves
                by Thin Layer Chromatography and Column
                Chromatography   92

Chapter 4 Fundamentals of Crystallography   99

Experiment 19 Ball Models of Atomic Arrangements in Pure
                Metal Crystals   99
Experiment 20 Construction and Analysis of Crystal Structure Models   105

Part III Material Properties   111

Chapter 5 Mechanical Properties of Materials   113

Experiment 21 Tensile Test of Materials   113
  Experiment 21.1 Tensile Properties of Metals and Work Hardening   113
  Experiment 21.2 Precise Measurement of Tensile Properties   118
  Experiment 21.3 Influence of Molecular Structure and Strain
                    Rate on the Tensile Properties of Polymers   121
Experiment 22 Hardness Test of Materials   124
  Experiment 22.1 Heat Treatment and Hardness Test of Metals   124
  Experiment 22.2 Shore Hardness Test of Polymers   129

Experiment 23	Charpy Impact Test of Polymers	132
Experiment 24	Compression Test of Materials	135
Experiment 25	Three-Point Bending Test of Materials	139
Experiment 26	Torsion Test of Metals	143

Chapter 6 Physical Properties of Materials — 149

Experiment 27	Determination of Ionization Rate of Plasma in Direct Current Sputtering	149
Experiment 28	Coprecipitation Synthesis and Magnetic Properties of Magnetite Nanoparticles	154
Experiment 28.1	Synthesis of Magnetite Nanoparticles by Coprecipitation Method	155
Experiment 28.2	Characterization of Structure and Magnetic Properties of Magnetite Nanoparticles	157
Experiment 29	Nanoemulsion Synthesis and Optical Properties of Iron-Gold Nanocrystals	161
Experiment 30	Preparation of ZnO Thin Film by Sol-Gel Method and Determination of Energy Bandgap	166
Experiment 31	Preparation and Characterization of $BaTiO_3$ Based Piezoelectric Ceramics	170

Part IV Material Preparation and Treatment — 179

Chapter 7 Forming and Heat Treatment — 181

Experiment 32	Microstructure Observation of Crystallization and Aluminum Alloy Ingots	181
Experiment 33	Preparation of Stainless Steel Pellets by Powder Metallurgy Method	184
Experiment 34	Construction of Sn-Bi and Sn-Zn Binary Phase Diagrams	187
Experiment 35	Microstructures and Diffusion of Iron-Carbon Alloys	193
Experiment 35.1	Observation of Equilibrium and Non-Equilibrium Microstructures of Iron-Carbon Alloys	194
Experiment 35.2	Surface Carburizing Heat Treatment of Low Carbon Steel	199
Experiment 36	Heat Treatment of Aluminum Alloy for Precipitation Hardening	203

Experiment 37 Forming Process of Polymers and Application in
             3D Printing   206
  Experiment 37.1 Determination of Melt Flow Index of Thermoplastics   206
  Experiment 37.2 Extrusion and Injection Molding of Polymers   209
  Experiment 37.3 3D Modeling, Printing, and Tensile Properties   213

Chapter 8 Organic and Polymeric Materials   219

Experiment 38 Preparation of Copolymer with LCST Phase
             Transition by Atom Transfer Radical
             Polymerization   219
Experiment 39 Synthesis of Aspirin by Esterification Reaction   222
Experiment 40 Synthesis of Polystyrene Microspheres by
             Dispersion Polymerization   228
Experiment 41 Synthesis of Organic Semiconductor and
             Determination of Frontier Orbital Energies   231
  Experiment 41.1 Synthesis of Organic Semiconductor   232
  Experiment 41.2 Determination of Frontier Orbital Energies by
                 Cyclic Voltammetry   238

Chapter 9 Nanomaterials   243

Experiment 42 Synthesis of Metal Nanoparticles   243
  Experiment 42.1 Citrate Synthesis of Gold Nanoparticles   243
  Experiment 42.2 Synthesis of Silver Nanoparticles by Chemical
                 Reduction Method   246
  Experiment 42.3 Synthesis and Characterization of Au@Ag
                 Core-Shell Composite Nanoparticles   249
Experiment 43 Synthesis of Cadmium Sulfide Nanoparticles by
             Microemulsion Method   252
Experiment 44 Preparation of Nickel Nanowires by Template
             Electrodeposition   255
Experiment 45 Synthesis and Characterization of Rod-like Silica
             Nanoparticles   258
Experiment 46 Preparation of Surface Conductive Glass by Spray
             Pyrolysis Method   260
Experiment 47 Preparation and Characterization of Porous Thin
             Films   263

Part V Material Applications 267

Chapter 10 Energy Materials 269

Experiment 48 Fabrication and Characterization of Lithium-Ion
　　　　　　　　Batteries 269
　Experiment 48.1 Synthesis and Characterization of Cathode
　　　　　　　　　　Material LiCoO$_2$ 269
　Experiment 48.2 Fabrication and Performance Test of Lithium-Ion
　　　　　　　　　　Batteries 273
Experiment 49 Fabrication and Characterization of Polymer Solar Cell 281

Chapter 11 Electronic Materials and Devices 287

Experiment 50 Magnetron Sputtering Deposition of ITO Thin Film
　　　　　　　　and Hall Effect Test 287
Experiment 51 Fabrication and Characterization of In-Ga-Zn-O Thin
　　　　　　　　Film Transistor 293
Experiment 52 Micro-nano Processing by Lithography 299
　Experiment 52.1 Preparation of a Logo Pattern by Photolithography 299
　Experiment 52.2 Fabrication of DVD Microstructure and a Simple
　　　　　　　　　　Microfluidic Device by Soft Lithography 301
　Experiment 52.3 Pattern Transfer by Nanoimprint Lithography 305
Experiment 53 Fabrication of Enzyme-Based Glucose Biosensor 308

Chapter 12 Biomaterials 313

Experiment 54 Preparation and Antibacterial Property of Silver-Loaded
　　　　　　　　Activated Carbon Composites 313
　Experiment 54.1 Preparation and Structure Characterization of
　　　　　　　　　　Antibacterial Silver-loaded Activated Carbon
　　　　　　　　　　Composites 313
　Experiment 54.2 Antibacterial Test of Silver-Loaded Activated
　　　　　　　　　　Carbon Composites 316
Experiment 55 Preparation and In Vitro Cytotoxicity of Pluronic
　　　　　　　　F127-Encapsulated Curcumin Micelles 323
　Experiment 55.1 Preparation and Characterization of Pluronic
　　　　　　　　　　F127-Encapsulated Curcumin Micelles 323

Experiment 55.2 In Vitro Cytotoxicity of Pluronic F127-Encapsulated
Curcumin Micelles 329
Experiment 56 Preparation and Bioactivity of Polylactic Acid/Nano-
Hydroxyapatite Composite 335
Experiment 56.1 Preparation and Characterization of Polylactic
Acid/Nano-Hydroxyapatite Composite 335
Experiment 56.2 Bioactivity Evaluation of Polylactic Acid/Nano-
Hydroxyapatite Composite 338

Chapter 13 Materials and Environment 343

Experiment 57 Passivation and Corrosion Behavior of Ferrous Alloys 343
Experiment 58 Salt Spray Test of Metals 349
Experiment 59 Synthesis and Photocatalytic Activity of TiO_2 Powder 352

Appendices 359

A1 Experimental Safety 359
A2 Experimental Errors 361
A3 Data Presentation 363
A4 Experimental Report 365

Part I
Preliminary Exploration of Materials Science and Engineering

Chapter 1 Practice for Materials Science and Engineering

Experiment 1 Practice for Optical Metallography—Preparation and Observation of Metallographic Sample

Type of the experiment: Cognitive
Recommended credit hours: 4
Brief introduction: Optical metallography is a widely used technique to examine the microstructures of metal alloys using visible light. In this experiment, the microstructure of a commonly used steel grade, No. 45 carbon steel, is observed by the metallographic microscope. The basic principle and operation of the microscope and the general preparation procedure of the metallographic sample are introduced.

1 Objectives

(1) To understand the basic knowledge of metallography.
(2) To understand the principle and operation of the metallographic microscope.
(3) To learn the procedure of metallographic sample preparation.
(4) To observe the microstructure of No. 45 carbon steel.

2 Principles

2.1 Metallography

Many important macroscopic properties of materials are highly sensitive to microstructures. The understanding of the relationship between the microstructure and macroscopic properties plays a crucialrole in the development and manufacture of materials. For example, the mechanical properties of materials strongly depend upon grain size, grain shape, and the amount and distribution of second phases.

Metallography is the study of the microstructures of metallic alloys. These microstructure features include the chemical and atomic structure, spatial distribution of grains, inclusions, or phases in metallic alloys. The metallographic analysis can be used to help identify a metal or alloy, examine multiple phases within a material, characterize defects, or determine whether an alloy was processed correctly. The method most often used in these evaluations is optical metallography.

2.2 Metallographic microscope

2.2.1 Imageforming principle

The metallographic microscope is most widely used in metallography. Comparing with the transmitting microscope, it observes light reflected from highly polished samplessince metals are not transparent to visible light. Most materials require etching before observation, which is a process of chemical corrosion to induce some irregularities on the polished surfaces. Light is collected, which forms contrasting features of the microstructure. Flat areas that reflect lightinto the objective lens appear bright. Second phases and non-metal inclusions with reflectivity different from the matrix material appear as darker areas. Thefeatures which do not reflect light, such as interfaces and scratches, also appear dark.

Figure 1.1 shows the structure and light pathway of a typical optical microscope. An optical microscope creates a magnified image of an object sample with an objective lens. The image is further magnified with an eyepiece to allow the user to observe it with naked eyes. The image can also be observed on a monitor

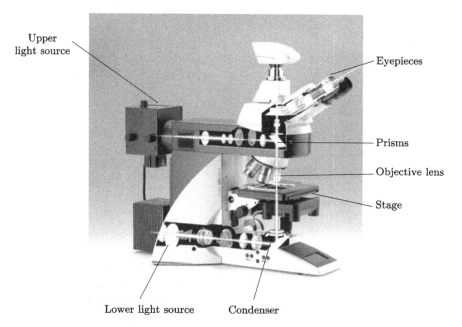

Fig. 1.1 Light pathway of a metallographic microscope.

when it is directly captured with a CCD camera.

In the microscope shown in Fig. 1.1, the sample is under the objective lens, and we look down to see the image. Another type of microscope, the inverted microscope, is commonly used in metallurgy, by which we look up to see the image. An inverted model can allow the large sample to be observed since its working distance is not limited as that in Fig. 1.1.

2.2.2 Performance of a microscope

Since the magnified image of an object is created by both the objective lens and the eyepiece, the final magnification, M, can then be calculated by the product of the magnifying power of the objective lens, M_{OL}, and the magnifying power of the eyepiece, M_{EP}. Thus, M is expressed as

$$M = M_{OL} \times M_{EP}. \tag{1.1}$$

Generally, the magnification of the objective lens is up to 100 times (denoted as 100×), and that of the eyepiece is 10×. Hence, the magnification of a microscope is up to 1000×.

The resolution of a microscope is the shortest distance between two points that can be distinguished. It is mainly determined by the objective lens and expressed as

$$d = 0.61 \frac{\lambda}{\text{NA}}, \tag{1.2}$$

where λ is the wavelength of the light and NA is the numerical aperture value of the objective lens. NA is calculated by

$$\text{NA} = n \sin \alpha, \tag{1.3}$$

where n is the refractive index of the medium between lens and object. Angle α is subtended by lens diameter at the object. The magnification and NA of a lens can be found on its shell.

From Equation (1.2), we can state a general rule that a higher NA means a better light-gathering property of the lens and a better resolution. A higher NA also means a shorter working distance of the objective lens. Therefore, the objective lens gets closer to the sample when the magnification of the microscope increases.

2.3 Preparation of metallographic sample

A properly prepared sample is the key to obtain an accurate interpretation of a microstructure. It must be truly representative of the material being examined. The surface of a proper metallographic sample must be flat and free from scratches, stains, and other imperfections which tend to mar the surface. It contains all non-metallic inclusions and intermetallic compounds and shows no chipping or galling.

Proper preparation of metallographic samples requires a rigid step-by-step process to be followed. In sequence, the process generally includes the following steps:

(1) Sectioning.

The region of the sample that is of interest must be sectioned from the component since the size of the sample is limited under microscopes. It is representative of the material to be evaluated. Abrasive cutting and electric discharge machining are widely used methods.

(2) Mounting.

Because large samples are generally more challenging to be prepared than small ones, and sometimes the material to be examined is precious, the sample size should be minimized. In this case, the sample can be embedded in a hot thermosetting powder or cold castable mounting material for ease of manipulation. Besides, it is also used for other factors such as fragility, edge preservation. This step is omitted for the samples with a suitable size or prepared with specific methods.

(3) Grinding.

Grinding is generally considered as the most important step in the process of sample preparation. The sample surface is ground against a series of successively finer waterproof abrasive papers. Proper grinding involves the rotation of the sample after one abrasive paper is replaced by another finer one. The grinding angle must be held constant during the grinding on one abrasive paper. The sample must be washed thoroughly between grinding operations on different abrasive papers.

(4) Polishing.

The sample must be washed before proceeding to the polishing stage. Polishing generally involves rough polishing and fine polishing. In rough polishing, a piece of polishing cloth covered on a wheel or disc is impregnated with diamond paste or a slurry of silica or alumina powder. The particle size of the paste or slurry is in the range of 30 to 3 μm. The sample is held against the cloth. Fine polishing is conducted similarly but with finer abrasives down to 0.05 μm.

(5) Etching.

Because many microstructural details are not observable on the as-polished sample, the sample surface must be etched to reveal these structural features, such as grain boundaries, phase boundaries, twins, and slip lines. Etchants attack at different rates on different crystal orientations, imperfections, or compositions. The resulting surface irregularities differentially reflect the incident light, resulting in the contrast of the final metallographs.

3 Experimental

3.1 Preparation

(1) Material: No. 45 carbon steel (Fe-0.45wt%C) with a size of $\Phi 15$ mm × 20 mm.

(2) Chemical: 4% nitric acid ethanol solution.

(3) Apparatuses: grinding machine, polishing machine, metallographic microscope with a digital camera.

(4) Tools: tweezers, air blower.

(5) Others: metallographic abrasive papers from 240 to 2000 grits, polishing cloth, diamond metallographic polishing agents with sizes of 7 μm and 1 μm, cotton balls.

3.2 Procedure

3.2.1 Preparation of metallographic sample

(1) Grind the sample using the grinding machine with a rotating disc covered with a series of abrasive papers from 240 to 2000 grits. The direction of the sample needs to rotate about 90° after replacing the abrasive papers.

(2) Rough polish and fine polish the sample using the polishing machine with a rotating disc covered by the polishing cloth. Add a small amount of the polishing agents during polishing.

(3) Clean the sample with water and ethanol, and then dry the sample.

(4) Etch the sample using the etchant (4% nitric acid ethanol solution) by immersion method or wiping method. That is, the sample can be submerged in the etchant or wiped using a cotton ball. The sample should be cleaned again after etching.

3.2.2 Microstructure observation

(1) Switch on the microscope.

(2) Select the desired objective lens. Usually, start from lower magnification to find the area for observation. Focus the sample using the coarse focus knob first, then the fine focus knob.

(3) Go up to a higher magnification. Then use the CCD camera and computer to capture the metallographic image.

4 Requirements for experimental report

(1) Describe the preparation process of the metallographic sample in detail.

(2) Present the microstructure image. Give the magnification in the figure caption.

(3) Schematically draw the observed microstructure.

5 Questions and further thoughts

(1) How can we avoid scratches, stains, and other imperfections that tend to mar the surface during the sample preparation?

(2) Why do grain boundaries and phase boundaries exhibit dark lines under the metallographic microscope?

6 Cautions

(1) The preparation of metallographic samples involves using etchants that can be very corrosive and poisonous. Therefore, the etchants must be used in a well-ventilated area and must not directly contact skin or eyes.

(2) It is necessary to keep the optical parts of the microscope clean. Do not touch lenses with bare hands.

Experiment 2   Practice for Material Processing — 3D Printing

Type of the experiment: Cognitive
Recommended credit hours: 4
Brief introduction: Processing materials into parts is an important way to use materials. In this experiment, an advanced processing method, 3D printing, is introduced. The experiment aims to provide a preliminary understanding of material processing technology through 3D printing a simple object. The object will be modeled by Solidworks, a well-known 3D design software. Then, the model is sliced into 2D layers by the Flashprint software and printed by a fused deposition modeling (FDM) 3D printer with polylactic acid (PLA) filament.

1 Objectives

(1) To understand the principle of 3D printing.
(2) To understand several widely used 3D printing techniques.
(3) To learn the basics of Solidworks and Flashprint software.
(4) To design and print an object using an FDM 3D printer.

2 Principles

2.1 3D printing

3D printing, also called additive manufacturing, is a process that produces objects by successively adding material in layers. These layers correspond to a series of cross-sections of 3D objects. From sand to living tissue, nearly anything can be used, and plastics and metal alloys are the most commonly used materials for 3D printing.

Compared with traditional subtractive or formative manufacturing technologies,3D printing is a fundamentally different way of producing parts. In 3D printing, parts are manufactured directly onto the built platform layer by layer, which leads to a unique set of benefits and limitations. It is the best option for manufacturing a single or only a few parts at a quick turnaround time and a low cost. It is also used for parts with special geometries that cannot be produced with traditional manufacturing technologies.

2.2 3D printing techniques

Various 3D printing techniques build objects based on the same main principle, that is, a digital model is turned into a physical three-dimensional object by adding material layer by layer. This is where the alternative term "Additive Manufacturing" comes from. The process of 3D printing always begins with a digital 3D model, which is the blueprint of the physical object. This model is sliced by software into 2D layers and then turned into a set of instructions in the machine language of3D printers.After that, the model will be printed by 3D printers that work based on various techniques.

This experiment will use the well-known 3D design software, Solidworks, to design a simple model. The model will then be sliced by the Flashprint software and printed with PLA filament. Since this experiment aims to give students a preliminary understanding of 3D printing, only the basics of the software and 3D printers will be learned during the experiment. The details will be introduced in Experiment 37.3. Instead,several well-developed 3D printing techniques are introduced here.

2.2.1 FDM technique

FDM is the most cost-effective way of producing thermoplastic parts and prototypes (Fig. 1.2). Various thermoplastic materials with different trade-offs between mechanical and thermal properties are available, such as PLA, acrylonitrile butadiene styrene (ABS), polycarbonate (PC), polyamide (PA), and polystyrene (PS). The printer may use multiple materials for one model to achieve different goals. For example, one material is used to build up the model, and another is used as a soluble support structure, or multiple colors of the same thermoplastic material are used on the same model.

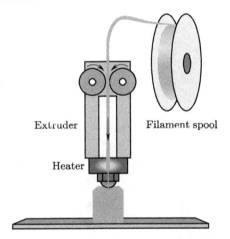

Fig. 1.2 Schematic illustration of the FDM process. (See color illustration at the back of the book)

2.2.2 Stereolithography

Stereolithography (SLA) positions a perforated platform just below the surface of a liquid photocurable polymer (Fig. 1.3). An ultraviolet (UV) laser beam then traces the first slice of an object on the surface of this liquid, causing a thin layer of photopolymer to harden. The perforated platform is then lowered slightly, and another slice is cured by the UV laser. A complete object can be printed by repeating the process. After printing, the object is cleaned from the photopolymer and exposed to a UV source to improve its strength. The support structures are removed, and additional post-processing steps are carried out if a high-quality surface is required.

2.2.3 Selective laser sintering

The selective laser sintering (SLS) process builds objects using a laser to selectively

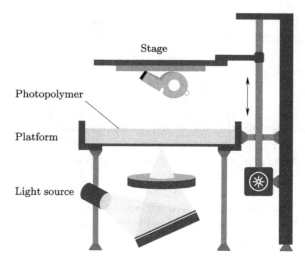

Fig. 1.3 Schematic illustration of the SLA process.

fuse successive layers in a bin of powdered metal, ceramic, or polymer materials. As shown in Fig. 1.4, when the entire cross-section is scanned, the building platform moves down one layer, and the recoating roller or blade then deposits a thin layer of powder with a thickness of about 0.1 mm onto the build platform. After repeating the process, the result is a bin filled with an object surrounded by unsintered powder. The object is then removed from the unsintered powder and cleaned. Post-processing steps, such as polishing or dying, can be employed to improve the visual appearance of the parts.

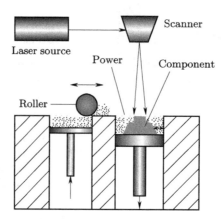

Fig. 1.4 Schematic illustration of the SLS process.

SLS parts are ideal for prototypes because they have excellent, almost-isotropic mechanical properties. No support structures are required since the unsintered powder acts as support. Thus, designs with very complex geometries can be easily manufactured.

2.2.4 Multi-jet modeling

Multi-jet modeling (MJM) works in a similar way to an inkjet printer. However, instead of printing a single layer of ink on a piece of paper, multiple layers of printing material are deposited successively upon each other to create a solid object (Fig. 1.5). Multiple print heads jet a large number of tiny droplets of photopolymer onto the build platform, which is then cured by the UV light source. After a layer is complete, the build platform moves down slightly, and the process repeats for another layer.

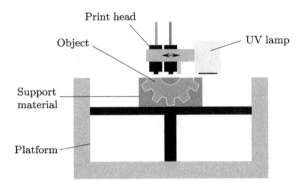

Fig. 1.5 Schematic illustration of the MJM process.

MJM is a precise 3D printing technology. Jetted objects have a very smooth surface and very high dimensional accuracy. Thus, this technique is ideal for realistic prototypes and parts that need an excellent visual appearance.

3 Experimental

3.1 Preparation

(1) Material: PLA filament.

(2) Apparatuses: 3D printer (FlashForge Corporation, Dreamer Dual Extruder), computer installed with the Solidworks and Flashprint software.

3.2 Procedure

(1) Design a simple object, such as a cylinder, a polyhedron, a cup, or a pen container, using the Solidworks software. Save the model as a ".stl" file.

(2) Slice the model using the Flashprint software and save it as a ".gx" file on an SD card.

(3) Switch on the 3D printer. Plug the SD card in the 3D printer.

(4) Level the print platform, and adjust the gap between the nozzle and the print platform.

(5) Select "Print from SD card" on the operation panel. Select the model on the SD card and start the print. It may take several hours to complete the print.

(6) Take the printed object out from the printer.

12　Chapter 1　Practice for Materials Science and Engineering

4 Requirements for experimental report

(1) Describe the general procedure of 3D printing in detail.
(2) Describe the main steps of using the Solidworks and Flashprint software.

5 Questions and further thoughts

(1) Compare the advantages and disadvantages of 3D printing with traditional manufacturing technologies.
(2) What is post-processing in 3D printing? What is it used for?

Experiment 3　Practice for Energy Materials and Devices —Preparation of Proton Exchange Membrane Fuel Cell

Type of the experiment: Comprehensive
Recommended credit hours: 4
Brief introduction: The proton exchange membrane fuel cell (PEMFC) is a type of fuel cell which operates at relatively low temperatures and can tailor the electrical output to meet dynamic power requirements. It is currently the leading technology for vehicles and, to a lesser extent, for stationary and other applications. This experiment will prepare a simple PEMFC and test its power generation behavior. A mini electric fan will be driven by the as-prepared PEMFC. Through the experiment, the students will understand the basic structure and working principle of fuel cells.

1 Objectives

(1) To understand the working principle and components of fuel cells.
(2) To understand the advantages and disadvantages of PEMFC.
(3) To prepare a simple PEMFC.

2 Principles

2.1　Fuel cells

Fuel cells are a class of device that converts the chemical energy from a fuel into electricity through a chemical reaction of the fuel with oxygen or oxidizing agent. Fuel cells are different from general batteries. They require a continuous source of fuel and oxygen to sustain the chemical reaction and generate an electromotive force.

　　Because fuel cells can produce electricity continuously as long as the fuel and oxygen are supplied, they have been used for decades in space probes, satellites, and spacecraft. Besides, thousands of stationary fuel cell systems have been installed worldwide for both primary and backup power. They are also used in waste-treatment plants to generate power from the methane gas produced by decomposing garbage. Fuel cell vehicles are also coming to market.

2.2 Working principle of fuel cells

The basic principle of fuel cells is to decompose a chemical reaction into several steps. The free exchange of electrons in the oxidation reaction is prohibited. Instead, the electrical transfer is divided into an electronic and an ionic transfer. Consequently, the electronic transfer can be utilized as an external electrical current.

This principle can be readily understood based on the structure of fuel cells. A fuel cell has essentially the same kinds of components as a battery. It has a matching pair of electrodes, including an anode, which supplies electrons, and a cathode, which absorbs electrons. Both electrodes must be separated by an electrolyte, which may be a liquid or a solid, and must conduct ions between the electrodes to complete the chemical reaction of the system. Figure 1.6 illustrates the structure of a hydrogen fuel cell as an example. When the fuel, i.e., hydrogen, in this case, is supplied to the anode, it is oxidized, producing hydrogen ions and electrons. An oxidizer, i.e., oxygen or air, is supplied to the cathode, where the hydrogen ions from the anode absorb electrons and react with the oxygen to produce water. The reactions at the electrodes are as follows:

$$\text{Anode}: 2H_2 \rightarrow 4H^+ + 4e^-, \tag{1.4}$$

$$\text{Cathode}: 4e^- + 4H^+ + O_2 \rightarrow 2H_2O. \tag{1.5}$$

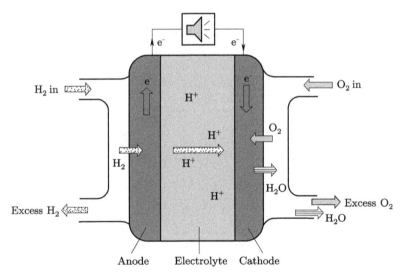

Fig. 1.6 The working principle of the hydrogen fuel cell.

The difference between the respective electrode potentials is the electromotive force, which produces the voltage per unit cell. The amount of electric current available to the external circuit depends on the chemical activity and amount of the supplied fuel.

2.3 Structure of PEMFC

A variety of fuel cells have been developed. They are generally classified based on the electrolyte because the electrolyte is the key factor that determines the operating temperature of a system and the kind of fuel that can be employed. These include alkaline fuel cell, phosphoric acid fuel cell, molten carbonate fuel cell, solid oxide fuel cell, and PEMFC.

PEMFC is a type of hydrogen-powered cell that offers high power density and can operate at low temperatures enabling it to be incorporated in various stationary and transport applications. A cell of this sort is built using an ion-conducting membrane as the electrolyte. The electrodes are catalyzed carbon. A noble metal such as platinum or one of its alloys is used as the catalyst to increase the reaction rate at electrodes. The platinum is formed into nanoparticles to increase the surface area contacting with the reactants. Thus, its catalytic effect is maximized. They are spread onto the surface of large particles of carbon support. The catalyst and carbon support particles are affixed to a porous and electrically conductive layer, such as carbon cloth or carbon paper. The reactant gases diffuse to the surfaces of the catalysts, and the water produced at the cathode diffuse away through the carbon cloth or carbon paper.

3 Experimental

3.1 Preparation

(1) Materials: proton exchange membrane (Dupont, Nafion film 117) with a size of 20 mm × 20 mm, carbon paper, conductive cloth.

(2) Chemicals: Pt/C catalyst (Sunlaite News Energy Company, HyCa-PT60), Nafion dispersion (Dupont, type D520), isopropyl alcohol, deionized (DI) water. The Pt content in the Pt/C catalyst is about 60%, and the particle size is about 4 nm.

(3) Apparatuses: spray gun and mini air compressor, electronic balance, ultrasonic cleaner, ammeter and voltmeter, water electrolysis device, mini electric fan.

(4) Tools: customized sealant clamp (consists of PMMA clamps, PTFE screws and nuts, leakage proof sheets), 20 mL glass bottle, pipette, tweezers, plastic pipes.

3.2 Procedure

3.2.1 Assembly of the PEMFC

(1) Prepare catalyst slurry by ultrasonically mixing 20mg Pt/C catalyst, 200 μL Nafion dispersion, 4 mL isopropyl alcohol, and 1 mL DI water in a 20 mL glass bottle for 30 min.

(2) Take a piece of Nafion film and spray the catalyst slurry on both sides evenly.

(3) Assemble the PEMFC with the sealant clamp. From the middle to both sides, the order is the Nafion film with catalyst, carbon paper, conductive cloth, and leakage proof sheet.

3.2.2 Test of the as-prepared PEMFC

(1) Produce hydrogen by the water electrolysis device and import the hydrogen to the anode side of the fuel cell through a plastic pipe. The cathode side contact with air directly.

(2) Connect the fuel cell to the mini electric fan. The fan will start to rotate.

(3) Use the ammeter and voltmeter to measure the current and voltage of the fuel cell during working.

4 Requirements for experimental report

(1) Describe the working principle of PEMFCs in detail.

(2) Compare the performance of the PEMFC prepared in this experiment with those reported in the literature. Discuss the possible factors leading to the difference.

5 Questions and further thoughts

(1) What are the advantages and disadvantages of PEMFCs?

(2) Please illustrate the role of materials in the development of fuel cells with several examples.

Experiment 4 Practice for Electronic Materials and Devices — Preparation and Characterization of Organic Photodetector

Type of the experiment: Comprehensive
Recommended credit hours: 4
Brief introduction: An organic photodetector with the bulk heterojunction structure is prepared in this experiment to provide an intuitive understanding of the preparation and application of the electronic materials and devices. The spin coating technique is used to prepare the organic active layer and other layers in the photodetector. The performance of the organic photodetector is then characterized using light-emitting diode (LED) light.

During the experiment, the preparation of several solutions for spin coating takes a long time. Since this experiment is designed as a preliminary experiment, the solutions can be prepared before class.

1 Objectives

(1) To understand the principle and the structure of organic photodetectors.

(2) To learn the spin coating method.

(3) To understand the characterization method of photodetectors.

2 Principles

2.1 Organic electronics

Organic electronics is a discipline that concerns the design, synthesis, characterization, and application of organic molecules or polymers with desirable electronic properties. Unlike conventional electronics, organic electronic materials are constructed from organic conjugated small molecules or polymers. They show advantages of low cost, solution processability, and mechanical flexibility.

Organic electronics have wide applications, such as organic conductors, organic semiconductors, organic insulators, and other types of organic electronic devices. Notably, organic semiconductors have attracted significant attention. Organic light-emitting diodes (OLEDs), organic transistors, and organic photovoltaics have been pushed to commercial markets.

2.2 Organic photodetector based on the photovoltaic effect

A photodetector is an optoelectronic device that absorbs optical energy and converts it into electrical signals. Active materials in the photodetectors possess the function of photoelectric conversion. Photodetectors have been widely applied, such as to environmental monitoring, data communication, image sensing, industrial automation, photometric, radiation measurement, and detection.

Photodetectors can be classified according to the detection mechanism, including photoemission or photoelectric effect, photovoltaic effect, thermal effect, and photochemical effect. In this experiment, an organic photodetector based on the photovoltaic effect will be prepared. The photovoltaic effect refers to photons of light exciting electrons into a higher state of energy. These electrons act as charge carriers for an electric current. In this way, the photons can be converted into electric current signals. The well-known solar photovoltaic cells, which generate electric power by converting light radiation into direct current electricity, are based on this mechanism.

The organic photodetector with the bulk heterojunction (BHJ) structure is shown in Fig. 1.7. In this structure, a two-way channel must be built since the balance of transmission between photoelectrons and holes is very important. The BHJ structure aims at solving the restrictions of exciton separation and carrier transport by extending the interface to the inner body. Such continuous organic semiconductor structure is beneficial to the transportation of carriers. A disadvantage is that phase separation tends to occur because of the poor solid compatibility of the BHJ structure.

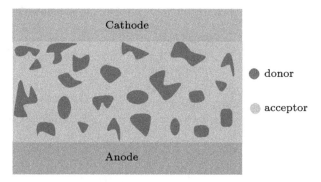

Fig. 1.7 Schematic illustration of the BHJ structure. (See color illustration at the back of the book)

Based on the BHJ structure, an organic photodetector is designed as shown in Fig. 1.8 and will be prepared in this experiment. The active layer is a conjugated polymer, poly (3-hexylthiophene-2, 5-diyl) (P3HT), blended with a fullerene derivative, [6,6]-phenyl-C_{61}-butyric acid methyl ester ($PC_{61}BM$). The preparation starts from a glass substrate coated with indium tin oxide (ITO) as the anode. The hole transport layer (HTL), poly (3, 4-ethylene dioxythiophene)-poly (styrene sulfonate) (PEDOT: PSS (AI 4083)), the active layer, the electron transport layer (ETL), perylene diimides functionalized with amino N-oxide (PDINO), and the cathode layer, PEDOT: PSS (PH 1000), are deposited in order by the spin coating method.

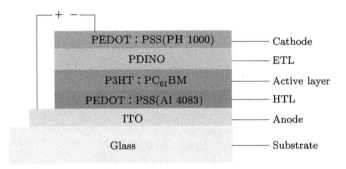

Fig. 1.8 The structure of an organic photodetector.

3 Experimental

3.1 Preparation

(1) Chemicals: PEDOT: PSS (AI 4083), P3HT, PC61BM, PEDOT: PSS (PH1000), PDINO, isopropyl alcohol, acetone, ethanol, ethylene glycol, chlorobenzene, surfactant (Toynol Superwet-340), colloid silver paste.

(2) Substrate: ITO coated conductive glass with a size of 1.5 cm × 1.5 cm.

(3) Apparatuses: electromagnetic stirrer, ultrasonic cleaner, Ultraviolet (UV) ozone cleaner, drying oven, heating plate, spin coater (SETCAS Electronics, KW-4A), LED light sources (with light colors of yellow, green, blue, red, purple, and white), electrochemical workstation (CH Instruments, CHI 604E).

(4) Tools: pipettes, 2mL bottles.

3.2 Procedure

3.2.1 Cleaning of the ITO substrate

(1) Determine which side of the substrate is coated with the ITO layer by comparing the light reflection or the electrical resistivity of the two sides. The electrical resistivity can be measured with a multimeter. Then mark on the side of the substrate without ITO.

(2) Ultrasonically clean the ITO substrate in acetone, ultrapure water, and isopropyl alcohol for 20 min sequentially.

(3) Put the substrate into the drying oven at 80 °C for 2 h to remove the

solvent.

(4) Treat the substrate with the UV ozone cleaner for 15 min to remove the organic moieties.

3.2.2 Deposition of the HTL layer

(1) Switch on the spin coater. Put the ITO substrate on the stage. Switch on the vacuum pump to hold the ITO substrate.

(2) Spread 70 μL PEDOT: PSS (AI 4083) solution on the substrate. Spin at 500 r/min for 10 s and then at 2500 r/min for 30 s.

(3) Switch off the vacuum pump and the spin coater. Take out the sample.

(4) Heat the substrate on the heating plate at 150 °C for 10 min to dry the HTL layer.

3.2.3 Deposition of the active layer

(1) Weigh 3 mg P3HT and 1.8 mg $PC_{61}BM$ in a 2 mL bottle. Add 200 μL chlorobenzene.

(2) Stir the solution by the electromagnetic stirrer for 2 h at 80 °C.

(3) Deposit the active layer by spin coating. Spread 40 μL solution on the substrate and spin at 800 r/min for 60 s.

3.2.4 Deposition of the ETL layer

(1) Weigh 1.5 mg PDINO and disperse it in 1 mL ethanol in a 2 mL bottle.

(2) Stir the solution by the electromagnetic stirrer for 12 h.

(3) Deposit the ETL layer by spin coating. Spread 40 μL solution on the substrate and spin at 500 r/min for 10 s and then at 3000 r/min for 30 s.

3.2.5 Deposition of the cathode layer

(1) Weigh 6 mg surfactant, 3 mg ethylene glycol, and 600 mg PEDOT: PSS (PH 1000), respectively. Put them into a 2 mL bottle.

(2) Stir the solution by the electromagnetic stirrer for 12 h.

(3) Deposit the cathode layer by spin coating. Spread 60 μL solution on the substrate and spin at 500 r/min for 10 s and then at 1000 r/min for 10 s.

(4) Heat the sample on the heating plate at 50 °C for 10 min.

3.2.6 Characterization of the photodetector

(1) Scratch one edge of the sample to reveal the ITO film with a width of about 4 mm. Brush the colloid silver paste on this area.

(2) Brush the colloid silver paste on the edge of the cathode with a width of about 4 mm. Dry the silver paste on the heating plate at 60 °C for 5 min.

(3) Switch on the electrochemical workstation and a LED light source.

(4) Connect the sample to the electrochemical workstation by clamping the silver electrode. Make the sample surface face the LED light.

(5) Select the linear sweep voltammetry technique. Set the voltage toincrease from −1 to 1 V at a scan rate of 50 mV/s.

(6) Start test. An I–V curve will be obtained. Export the data as a ".txt" file.

(7) Repeat steps (4) to (6) for the LED light sources with different colors.

4 Requirements for experimental report

(1) Describe the structure of the organic photodetector prepared in this experiment in detail.

(2) Plot the I–V curves recorded in the experiment. Analyze the effect of light colors.

5 Questions and further thoughts

(1) What are the advantages and disadvantages of organic photodetectors?

(2) Please describe the working principles of photodetectors based on photoemission or photoelectric effect and photovoltaic effect with typical examples.

Experiment 5   Practice for Biomaterials — Observation of Biological Samples

Type of the experiment: Comprehensive
Recommended credit hours: 6
Brief introduction: In this experiment, biological samples are observed by both scanning electron microscope (SEM) and fluorescence microscope, which areboth commonly used to study the morphology of various materials. This experiment is divided into two experiments that can be conducted respectively. The first experiment introduces the fundamentals of SEM, and describes the preparation method of biological samples for SEM observation. Then, the MC3T-E1 cells seeded on an artificial bone composite are prepared and observed by SEM. The second experiment introduces the fundamental of fluorescence microscopy, and the pollen and potato tuber slices are observed.

Experiment 5.1   Preparation and Observation of Biological Samplesby Scanning Electron Microscope

Type of the experiment: Cognitive
Recommended credit hours: 4

1 Objectives

(1) To understand the principle of SEM.
(2) To learn the preparation method of biological samples for SEM observation.
(3) To understand the principle of freeze-drying and the operation of the freeze-drying machine.

2 Principles

2.1   Principle of SEM

The SEM uses a focused beam of high-energy electrons to generate various signals at the surface of solid samples. As shown in Fig. 1.9, these signals consist of secondary electrons (SEs), backscattered electrons (BSEs), X-rays, and visible lights. The signals produced from transmitted electrons are also shown, which are generally used in transmission electron microscope (TEM). The signals from electron-sample interactions reveal information about the samples, including surface morphology, chemical composition, crystalline structure, and orientation.

SEs and BSEs are most commonly used for imaging samples. Specifically, SEs are used to show the morphology and topography of samples, and BSEs are used to illustrate contrasts in composition in multiphase samples.

X-rays are generated by the inelastic collision of the incident electrons with electrons in discrete orbitals or shells of atoms. As the excited electrons return to lower energy states, characteristic X-rays are yielded, which have a fixed wavelength related to the difference in energy levels of electrons in different shells of a given element. The chemical microanalysis technique named X-ray energy–dispersive spectroscopy (EDS) is based on the detection of the characteristic X-rays. It isa measurement of the relative abundance of emitted X-rays versus their energy.

It is noted that the SEM can perform analyses of selected regions on samples. This is especially useful in qualitatively determining chemical compositions using EDS and crystal structures and orientations using electron backscattering diffraction (EBSD) analysis.

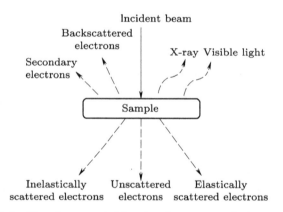

Fig. 1.9 Signals generated by the electron-sample interaction.

2.2 SEM instrumentation

SEM is a complicated instrument. Figure 1.10 is a schematic illustration of the electro-optical system, which comprises the following essential components: an electron gun, electron lenses, various detectors, and display or data output devices.

The electron gun acts as an electron source that emits electrons. The electrons can be created by heating a tungsten filament. The field emission gun is also widely used, which provides a more efficient electron beam.

There are a series of electromagnetic lenses in SEM. They create an electromagnetic field that acts as a convex lens bringing off-axis electrons back to focus. The condenser lens and the objective lens are probably the most important ones. The former narrows the electron beam from the electron gun, and the latter focuses

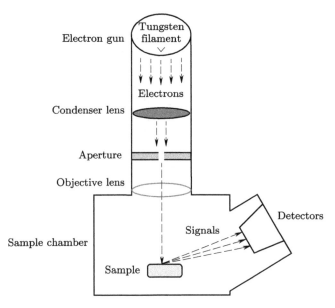

Fig. 1.10 Schematic illustration of the essential components of SEM.

the electron beam onto the sample.

Between the condenser lens and the objective lens, there is an aperture, which is a metal rod with a series of holes in different sizes. The diameter of the electron beam is controlled by changing the holes.

The electron beam enters the sample chamber after passing through the objective lens. The sample itself needs to be conductive to prevent charging. If the sample is an insulating material, it can be coated with a thin conductive layer of carbon or metals (Au, Pt, W, etc.). Carbon is recommended if X-ray analysis is required.

There are detectors that collect signals from electron-sample interaction. They can create magnified images, EDS spectra, and EBSD patterns on display or other data output devices for different analysis purposes.

2.3 Preparation of biological samples for SEM observation

Conventional SEM is operated at a very high vacuum to avoid gas molecules interfering with the electron beam from the electron gun and the secondary or backscattered electrons emitted from the sample. This means that the sample for SEM analysis must be completely dry and free of any organic contaminants that may potentially outgas in a high vacuum environment. This request poses a problem for biological samples, which are primarily composed of water. Compared with inorganic materials, additional preparation steps are required to retain the native structure of the organism.

The preparation steps strongly depend on sample types. Some biological tissues with hard exoskeletons, such as insects, require less stringent processing to preserve their structures. Other more delicate samples, such as cells and viruses, take time and care to avoid the introduction of drying artifacts. In this case, the samples should be processed sequentially by fixation, dehydration, and freeze-drying.

Fixation is the most crucial step in preparation. It is a reaction between protein and fixative, leading to the formation of a less permeable gel. Consequently, cells or tissue components are preserved as close to their in vivo state as possible.

Dehydration is simply the removal of water from aqueous-fixed tissue. Common dehydrating fluids are ethanol and acetone. Dehydration must be conducted rapidly to prevent excessive extraction of ethanol and acetone-soluble compounds and slowly to prevent plasmolysis.

Freeze drying is the removal of ice or other frozen solvents from a material through the process of sublimation. Sublimation is a phenomenon that solid ice transforms directly to a vapor state without going through a liquid phase. As shown in the phase diagram of water in Fig. 1.11, low pressure is required for sublimation.

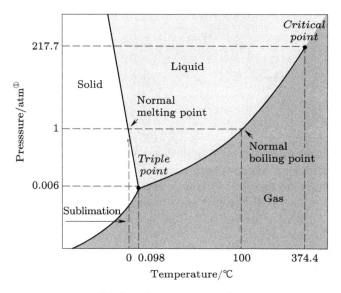

Fig. 1.11 Phase diagram of water.

The freeze-drying process generally involves the following steps:
(1) Freeze: The sample is completely frozen.
(2) Vacuum: The sample is then placed in a deep vacuum below the triple point of water, as shown in Fig. 1.11.
(3) Dry: Heat is applied to the sample causing the sublimation of ice.

This process is generally accomplished by the freeze-drying machine. As illustrated in Fig. 1.12, it comprises five main components: refrigeration system, vacuum system, control system, chamber, and condenser. The refrigeration system cools the chamber located inside the freeze dryer. The vacuum system consists of a separate vacuum pump connected to the airtight chamber and the attached

---

① 1 atm = 1.01325 ×10^5 Pa, the same below.

condenser. The control system usually includes temperature and pressure sensing abilities. The chamber is typically a large space with shelves to place the samples. The sublimated ice accumulates in the condenser and is removed at the end of the freeze-drying cycle.

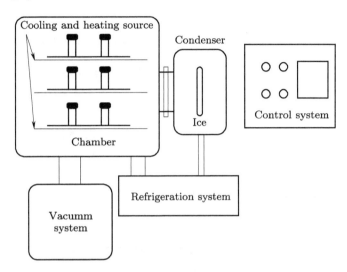

Fig. 1.12 Schematic illustration of the freeze-drying machine.

In this experiment, MC3T-E1 cells seeded on a polymeric nanocomposite, an artificial bone composite, are observed. The osteoblastic cell line MC3T3-E1 is established from a mouse calvaria. Diluted glutaraldehyde is used as the fixative, and dehydration is conducted using a series of aqueous solutions of ethanol with gradually increased concentration. Since the sample is insulated, it is coated with a conductive gold layer by sputter coater after freeze-drying.

3 Experimental

3.1 Preparation

(1) Materials: MC3T3-E1 cells seeded on a polymeric nanocomposite. The composite was prepared by incorporating nanohydroxyapatite and carbon nanotube into polymeric polyetheretherketone.

(2) Chemicals: 2.5%–3% glutaraldehyde, phosphate-buffered saline (PBS) solution, ethanol.

(3) Apparatuses: sputter coater, freeze-drying machine (EYELA FDU-1200), SEM (TECAN Vega3 LM), refrigerator.

(4) Tools: 96 wells plates, vials, Pasteur pipets, pipettes, tweezers.

3.2 Procedure

3.2.1 Preparation of biological sample for SEM observation

(1) Wash the sample with normal saline or buffer solution (PBS solution) to remove blood, mucus, or tissue fluid before fixation.

(2) Fix the sample with 2.5%–3% glutaraldehyde for 15 min at 4 °C in the refrigerator.

(3) Wash the sample three times with the PBS solution. Then wash the sample with 30%, 50%, 70%, and 90% aqueous solutions of ethanol and absolute ethanol twice for each solution and 5 min for each time.

(4) Freeze dry the sample in the freeze-drying machine. The chamber pressure and temperature are 0.035 mbar and –56 °C, respectively. It takes about 1 h to dry the sample completely.

(5) Mount the dried sample on an SEM sample holder using conductive double-sided adhesive tape.

3.2.2 SEM observation

(1) Coat the sample with a gold layer of 100–200 Å in thickness.
(2) Observe the morphology and distribution of the cells on the composite.
(3) Capture images at magnifications of 500× and 1000×.

4 Requirements for experimental report

(1) Present the SEM images of the sample. Label the cells and the polymeric nanocomposite.

(2) Describe the morphology and distribution of the cells on the composite.

5 Questions and further thoughts

(1) Why do biological samples need to be fixed, dehydrated, dried, and coated before SEM observation?

(2) What is biocompatibility? Analyze the biocompatibility of the polymeric composite based onthe morphology of the MC3T3-E1 cells.

6 Cautions

(1) Do not touch the sample or anything inside the SEM chamber with bare hands to avoid potential contamination.

(2) Do not allow the sample surface to be exposed to air before freeze-drying to prevent air-drying.

(3) The as-prepared SEM sample should be stored in a drying box if it will not be observed in a short time.

Experiment 5.2 Observation of Biological Samples by Fluorescence Microscope

Type of the experiment: Cognitive
Recommended credit hours: 2

1 Objectives

(1) To understand the principle of fluorescence microscopy.
(2) To learn the operation of the fluorescent microscope.

2 Principles

2.1 Principle of fluorescence

Fluorescence is the emission of visible light by a substance that has absorbed light or other electromagnetic radiation. For instance, the absorbed ultraviolet radiation is invisible to human eyes, while the emitted fluorescence is visible. The most common application of fluorescence in daily life is fluorescent lamps. Besides, it has been widely used in mineralogy, gemology, chemical sensors, fluorescent labeling, dyes, and biological detectors.

Fluorescence occurs when an orbital electron of a molecule, atom, or nanostructure relaxes to its ground state by emitting a photon from an excited singlet state. As shown in Fig. 1.13, the process of fluorescence has three stages:

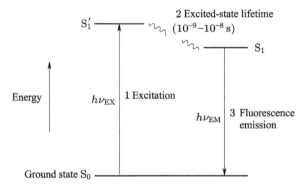

Fig. 1.13 The three-stage process of fluorescence.

(1) Excitation. A photon of energy ($h\nu_{EX}$) is supplied by an external source, such as an incandescent lamp or a laser, and absorbed by the fluorophore, creating an excited electronic singlet state (S_1').

(2) Excited-state lifetime. The excited state exists for a finite time (typically 1–10 ns). The fluorescent substance undergoes conformational changes and is also subject to possible interactions with its molecular environment. Consequently, the energy of S_1' is partially dissipated, yielding a relaxed singlet excited state (S_1). The energy difference between the S_1' and S_1 states is called the Stokes shift.

(3) Fluorescence emission. A photon of energy ($h\nu_{EM}$) is emitted, returning the fluorophore to its ground state S_0. Due to energy dissipation during the excited-state lifetime, the energy of this photon is lower, than the excitation photon ($h\nu_{EX}$).

2.2 Fluorescence microscopy/microscope

Fluorescence microscopy is an imaging technique that visualizes possible fluorescence from the analyzed materials. The technique has become an essential tool in biology and materials science because it shows attributes that are not readily available with conventional optical microscopy. Fluorescence microscopy can detect particles below the resolution of a conventional optical microscope. The application of fluorochromes or fluorescent dyes has made it possible to identify cells and sub-microscopic cellular components with a high degree of specificity amid non-fluorescing material.

The basic function of a fluorescence microscope is to irradiate the sample with a specific band of wavelengths and then to separate the emitted fluorescence from the excitation light. Figure 1.14 schematically illustrates the structure of the fluorescence microscope. Optical filters are critical components in a fluorescence microscope. A typical system has three basic filters: an excitation filter, a dichroic mirror, and an emission filter (or barrier filter).

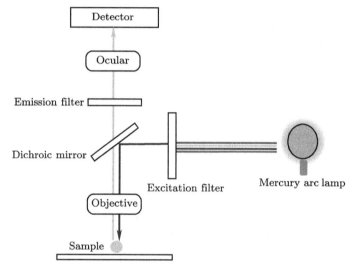

Fig. 1.14 Schematic illustration of the fluorescence microscope. (See color illustration at the back of the book)

Multispectral light from the high-intensity mercury arc lamp is filtered by the exciter to select short-wavelength excitation light. This light is reflected to the sample by the dichroic mirror. The sample is previously stained with a fluorescent dye. The portions, which retain the fluorescent dye, emit light with a longer wavelength. The emitted light gathered by the objective passes back through the dichroic mirror and is subsequently filtered by the emission filter, which blocks the unwanted excitation wavelengths.

3 Experimental

3.1 Preparation

(1) Samples: fluorescence slices of pollen, potato tuber.
(2) Apparatus: fluorescence microscope (Olympus IX-53).

3.2 Procedure

(1) Switch on the microscope, the halogen lamp, and the mercury arc lamp. Load the samples onto the stage.
(2) Select the desired objective lens. Usually, start with a lower magnification.

(3) Remove any filters. Find and focus the samples using the transmitted light from the halogen lamp.

(4) Dim or turn off transmitted light. Select desired filters. Open the baffle of the mercury arc lamp. Capture the images.

(5) Switch off the mercury arc lamp and the microscope.

4 Requirements for experimental report

(1) Present the images of the biological samples.

(2) Describe the features of the biological samples observed using the fluorescence microscope.

5 Questions and further thoughts

(1) What are the advantages of the fluorescence microscope comparing with the conventional optical microscope?

(2) Please discuss the factors that influence the quality of the fluorescence images.

6 Cautions

(1) The fluorescence images should be captured as soon as possible after switching to the mercury lamp to avoid fluorescence quenching.

(2) In order to prolong the life of the mercury lamp, frequent turn the lamp on and off should be avoided.

Extended readings

Bradbury S. Scanning electron microscope. Available at: https://www.britannica.com/technology/scanning-electron-microscope. Accessed: 2 March 2021.

Callister W D, Rethwisch D G. Fundamentals of Materials Science and Engineering An Integrated Approach. 5th Edition. John Wiley & Sons Inc., 2001.

Herring A M, Zawodzinski T A, Hamrock S J. Fuel Cell Chemistry and Operation. American Chemical Society, 2010.

Louthan M R. Optical metallography. In: Whan R E, editor, ASM Handbook, Volume10: Materials Characterizations. ASM International, 1986.

Page K A, Soles C L, Runt J. Polymers for Energy Storage and Delivery: Polyelectrolytes for Batteries and Fuel Cells. American Chemical Society, 2012.

Reimer L. Scanning Electron Microscopy: Physics of Image Formation and Microanalysis. Springer, 2010.

Solidworks Tutorials. Available at: https://www.solidworks.com/sw/resources/solidworks-tutorials.htm. Accessed: 7 April 2020.

Srinivasan S. Fuel Cells: From Fundamentals to Applications. Springer, 2006.

The Editors of Encyclopaedia Britannica. Fluorescence. Available at: https://www.britannica.com/science/fluorescence. Accessed: 1 March 2021.

Part II
Fundamentals of Chemistry and Crystallography

Chapter 2 Fundamentals of Physical Chemistry

Experiment 6 Determination of Combustion Heat

Type of the experiment: Cognitive
Recommended credit hours: 4
Brief introduction: Combustion heat is the heat released when matter reacts entirely with oxygen. It is generally measured in terms of the energy released by the combustion of per unit mole, mass, or volume of matter. In this experiment, the combustion heats of naphthalene and benzoic acid are measured using the oxygen bomb calorimeter.

1 Objectives

(1) To understand the concept of combustion heat.
(2) To understand the structure and usage of the oxygen bomb calorimeter.
(3) To measure the combustion heat of naphthalene and benzoic acid using the bomb calorimeter.
(4) To calibrate the variation of the temperature by the Renolds correction diagram.

2 Principles

2.1 Combustion heat

The combustion heat of a compound is the heat released when one mole or one gram compound is completely oxidized to $CO_2(g)$, $H_2O(l)$, and $N_2(g)$ without doing non-volume work. Under the measurement condition of constant volume or constant pressure, we can get the combustion heat expressed as Q_v or Q_p. According to the first law of thermodynamics, Q_v equals the change in internal energy ΔU, and Q_p equals the change in enthalpy ΔH. Therefore, if the gas involved or generated in the reaction is treated as an ideal gas, from

$$\Delta H = \Delta U + \Delta(pV), \tag{2.1}$$

the relationship between Q_v and Q_p can be expressed as

$$Q_p = Q_v + \Delta n RT, \tag{2.2}$$

where Δn is the change in the mole number of the gas according to the chemical equation, R is the gas constant (8.314 J/(mol · K), T is the temperature during the reaction.

Because the thermal effect of a chemical reaction is usually expressed under constant pressure, one can use the equation above to calculate Q_p with the Q_v, and Q_v can be measured under the experimental conditions.

2.2 Oxygen bomb calorimeter

Since the basic principle of calorimetry obeys the law of energy conservation, the combustion heat Q_v can be measured by burning a certain amount of sample in the oxygen bomb calorimeter and measuring the increase in the system temperature. The main parts of the calorimeter are shown in Fig. 2.1. The bomb is airtight and filled with high-pressure oxygen to ensure the sample can be burned completely.

If water is used as the measuring medium, Q_v can be obtained from

$$-n_{\text{sample}} Q_v - l Q_l = (m_{\text{water}} C_{\text{water}} + C_{\text{calorimeter}}) \Delta T, \tag{2.3}$$

where n_{sample} is the mole number of the sample, l and Q_l are respectively the length and the combustion heat per unit length of the fuse wire, m_{water} and C_{water} are the mass and heat capacity of water, $C_{\text{calorimeter}}$ is the heat capacity of the calorimeter exclusive of water. One can obtain the $C_{\text{calorimeter}}$ by conducting the experiment with a material that has a known combustion heat at constant volume. ΔT is the

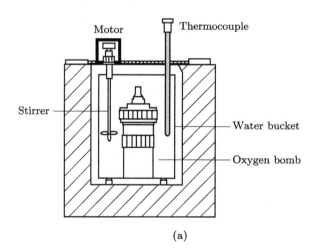

(a)

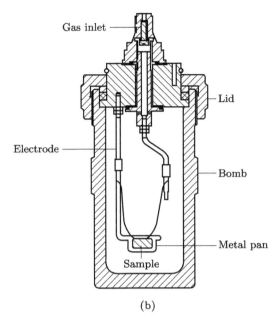

Fig. 2.1 Cross-sectional illustrations of (a) the oxygen bomb calorimeter and (b) the oxygen bomb.

temperature increase of the water and other parts of the testing system when the sample is completely oxidized.

2.3 Renolds temperature correction diagram

Actually, the experiment is not carried out under ideal adiabatic conditions since heat exchange between the system and the outside environment is inevitable. The Renolds temperature correction diagram is generally used to calibrate the temperature deviation caused by this effect.

Figure 2.2 illustrates Renolds temperature correction diagrams plotted using the recorded values of the water temperature against the measurement time. In this figure, point H is the starting of the reaction, point D is the maximum temperature value of the reaction process, and point J is the middle temperature between H and D. Point I is obtained by drawing a horizontal line from point J to the curve. Draw a vertical line ab through point I. Then, extend FH and GD lines to the vertical line ab, and the points A and C can be obtained. The temperature difference between points A and C is the calibrated ΔT.

Besides, points A' and C' are obtained by drawing a horizontal line from points H and D to the vertical line ab. The AA' segment is the temperature increase due to the stirring and the heat irradiation from the surrounding. The CC' segment is the temperature decrease due to the heat released from the apparatus to the environment.

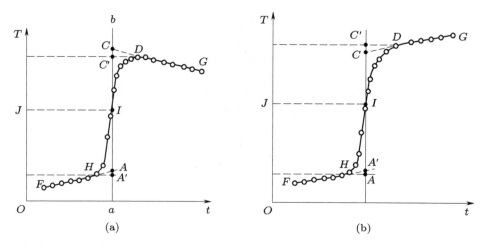

Fig. 2.2 Renolds temperature correction diagrams for the data obtained under (a) relatively poor adiabatic condition and (b) under relatively well adiabatic condition.

3 Experimental

3.1 Preparation

(1) Material: iron fuse wire.

(2) Chemicals: naphthalene, benzoic acid.

(3) Apparatuses: oxygen bomb calorimeter (Nanjing Sangli Electronic Equipment Factory, SHR-15B), oxygen cylinder with a regulator, press machine, counter scale (15 kg), electronic balance.

(4) Tools: pellet press die, multimeter, ruler.

3.2 Procedure

3.2.1 Preparation and assembly of the sample

(1) Weigh about 1 g of naphthalene or benzoic acid using the electronic balance. Press it into a pellet using the press machine and the pellet press die. Get rid of the powder on the pellet surface and weigh the sample precisely to 0.1 mg.

(2) Cut a piece of iron fuse wire. Measure the length of the wire.

(3) Tie up the sample and make the wire in good contact with the sample. Place the sample pellet in the center of the metal sample pan in the bomb. Fix the ends of the wire to the electrodes (see Fig. 2.1(b)). Measure the electrical resistance using the multimeter to ensure that the fuse wire does not contact the sample pan and the inner wall of the bomb.

(4) Screw the lid of the bomb on tightly. Measure the resistance between the electrodes using the multimeter to ensure the sample can be fired up. Fill the bomb with oxygen gas to a pressure of 2 MPa.

3.2.2 Temperature difference measurement

(1) Weigh 2.5 kg water accurately in the water bucket. The temperature is

adjusted to about 1–2 °C below the room temperature.

(2) Place the bomb in the water bucket. Connect the two electrodes of the bomb to the ignition voltage supplier.

(3) Switch on the stirrer.

(4) Measure the water temperature every 1 min until the water temperature changes regularly and slightly for about 10 min.

(5) Press the ignition button.Record the water temperature every 15 s until the difference between two successive data is within 0.005 °C.

(6) Record 10 more temperature data every 1 min.

(7) Switch off the apparatus and take out the bomb. Depressurize the residual gas and open the bomb.

(8) Measure the length of the residual fuse wire, and calculate the actual length of the burnt fuse wire.

(9) Clean and dry all parts of the bomb.

4 Requirements for experimental report

(1) Plot the Renolds correction curves for benzoic acid and naphthalene. Determine the calibrated ΔT from the curves.

(2) Calculate the $C_{\text{calorimeter}}$ using the known combustion heat of benzoic acid (–3226.9 kJ/mol or –26460 J/g). Q_1 is calculated by the combustion heat of pure iron (–367.7 kJ/mol) for the formation of Fe_3O_4.

(3) Calculate the combustion heats Q_v and Q_p for naphthalene.

(4) Compare the combustion heat Q_p of naphthalene with that in the literature (–5153.8 kJ/mol or –40205 J/g). Point out the primary error sources.

5 Questions and Further Thoughts

(1) Why should the solid sample be pressed into pellets?

(2) Can you suggest improvements to make the experiment easier and faster?

Experiment 7 Determination of Saturated Vapor Pressure of Pure Liquid

Type of the experiment: Cognitive

Recommended credit hours: 4

Brief introduction: In this experiment, the saturated vapor pressures of a widely used liquid, ethanol, are measured by the static method. This method is appropriate for liquids with relatively high vapor pressures. The saturated vapor pressures in a temperature range from 30 to 70°C will be measured. Consequently,the relationship between the saturated vapor pressure and temperature can be understood and expressed by the Clausius-Clapeyron equation.

1 Objectives

(1) To understand the concepts of saturated vapor pressure and the equilibrium between gas and liquid.

(2) To determine the saturated vapor pressure of ethanol at different temperatures by the static method.

(3) To calculate the average molar enthalpy of vaporization and the boiling point of ethanol.

2 Principles

2.1 Saturated vapor pressure

In a closed container, the vapor pressure of a liquid or a solid in phase equilibrium is called saturated vapor pressure at a specific temperature. Generally, the saturated vapor pressure increases with temperature for the same substance. Moreover, the saturated vapor pressure of a solid is less than that of the corresponding liquid. The saturated vapor pressure of a pure solvent is greater than that of the corresponding solution.

Take pure water as an example. The water in a cup will become less and less by evaporation. If we fill pure water in a closed container and remove the air from the top, the pressure of the vapor phase above the water surface, that is, the pressure of the steam of the water increases as the water evaporates. If the temperature is kept constant, the vapor pressure will eventually be stabilized at a fixed value, and it is called the saturated vapor pressure of water at that temperature. In this state, the liquid and the gas reach a dynamic equilibrium. In other words, the water molecules in the liquid phase are still vaporized, and the water molecules in the vapor phase are continuously condensing into liquid. Moreover, the vaporization rate of liquid water is equal to the condensation rate of water vapor.

2.2 Static method

There are many methods to measure the vapor pressure, including the static method, the dynamic method, and the saturated flow method. The static method is appropriate for liquid with relatively high vapor pressure. The liquid is placed in a sealed system, and then its vapor pressures at different temperatures are measured. In the present experiment, this method is used to measure the saturated vapor pressures of ethanol at different temperatures.

The apparatus of the static method is shown in Fig. 2.3. The core of the apparatus is the balance tube, in which the liquid for the measurement is filled in. The balance tube comprises three interconnected glass tubes, denoted as tubes a, b, and c. Tube b is connected to the pressure gauge. If only pure vapor exists above the liquid in tubes a and c, and the liquid levels in tubes b and c are equal, the numerical reading of the pressure gauge is the saturated vapor pressure of the liquid.

2.3 The Clausius-Clapeyron equation

The relationship between the vapor pressure of a pure liquid and the temperature can be expressed as the Clausius-Clapeyron equation

$$\frac{\mathrm{d}\ln p^*}{\mathrm{d}T} = \frac{\Delta_\mathrm{v} H_\mathrm{m}}{RT^2}, \tag{2.4}$$

where p^* is the vapor pressure, T is the thermodynamic temperature, $\Delta_\mathrm{v} H_\mathrm{m}$ is

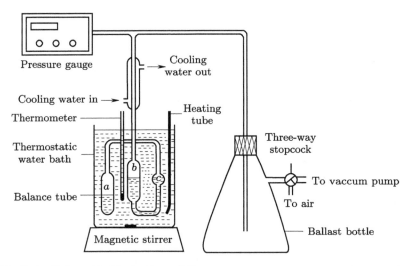

Fig. 2.3 Schematic illustration of the vapor pressure measurement apparatus.

the molar enthalpy of vaporization, and R is the ideal gas constant. Within a moderate range of temperature, $\Delta_v H_m$ can be considered as a constant, namely, the average molar enthalpy of vaporization. Integrating the Equation (2.4), one can obtain

$$\ln p^* = -\frac{\Delta_v H_m}{RT} + C, \qquad (2.5)$$

where C is a constant. A straight line can be fitted by plotting $\ln p^*$ against $1/T$, and the slope of the line is $-\Delta_v H_m/R$. Then the average molar enthalpy of vaporization of the liquid $\Delta_v H_m$ can be obtained. For instance, it is 40.7 kJ/mol for pure water.

One can also calculate the boiling point by input the ambient pressure (101.325 kPa) into the equation since the boiling point of a substance is the temperature at which the vapor pressure of the liquid equals the ambient pressure.

3 Experimental

3.1 Preparation

(1) Chemical: absolute ethanol.

(2) Apparatuses: vapor pressure measurement apparatus, thermostatic water bath, vacuum pump.

3.2 Procedure

(1) Fill ethanol into the balance tube and set up the apparatus as shown in Fig. 2.3. Then, conduct the leak testing.

(2) Switch on the water bath. Set the temperature to 30 °C.

(3) Switch on the vacuum pump and vacuum the system to a pressure lower than −90 kPa. The air in tubes a and c will be pumped out through the liquid, and bubbles can be seen in tube b. Hold the pressure by adjusting the stopcock

for about 5 min to remove all remained air between tubes a and c. The amount of the bubbles will decrease and disappear finally. Then turn off the stopcock to the pump.

(4) Adjust the liquid levels in tubes b and c, and record the pressure more than three times. When the liquid levels in tubes b and c are the same, the pressure in tube b read from the pressure gauge is the saturated vapor pressure. If the liquid level in tube b is higher than tube c, open the stopcock connected to the air; if the liquid level in tube c is higher than tube b, open the stopcock connected to the vacuum pump.

(5) Set the temperature of the water bath from 35 to 70 °C at intervals of 5 °C. Repeat step (4) to measure the saturated vapor pressures at these temperatures.

4 Requirements for experimental report

(1) Tabulate the experimental data. Calculate the average value of the saturated vapor pressure at each temperature.

(2) Plot $\ln p^*$ against $1/T$. Calculate the $\Delta_v H_m$ and the normal boiling point of ethanol. Compare them with the data in the literature (41.50 kJ/mol and 78.3 °C).

5 Questions and further thoughts

(1) The random errors in the measurement of pressures and temperatures are inevitable. Please derive the expression for error propagation in $\Delta_v H_m$.

(2) What do we need to pay attention to when heating flammable liquids?

6 Cautions

(1) The numerical reading of the differential pressure gauge is the difference from the atmospheric pressure. Thus, one must record the atmospheric pressure before the measurement.

(2) The stopcock should be opened evenly and slowly, and pay attention to the pressure at the same time.

Experiment 8   Construction of Binary Gas-Liquid Phase Diagram

Type of the experiment: Cognitive
Recommended credit hours: 4
Brief introduction: The relationship between the boiling point and composition of a liquid-liquid system can be expressed by the gas-liquid phase diagram. In this experiment, the binary gas-liquid phase diagram of a completely miscible liquid-liquid system, the cyclohexane-ethanol system, is constructed. The principles and usage of the ebulliometer and Abbé refractometer are introduced.

1 Objectives

(1) To understand the concepts of the completely miscible liquid-liquid system and the equilibrium gas-liquid phase diagram.

(2) To understand the principles and usage of the ebulliometer and Abbé refractometer.

(3) To learn the measurement method of the normal boiling point and the solution composition.

(4) To construct the gas-liquid phase diagram of the cyclohexane-ethanol system.

2 Principles

2.1 Equilibrium gas-liquid phase diagram

A completely miscible liquid-liquid system is that the two components are miscible in all proportions. The relationship between the boiling point and the solution composition under constant pressure can be expressed by the boiling point-composition diagram, which is also called the gas-liquid phase diagram.

The phase diagram of the completely miscible liquid-liquid system can be classified into three types. The simplest one is that the boiling point of the solution is between the boiling points of the two pure components. Figure 2.4 illustrates the phase diagram of the benzene-toluene system as an example. The vertical coordinates in Fig. 2.4 represent the temperatures, and the transverse coordinates indicate the molar fractions of the components. The gaseous line is located between the gas phase region and the gas-liquid two-phase region, above which the gas phase presents at all temperatures and compositions. The liquid us line is located between the liquid phase region and the gas-liquid two-phase region, below which only the liquid phase exists. The two lines also represent the equilibrium compositions of the gas and liquid phases at different temperatures, respectively.

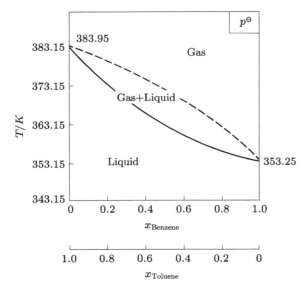

Fig. 2.4 The benzene-toluene gas-liquid phase diagram.

The other two types are those with an azeotropic point, as illustrated in Fig.

2.5. Examples include ethanol-water and nitric acid-water mixtures. In these phase diagrams, the gas line and the liquid line are tangent at the azeotropic point. When the azeotropic mixture reaches gas-liquid equilibrium at the azeotropic point, the gas and liquid phases have the same composition. Obviously, the temperature and composition of the azeotropic point of the same liquid-liquid system vary with ambient pressure.

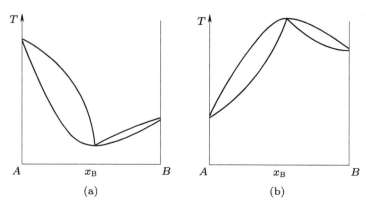

Fig. 2.5 The gas-liquid phase diagrams with azeotropic points at (a) lower and (b) higher temperatures.

2.2 Construction of binary gas-liquid phase diagram

2.2.1 Ebulliometer

The method for the construction of a binary gas-liquid phase diagram is to boil a series of solutions with different compositions in an ebulliometer. It is designed to obtain the boiling point and samples conveniently, and prevent samples from overheating or fractionating.

The ebulliometer used in this experiment is shown in Fig. 2.6, which is a long-necked flask with a reflux condenser. The condenser has a small hemisphere at the bottom for collecting the condensed sample.

The boiling temperature can be read directly from the thermometer or thermocouple inserted into the liquid. In order to construct the phase diagram under the standard atmospheric pressure $p^{\ominus}$, i.e., 101325 Pa, the normal boiling point should be used. Since the ambient pressure is not exactly the standard atmospheric pressure, the calibration of the measured boiling points is required. The calibration equation can be derived from the Trouton rule and the Clausius-Clapeyron equation, which is expressed as

$$t_B = t_A + \frac{273.15 + t_A}{10} \frac{101325 - p}{101325}, \tag{2.6}$$

where t_A and t_B are respectively measured and calibrated boiling temperatures in a unit of °C, p is the ambient pressure in Pa.

After recording the boiling point, it is also necessary to determine the corresponding compositions of the gas phase and the liquid phase. Obviously, they are

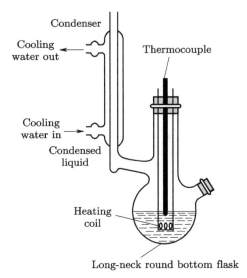

Fig. 2.6 Schematic illustration of the ebulliometer.

the compositions of the condensed liquid in the hemisphere and the residual liquid in the flask. In fact, the measured compositions are also affected by the pressure. Considering that the effect is relatively small, the compositions are not calibrated in this experiment to simplify the experiment.

2.2.2 Measurement of liquid composition by refractometry

In this experiment, the compositions of the liquids are measured by their refractive indices. A refractive index is a characteristic value of a substance, which is related to the concentration and temperature of the substance. It is necessary to measure the refractive indices of a series of solutions with known compositions and draw a standard curve about the refractive index versus the concentration. Since the variation of the refractive index with the concentration appears to be linear in the solubility range, the composition of the liquids can be calculated by fitting the standard curve into a linear equation.

The refractive index can be measured by the Abbé refractometer. Its structure and working principle are schematically shown in Fig. 2.7. A sample is put between two prisms known as measuring prism and illuminating prism. Light enters the sample from the illuminating prism and gets refracted at a critical angle at the bottom surface of the measuring prism. Then the telescope is used to measure the position of the border between bright and dark areas. Knowing the angle and the refractive index of the measuring prism, it is easy to calculate the refractive index of the sample, which can be read directly from the scale. The surface of the illuminating prism is matted so that the light enters the sample at all possible angles, including those almost parallel to the surface.

The refractive index of a substance is a function of wavelength. If the light source is not monochromatic, the light gets dispersed, and the shadow boundary

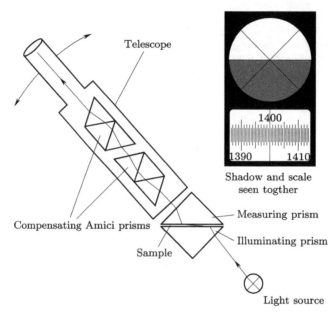

Fig. 2.7 The working principle of the Abbé refractometer.

is not well defined. A blurred blue or red border instead of a sharp edge between the bright and dark areas will be observed. This means measurements are either very inaccurate or even impossible. To prevent the dispersion, two compensating Amici prisms are added. Not only the telescope position can be changed to measure the angle, but the Amici prisms position can also be adjusted to correct the dispersion. Consequently,the edge of the shadow is well defined and easy to locate.

Abbé refractometer can be used to measure the refractive index of both liquids and solids. In both cases, the refractive index of the substance must be lower than the refractive index of the glass used to make the measuring prism.

2.2.3 Construction of the phase diagram

After measuring the boiling points of a series of solutions with different compositions in the ebulliometer and determining the corresponding compositions of the gas phase and the liquid phase, the gas-liquid phase diagram can be constructed. The compositions of the solutions are used as the transverse coordinates and the temperatures of the boiling points as the vertical coordinates. The gaseous line and the liquidus line are drawn by connecting data points using smooth curves.

3 Experimental

3.1 Preparation

(1) Chemicals: cyclohexane, absolute ethanol.

(2) Apparatuses: ebulliometer, Abbé refractometer (Shanghai Optical Instrument Factory, 2WAJ).

(3) Tools: glass funnel, 5 mL pipette, 50 mL and 250 mL glass breakers,

dropper.

3.2 Procedure

3.2.1 Plot the standard curve of the refractive index against concentration

(1) Prepare a series of cyclohexane-ethanol solutions, in which the molar fractions of cyclohexane are from 0 to 1 at intervals of 0.05. The volume of each solution is about 40 mL.

(2) Measure the refractive indices of the solutions by Abbé refractometer. The main operating steps are as follows:

(a) Place the Abbé refractometer in a bright place.

(b) Put two or three drops of a solution onto the surface of the illuminating prism with a dropper and close the prism.

(c) Focus the cross wire in the scope by adjusting the focal length of the eyepiece. Turn the dispersion control knob to sharpen the shadow boundary. Turn the measurement knob until the shadow boundary passes the cross point of the cross wire.

(d) Read the refractive index value to four decimal places three times and calculate the average value.

(3) Plot a standard curve of the refractive index against the concentration. Fit the curve into a linear equation.

3.2.2 Measure the data for constructing the phase diagram

(1) Assemble the ebulliometer according to Fig. 2.6.

(2) Add a solution into the flask from the branch tube using a glass funnel. Make sure that the heating coil is immersed in the solution completely.

(3) Switch on the heating electric power. Increase the output voltage gradually to heat the solution slowly for about 3 min. Measure the boiling point after the reading of the thermocouple is steady. Record the reading and calibrate it by equation (2.6).

(4) Switch off the heating electric power. Put the ebulliometer into a 250 mL beaker containing a mixture of ice and water to cool the system.

(5) Suck the condensed liquid in the hemisphere, and about 1 mL liquid from the flask, respectively. Measure their refractive indices. Calculate the cyclohexane concentrations, which can be considered as those of the gas and liquid phases.

(6) Repeat steps (1) to (5) for all the solutions. The ebulliometer is not required to be clean and dry when changing the solutions except the measurements for pure ethanol and cyclohexane.

4 Requirements for experimental report

(1) Tabulate the compositions and the refractive indices of the prepared solutions. Plot the standard curve and give the equation of curve fitting.

(2) Tabulate refractive indices and compositions of the gas and liquid phases, the corresponding measured boiling points, and the calibrated normal boiling points.

(3) Plot the gas-liquid phase diagram of the cyclohexane-ethanol system us-

ing the calibrated temperatures. The effect of pressure on the compositions is neglected. Determine the temperature and composition of the azeotropic point.

5 Questions and further thoughts

(1) What are the effects on the phase diagram if the solution is overheated or fractional distillation occurs during the experiment?

(2) Why is 95% ethanol often produced in the industry? Can absolute ethanol be obtained only by rectification of an aqueous solution of ethanol?

Experiment 9 Determination of Rate Constant for Hydrolysis Conversion of Sucrose

Type of the experiment: Cognitive
Recommended credit hours: 4
Brief introduction: Reaction rate is the speed at which a chemical reaction proceeds. In this experiment, the reaction rate of the hydrolysis conversion of sucrose is examined. The rate constant and half-life of the reaction at room temperature are determined by measuring the optical rotation of the sucrose solution using the polarimeter. The principles and the usage method are introduced.

1 Objectives

(1) To learn the basic principle and the operation of a polarimeter.

(2) To understand the relationship between the concentration and optical rotation.

(3) To master the method of determining the rate constant and the half-life of the hydrolysis conversion of sucrose.

2 Principles

2.1 Hydrolysis conversion of sucrose

Sucrose can be converted into glucose and fructose in water by the reaction

$$C_{12}H_{22}O_{11}(\text{sucrose}) + H_2O \xrightarrow{H^+} C_6H_{12}O_6(\text{glucose}) + C_6H_{12}O_6(\text{fructose}). \quad (2.7)$$

The conversion reaction is a second-order reaction, which is relatively slow in pure water. If there are H^+, the reaction can be catalyzed.

Since water exists in large quantities, the concentrations of water and H^+ may be considered as a constant. Hence, the conversion reaction of sucrose can be approximately regarded as a first-order reaction. The rate equation of the first-order reaction can be expressed as

$$-\frac{dc}{dt} = kc, \quad (2.8)$$

where c is the concentration of the sucrose at time t, and k is the rate constant. Integrating the equation, one can obtain

$$\ln c = \ln c_0 - kt, \quad (2.9)$$

where c_0 is the initial concentration of the sucrose. When $c = c_0/2$, time t is called the half-life of the reaction, which is denoted as $t_{1/2}$ and calculated by

$$t_{1/2} = \frac{\ln 2}{k} = \frac{0.693}{k}. \tag{2.10}$$

2.2 Analysis of the reactant concentration by the polarimeter

Generally, it is relatively difficult to analyze the reactant concentration in a chemical reaction instantaneously since the reaction proceeds continuously. However, in this experiment, both sucrose and its conversion products are optically active, and their optical activities are different. Therefore, the progress of the conversion reaction of sucrose can be followed by measuring the optical rotation change of the reaction system.

Optical activity is the ability of a compound to rotate the plane of polarized light, and the property is called optical rotation. This property arises from an interaction of the electromagnetic radiation of polarized light with the asymmetric electric field generated by the electrons in chiral molecules.

A polarimeter is a device for determining the polarization direction of light. Its structure and working principle are schematically illustrated in Fig. 2.8. It is made up of two Nicol prisms, named polarizer and analyzer respectively. The polarizer is fixed, and the analyzer can be rotated. The polarizer allows only those light waves which move in a single plane, creating a polarized beam. This beam is rotated as it passes through the sample. Then, the second polarizer, i.e., the analyzer, rotates such that all the light or no light can pass through. One can find the rotation angle, which is equal to the angle α (in a unit of °) in the former case or the angle $(90 - \alpha)$ in the latter case.

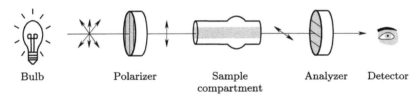

Fig. 2.8 The working principle of the polarimeter.

The optical rotation of a solution is associated with the optical activity of the optical active solute in the solution, solvent, concentration, the length of the sample compartment, and temperature. If all the other factors are fixed, the rotation α changes linearly with the solute concentration c (in a unit of g/10^2 mL), i.e.

$$\alpha = \beta c, \tag{2.11}$$

where β varies with the factors except the solute concentration.

The optical activity of a substance can be measured by the specific rotation expressed as

$$[\alpha]_D^{20} = \frac{100\alpha}{lc}, \tag{2.12}$$

where the superscript "20" indicates the experimental temperature is 20 °C, the subscript "D" represents the D line (589 nm) of the sodium lamp of the polarimeter, l is the length of the sample compartment in a unit of dm.

2.3 Determination of the rate constant by the polarimetric method

From Equation (2.9), it can be seen that the rate constant k can be determined from the slope of the $\ln c - t$ curve if the reactant concentrations at a series of reaction instants are measured. The sucrose is a dextrorotatory substance and its specific rotation $[\alpha]_D^{20}$ is 66.6°. The reaction product glucose is also a dextrorotatory substance whose $[\alpha]_D^{20}$ is 52.5°. The other product, fructose, is a levorotatory substance, and its specific rotation $[\alpha]_D^{20}$ is −91.9°. For $t = 0$, sucrose is not converted. The initial rotation of the system is

$$\alpha_0 = \beta_R c_0. \tag{2.13}$$

As the reaction proceeds, the dextrorotatory angle of the system decreases continuously. After the rotation reach zero, the system will further become levorotatory. For $t = \infty$, the sucrose is completely converted. The rotation reaches a maximum α_∞, which is expressed as

$$\alpha_\infty = \beta_P c_0. \tag{2.14}$$

β_R and β_P are the proportionality constants of reactant and product, respectively.

For $0 < t < \infty$, the relationship between the concentration of the sucrose c and the rotation α_t is expressed as

$$\alpha_t = \beta_R c + \beta_P (c_0 - c). \tag{2.15}$$

From Equations (2.13), (2.14), and (2.15), we can obtain

$$c_0 = \frac{\alpha_0 - \alpha_\infty}{\beta_R - \beta_P}, \tag{2.16}$$

and

$$c = \frac{\alpha_t - \alpha_\infty}{\beta_R - \beta_P}. \tag{2.17}$$

Substituting the Equations (2.16) and (2.17) into Equation (2.9), we can obtain

$$\ln(\alpha_t - \alpha_\infty) = -kt + \ln(\alpha_0 - \alpha_\infty). \tag{2.18}$$

Consequently, the rate constant k can be obtained from the slope of the $\ln(\alpha_t - \alpha_\infty) - t$ curve. It is noted that the specific rotation and the length of the sample tube are not included in the equation. This may facilitate the calculation of the rate constant at different temperatures.

The activation energy is the minimum amount of energy required to initiate a chemical reaction. The Arrhenius equation that relates the activation energy to the rate constant is

$$k = A e^{-\frac{E_a}{RT}}, \tag{2.19}$$

where E_a is the activation energy in a unit of kJ/mol, A is a pre-exponential factor, R is the universal gas constant, and T is the absolute temperature. In order to determine the activation energy, the equation is transformed to

$$\ln k = -\frac{E_a}{RT} + A. \tag{2.20}$$

Then, the activation energy can be obtained by linear fitting the $\ln k - 1/T$ curve if the rate constants at different temperatures have been determined.

3 Experimental

3.1 Preparation

(1) Chemicals: sucrose, 4 mol/L aqueous solution of HCl, deionized water.

(2) Apparatuses: polarimeter (Shanghai Optical Instrument Factory, WXG-4), thermostatic water bath, electronic balance.

(3) Tools: 50 mL stoppered test tube, 50 mL graduated cylinder, 50 mL Erlenmeyer flask.

3.2 Procedure

3.2.1 Zero-point calibration of the polarimeter

(1) Clean the sample compartment, which is a glass tube, with distilled water.

(2) Fill the tube with distilled water. Cover the ends of the tube with glass plates and screw on the caps. Do not overexert to avoid crushing the glass plates.

(3) Dry the outside of the tube with drinking paper and wipe the glass plates at both ends with lens paper. Place the tube in the light path of the polarimeter.

(4) Switch on the polarimeter and focus the ocular. Adjust the analyzer until the shade in the triple field of view is uniform. Record the rotation α of the analyzer.

(5) Repeat the measurement three times and average the data. This average value is the zero point, which is used to calibrate the systematic error.

3.2.2 Measurement of optical rotation during the reaction

(1) Weigh 5 g sucrose in a 50 mL Erlenmeyer flask. Add 25mL water to dissolve sucrose completely. Then add 25 mL HCl solution into the sucrose solution. At the same time, start recording the time. Shake the Erlenmeyer flask back and forth several times to mix the solution evenly.

(2) Rinse the sample compartment twice with a small amount of the solution. Then, fill the tube with the solution (about 10–15mL), and screw on the cap. The rest of the solution is filled in a 50 mL stoppered test tube and placed in a 55 °C water bath for further use.

(3) Place the tube in the polarimeter. Measure the optical rotation at intervals of 2 min. After 20 min, the time intervals may be lengthened to 5 min since the reaction slows down.Measure the rotation until the reaction time reaches 80 min.

3.2.3 Measurement of α_∞

(1) The reaction of the solution in the 55 °C water bath is fast. Take out the solution in the water bath after about 60 min when the reaction is completed.

(2) Cool the solution to room temperature.

(3) Measure the optical rotation of the solution three times and average the data, which is considered as α_∞ of the reaction.

4 Requirements for experimental report

(1) Tabulate the optical rotation α_t measured during the reaction at room temperature and the corresponding time t. Plot the $\alpha_t - t$ curve.

(2) Plot the $\ln(\alpha_t - \alpha_\infty) - t$ curve. Calculate the rate constant k by linear fitting the curve, and then calculate the half-life of the reaction $t_{1/2}$.

5 Questions and further thoughts

(1) Why is the initial rotation of the sucrose solution α_0 not determined in the experiment?

(2) Besides measuring the optical rotation using the polarimeter, please suggest other methods or techniques that can be used to determine the reactant concentration instantaneously during a reaction.

Experiment 10   Determination of Electromotive Force of Galvanic Cells

Type of the experiment: Cognitive
Recommended credit hours: 4
Brief introduction: The measurement of the electromotive force of galvanic cells is of great practical significance in the research of energy materials. In this experiment, the electromotive force of the Daniell cell, a Zn-Cu cell, and the electrode potentials of Zn and Cu will be determined using the potentiometer. Then, the electromotive of a Cu-Cu concentration cell is also measured. The principle and structure of galvanic cells will be understood by the experiment.

1 Objectives

(1) To understand the principle of galvanic cells.

(2) To determine the electromotive force of a Zn-Cu galvanic cell and the electrode potentials of Zn and Cu.

(3) To determine the electromotive force of a Cu-Cu concentration cell.

(4) To understand the principle and usage of the potentiometer.

2 Principles

2.1 Structure and reaction of galvanic cells

An electrochemical cell is a device that converts chemical energy into electrical energy. It is known as a galvanic cell when it acts as a source of electrical energy. It generally consists of two electrodes and an electrolyte. Oxidation and reduction occur at the anode and cathode, respectively. The overall cell reaction is the sum of these two reactions.

Take the Daniell cell, a Zn-Cu cell, as an example. Its representation is

$$\mathrm{Zn}|\mathrm{ZnSO_4}(m_{\mathrm{Zn}})||\mathrm{CuSO_4}(m_{\mathrm{Cu}})|\mathrm{Cu}, \tag{2.21}$$

where the single vertical line "|" represents an interface between the solid electrode (Zn or Cu) and liquid solution ($\mathrm{ZnSO_4}$ or $\mathrm{CuSO_4}$), the double line "||" represents a salt bridge between the two liquids, and m_{Zn} and m_{Cu} are the molalities of the solutions. During discharge, the reactions involve

$$\text{oxidation on anode}: \mathrm{Zn} \rightarrow \mathrm{Zn^{2+}} + 2e^-, \tag{2.22}$$

$$\text{reduction on cathode}: \mathrm{Cu^{2+}} + 2e^- \rightarrow \mathrm{Cu}, \tag{2.23}$$

$$\text{the cell reaction}: \mathrm{Zn} + \mathrm{Cu^{2+}} \rightarrow \mathrm{Zn^{2+}} + \mathrm{Cu}. \tag{2.24}$$

The cell reaction can also be expressed as

$$\mathrm{Zn} + \mathrm{CuSO_4} \rightarrow \mathrm{ZnSO_4} + \mathrm{Cu}. \tag{2.25}$$

2.2 Theoretical calculation of the electromotive force

Under constant pressure, constant temperature, and reversible conditions, the Gibbs free energy change, ΔG, of the cell reaction can be expressed as

$$\Delta G = -nFE, \tag{2.26}$$

where n is the number of electrons transferred in the reaction, F is the Faraday constant, E is the electromotive force (EMF). From this equation, it can be seen that the electrical energy of the galvanic cell is derived from the free energy change caused by the cell reaction.

For the Zn-Cu cell, the Gibbs free energy change of the cell reaction is

$$\Delta G = \Delta G^{\ominus} + RT \ln \frac{a(\mathrm{Zn^{2+}})a(\mathrm{Cu})}{a(\mathrm{Cu^{2+}})a(\mathrm{Zn})}, \tag{2.27}$$

where $\Delta G^{\ominus}$ is the free energy change under the standard conditions, a is the activity of each substance. It is known that the activity of pure solid is 1. In addition, under the standard conditions, the activity of each substance in solution is 1. That is,

$$a(\mathrm{Zn}) = a(\mathrm{Cu}) = a(\mathrm{Zn^{2+}}) = a(\mathrm{Cu^{2+}}) = 1. \tag{2.28}$$

Then

$$\Delta G = \Delta G^{\ominus} = -nFE^{\ominus}, \tag{2.29}$$

where $E^{\ominus}$ is the standard EMF of the cell. From Equations (2.26) to (2.29), the EMF of the cell can be expressed as

$$E = E^{\ominus} - \frac{RT}{nF} \ln \frac{a(\mathrm{Zn^{2+}})}{a(\mathrm{Cu^{2+}})}. \tag{2.30}$$

The EMF can also be expressed as the potential difference of the two electrodes, i.e.

$$E = \varphi_+ - \varphi_-, \tag{2.31}$$

where φ_+ and φ_- are the potentials of cathode and anode, respectively. For the Zn-Cu cell,

$$\varphi_- = \varphi^{\ominus}_{Zn^{2+},Zn} - \frac{RT}{2F} \ln \frac{1}{a(Zn^{2+})}, \quad (2.32)$$

$$\varphi_+ = \varphi^{\ominus}_{Cu^{2+},Cu} - \frac{RT}{2F} \ln \frac{1}{a(Cu^{2+})}, \quad (2.33)$$

where $\varphi^{\ominus}_{Zn^{2+},Zn}$ and $\varphi^{\ominus}_{Cu^{2+},Cu}$ are the standard electrode potentials of zinc and copper, respectively.

Although the ionic activities cannot be determined directly by experiment, they can be expressed as the product of mean ionic activity coefficient and molality, i.e.

$$a(Zn^{2+}) = \gamma_\pm(Zn^{2+}) \times m_{Zn}, \quad (2.34)$$
$$a(Cu^{2+}) = \gamma_\pm(Cu^{2+}) \times m_{Cu}, \quad (2.35)$$

where $\gamma_\pm$ is the mean ionic activity coefficient, which depends on the concentration, nature of ions, and temperature. The $\gamma_\pm$ values that will be used in this experiment are listed in Table 2.1.

Table 2.1 Parameters for calculating the activities and standard potentials of Zn and Cu electrodes

$\gamma_\pm$	$\varphi^{\ominus}_{298K}$, α and β
$\gamma_\pm(Zn^{2+}, 0.1 \text{ mol/L}) = 0.15$	$[Zn^{2+}, Zn]$: $\varphi^{\ominus}_{298K} = -0.7627$ V,
$\gamma_\pm(Cu^{2+}, 0.1 \text{ mol/L}) = 0.15$	
$\gamma_\pm(Zn^{2+}, 0.01 \text{ mol/L}) = 0.387$	$\alpha = 1 \times 10^{-4}$ V/K, $\beta = 6.2 \times 10^{-7}$ V/K
$\gamma_\pm(Cu^{2+}, 0.01 \text{ mol/L}) = 0.40$	(Cu^{2+}, Cu): $\varphi^{\ominus}_{298K} = 0.3419$ V,
	$\alpha = -1.6 \times 10^{-5}$ V/K, $\beta = 0$

It is noted that the value of electrode potential depends not only on the type of electrode and the solution concentration but also on the temperature. Generally, it is easy to find the standard potentials of various electrodes at 298 K from the literature or handbook. The standard potential at experimental temperature can be calculated by

$$\varphi^{\ominus}_T = \varphi^{\ominus}_{298K} + \alpha(T - 298) + \frac{1}{2}\beta(T - 298)^2, \quad (2.36)$$

where α and β are the temperature coefficients of the electrodes, T is the experimental temperature in a unit of K. These values for Zn and Cu electrodes are listed in Table 2.1. Then, the potentials of the electrodes and the EMF at the experimental condition can be calculated using Equations (2.31), (2.32), and (2.33).

2.3 Concentration cell

Because chemical reactions occur in the cells like the Daniell cell, these cells are also called chemical cells. In contrast, another type of cell without chemical reactions is the concentration cell.

A concentration cell is a limited form of a galvanic cell that comprises two half-cells with the same electrode but different concentrations. A concentration cell acts to dilute the more concentrated solution and concentrate the more dilute solution. A current is created by transferring the electrons from the half-cell with the lower concentration to the half-cell with the higher concentration until the cell reaches an equilibrium. A typical example of the concentration cell is

$$\text{Cu}|\text{CuSO}_4(0.01 \text{ mol/L})\|\text{CuSO}_4(0.1 \text{ mol/L})|\text{Cu}. \tag{2.37}$$

The standard EMF of a concentration cell is equal to zero because the electrodes are identical. Nevertheless, because the ion concentrations are different, there is a potential difference between the two half-cells.

2.4 Measurement of the electromotive force

In this experiment, the electrode potentials of Cu and Zn and the EMF of the Zn-Cu cell will be measured. The electrode potential cannot be measured singly. Instead, it can be obtained by measuring the EMF of a galvanic cell which comprises the test electrode and a reference electrode with a known potential. For instance, the standard hydrogen electrode (SHE) can be used as the reference electrode, and its potential at 298 K is defined as zero volts. Considering the inconvenience of using SHE in practice, other subsidiary reference electrodes are widely used, such as the saturated calomel electrode (SCE), which is a reference electrode based on the reaction between elemental mercury and mercury chloride.

The EMF of galvanic cells can be measured by the potentiometer with the compensation method. Figure 2.9 illustrates a compensation coupling by which the electromotive force of a voltage source can be measured. E_x is the studied voltage source, E_n is a standard element which is a voltage source with accurately known voltage, and R_e is a bias resistor.

When E_n and E_x are switched off, a current flow I appear in the main circuit. Switch E_x on to the circuit and find point C_x between A and B, for which there is no current flowing through the μA-meter. If R_x is the resistance of the interval AC_x, then

$$E_x = IR_x. \tag{2.38}$$

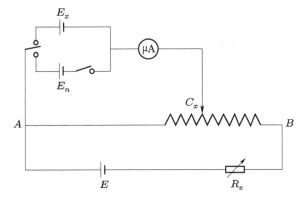

Fig. 2.9 The working principle of the potentiometer.

Correspondingly, by switching the standard element E_n on to the circuit, the point C_n is found for no current flow through the μA-meter. Then, we get

$$E_n = IR_n. \tag{2.39}$$

From Equations (2.38) and (2.39), the voltage to be measured is determined as

$$E_x = \frac{R_x}{R_n} E_n. \tag{2.40}$$

3 Experimental

3.1 Materials and apparatuses

(1) Chemicals: $CuSO_4 \cdot 5H_2O$, $ZnSO_4 \cdot 7H_2O$, 0.1 mol/L aqueous solution of HCl, KCl, deionized water.

(2) Apparatus: digital potentiometer (Sangli Electronic Factory, SDC-IIB).

(3) Tools: 100 mL volumetric flask, 100 mL beaker, 300 mL beaker, saturated calomel electrode (SCE), copper electrode, zinc electrode, two electrode tubes.

3.2 Procedure

3.2.1 Setting up the galvanic cells

(1) Prepare 0.1 mol/L $ZnSO_4$, 0.1 mol/L $CuSO_4$, and 0.01 mol/L $CuSO_4$ solution in 100 mL volumetric flask, respectively.

(2) Prepare saturated KCl solution by dissolving enough KCl (about 80 g) into 300 mL H_2O.

(3) Grind the zinc and copper electrodes with abrasive paper and dip them into 0.1 mol/L HCl solution for a few seconds. Rinse them with water.

(4) Assemble the electrode tubes with Zn and Cu electrodes as shown in Fig. 2.10.

(5) Add about 50 mL saturated KCl solution into a 100 mL beaker to act as the salt bridge. Immerse the tips of the side arms of the electrode tubes into the solution. The cell can be expressed as

$$Zn|ZnSO_4(0.100 \text{ mol/L})||CuSO_4(0.100 \text{ mol/L})|Cu. \tag{2.41}$$

(6) After measuring the EMF of this cell, assemble the following cells respectively in a similar way, which are

$$Zn|ZnSO_4(0.100 \text{ mol/L})||KCl(\text{saturated})|Hg_2Cl_2|Hg, \tag{2.42}$$

$$Hg|Hg_2Cl_2|KCl(\text{saturated})||CuSO_4(0.100 \text{ mol/L})|Cu, \tag{2.43}$$

$$Cu|CuSO_4(0.0100 \text{ mol/L})||CuSO_4(0.100 \text{ mol/L})|Cu. \tag{2.44}$$

3.2.5 Measurement of EMF

(1) Record the experiment temperature.

(2) Respectively measure the EMF of the above cells with the potentiometer. The main operation steps of the potentiometer are as follows:

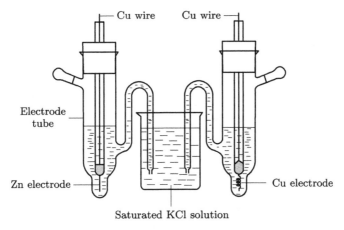

Fig. 2.10 Schematic illustration of the Zn-Cu cell.

(a) Select "Internal calibration". Press "Zero" to set the current to zero.
(b) Select "Measurement". Connect the cell to the potentiometer.
(c) Adjust the compensation buttons until the current becomes zero again. Read the potential, which is the EMF of the cell.

4 Requirements for experimental report

(1) Calculate the electrode potential of the SCE at the experimental temperature by
$$\varphi_{\text{SCE}} = 0.2415 - 7.61 \times 10^{-4} \times (T - 298). \tag{2.45}$$

(2) Calculate the electrode potentials, φ_T, $\varphi_T^\ominus$ and $\varphi_{298\text{K}}^\ominus$, of zinc and copper, respectively.

(3) The EMF of the Zn-Cu cell can be obtained in three ways: to be calculated theoretically, to be calculated from the potential difference of the electrodes which has been measured in the experiment, and to be measured directly. Compare these EMF values and discuss the sources of error.

5 Questions and further thoughts

(1) What factors should be considered in choosing a salt bridge?
(2) What factors may affect the EMF of the Zn-Cu galvanic cell?

Experiment 11 Determination of Surface Tension of Ethanol-Water Solutions

Type of the experiment: Cognitive
Recommended credit hour: 4
Brief introduction: Surfactants are widely used as decontamination agents, wetting agents, emulsifiers, and foaming agents in industry and daily life. Since their main role occurs at the surface, it is of great practical significance to study their

surface effects. In this experiment, the surface tension of a series of ethanol-water solutions is measured by the maximum bubble pressure method. Then, the surface concentration and the cross-sectional area of ethanol molecules are calculated.

■ Objectives

(1) To understand the concepts of surface tension, surface adsorption, and the cross-sectional area of solute molecules.

(2) To understand the principle of the maximum bubble pressure method for determining the surface tension.

(3) To determine the surface tension of ethanol-water solutions.

(4) To determine the surface concentration and the cross-sectional area of ethanol molecules.

■ Principles

2.1 Surface free energy and surface tension

The energy state of the surface molecules of a solution is different from the internal molecules. This is because the molecules inside the solution are attracted in all directions, and the vector sum of the forces is zero. However, the molecules near the surface are dominantly attracted by the internal molecules, and, hence, the net force is directed towards the interior liquid. In other words, the surface molecules are subjected to the internal pull force.

In order to increase the surface area of a liquid, it is necessary to overcome the internal force. Under a certain temperature and pressure, the increase of work, ΔW, on the surface of the liquid is proportional to the increase in the surface area, ΔA, i.e.

$$\Delta W = \sigma \cdot \Delta A, \tag{2.46}$$

where the coefficient σ is the surface free energy or surface tension in a unit of J/m or N/m. It reflects the tendency for the liquid to reduce its surface area spontaneously.

2.2 Measurement of surface tension

2.2.1 Excess pressure on the curved surface

The curved surface bears not only the external pressure p but also the excess pressure Δp caused by the surface tension of a liquid. The excess pressure depends on the curvature radius and surface tension of the liquid surface. It can be derived by a simple model shown in Fig. 2.11.

In this figure, a capillary tube is filled with a liquid. A spherical droplet of radius r is suspended at the end. Under the equilibrium conditions, the pressure inside the droplet is

$$p' = p + \Delta p. \tag{2.47}$$

When an infinitesimal force is applied on the piston at the top of the capillary tube, the droplet volume increases $\mathrm{d}V$, and its surface area increases $\mathrm{d}A$. In this process, the work done on the droplet by pushing the piston is

$$p'\mathrm{d}V = (p + \Delta p)\mathrm{d}V. \tag{2.48}$$

At the same time, the work done by the droplet expansion on the environment is $-p\mathrm{d}V$. Thus, the net volume work done by the piston on the droplet is

$$(p + \Delta p)\mathrm{d}V - p\mathrm{d}V = \Delta p \mathrm{d}V. \tag{2.49}$$

This work is used to overcome the energy increase with the increase in the surface area $\mathrm{d}A$. Thus, from Equation (2.49), one gets

$$\Delta p = \frac{\sigma \cdot \mathrm{d}A}{\mathrm{d}V}. \tag{2.50}$$

The area of the droplet $A = 4\pi r^2$, then $\mathrm{d}A = 8\pi r \mathrm{d}r$. The volume $V = \frac{4}{3}\pi r^3$, then $\mathrm{d}V = 4\pi r^2 \mathrm{d}r$. Thus, the excess pressure can be expressed as

$$\Delta p = \frac{\sigma 8\pi r \mathrm{d}r}{4\pi r^2 \mathrm{d}r}. \tag{2.51}$$

This equation is the simplified form of the Young-Laplace equation, which describes the equilibrium pressure difference sustained across the interface between two static fluids.

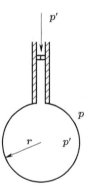

Fig. 2.11 Relationship between the excess pressure and the radius of a liquid droplet.

2.2.2 Maximum bubble-pressure method

There are many experimental techniques for measuring the surface tension of liquids. The maximum bubble pressure method is adopted in this experiment. Its principle is based on the expression of the excess pressure.

Figure 2.12 is a schematic illustration of the experiment apparatus. The bottom of a capillary tube is just in contact with the surface of the test liquid. When the gas-pumping bottle is slowly opened to discharge water from the bottle, the pressure in the test tube is reduced accordingly. The atmospheric pressure in the capillary expels the liquid from the capillary, leading to the formation of a bubble. When the bubble radius is the same as the capillary radius, it endures the maximum pressure difference between the atmospheric pressure and the pressure in the test tube. Obviously, this pressure difference can be expressed using Equation (2.50). When the pumping continues, the bubble will burst up as the pressure difference further increases. The maximum pressure difference can be read from the digital micro-pressure detector.

When the same capillary is used to test two different solutions with surface tensions of σ_1 and σ_2, from equation (2.50) we can get

$$\frac{\sigma_1}{\sigma_2} = \frac{\Delta p_1}{\Delta p_2}, \qquad (2.52)$$

and

$$\sigma_1 = \sigma_2 \frac{\Delta p_1}{\Delta p_2} = K_c \Delta p_1, \qquad (2.53)$$

where Δp_1 and Δp_2 are their maximum bubble pressures. K_c can be considered as the capillary constant. Hence, the surface tension σ_1 can be simply calculated after K_c is determined from the other liquid with known surface tension σ_2.

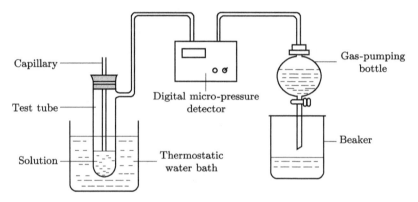

Fig. 2.12 Schematic illustration of the surface tension measurement apparatus.

2.3 Surface adsorption of solution

Since the solute can change the surface tension of the solvent, surface free energy can be altered by adjusting the solute concentration in the surface layer. The phenomenon that the solute concentration in the surface layer is different from that in bulk is called the surface adsorption of the solution. When the concentration in the surface layer is less than the bulk concentration, it is called negative adsorption. When the concentration in the surface layer is larger than the bulk concentration, it is called positive adsorption. For the aqueous solution, the inorganic solute usually can cause negative adsorption, and the soluble organic compound can cause positive adsorption. In this experiment, only the situation of positive adsorption will be considered.

The solute that decreases the surface tension of the solvent is called the surface-active substance. It possesses a typical asymmetric structure with both hydrophilic polar and hydrophobic nonpolar groups. Ethanol belongs to this category. The arrangement of the solute on the surface of the solution varies with their concentration, as shown in Fig. 2.13. The surface molecules and dangling bonds may lie down when the concentration of the solute is low. When the concentration is increased, the solute molecules will occupy the entire surface leading to the formation of a saturated layer.

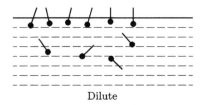

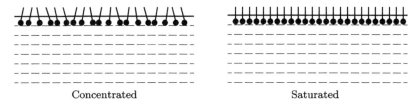

Fig. 2.13 The arrangements of solute molecules on the solution surface.

The relationship between the variation rate of the surface tension with concentration and the amount of surface adsorption can be expressed by the Gibbs adsorption equation

$$\Gamma = -\frac{c}{RT}\left(\frac{d\sigma}{dc}\right)_T, \quad (2.54)$$

where Γ is surface adsorption capacity (mol/m), c is the concentration of the dilute solution (mol/L), T is the thermodynamic temperature (K), R is the gas constant. From the $\sigma - c$ curve shown in Fig. 2.14, one can see that σ decreases rapidly with an increase in the solute concentration at the beginning, then it changes very slowly. From the slope $\left(\frac{d\sigma}{dc_i}\right)_T$ corresponding to concentration c_i at point i, the Γ value can be calculated at the concentration using the Equation (2.54) if the surface tension of the solution has been determined.

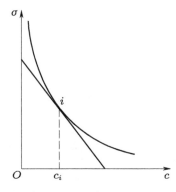

Fig. 2.14 Relationship between surface tension and bulk concentration.

2.4 Cross-sectional area of solute molecules

When saturated adsorption occurs, the solution surface is almost completely occupied by solute molecules. The saturated adsorption capacity Γ_∞ can be regarded as the number of solute molecules in a unit area, and the cross-sectional area S of the solute molecules can be calculated by the equation

$$S = \frac{1}{\Gamma_\infty N_A}, \qquad (2.55)$$

where N_A is the Avogadro constant.

The relationship between the surface adsorption capacity Γ and the solute concentration c can be expressed by the Langmuir adsorption equation

$$\Gamma = \Gamma_\infty \frac{K_A C}{1 + K_A C}, \qquad (2.56)$$

where K_A is a constant. Take the reciprocal of the equation, we have

$$\frac{c}{\Gamma} = \frac{c}{\Gamma_\infty} + \frac{1}{K_A \Gamma_\infty}. \qquad (2.57)$$

Then the Γ_∞ can be obtained from the slop of the $c/\Gamma - c$ curve, and the cross-sectional area S of the solute molecules can be calculated.

3 Experimental

3.1 Preparation

(1) Chemical: absolute ethanol.

(2) Apparatuses: surface tension measurement apparatus, Abbé refractometer, thermostatic water bath, electronic balance.

3.2 Procedure

(1) Prepare 50 mL of ethanol-water solutions with a series of concentrations from 5% to 40% at intervals of 5%.

(2) Measure the refractive indices of the solutions by Abbé refractometer. Plot a standard curve of the refractive index against the ethanol concentration and fit the curve into a linear equation.

(3) Adjust the temperature of the apparatus for surface tension measurement to 25 °C.

(4) Add pure water into the test tube until the liquid level is just in contact with the end of the capillary tube.

(5) Fill water in the gas-pumping bottle. Open the stopcock of the gas-pumping bottle slowly. Ensure that the capillary tube is vertical and the formation of the bubbles is stable, and then read the maximum pressure difference. A suitable rate of bubble formation is about 1or 2 bubbles per second.The readout of the digital micro-pressure detector, i.e., the instant maximum pressure difference, should be in the range of 700–800 Pa.

(6) Read the instant maximum pressure difference three times and calculate the average value.

(7) Repeat steps (5)and(6) to measure the surface tension of the ethanol-water solutions from low to high concentrations. During replace the solution, the test tube is not required to be cleaned. Instead, the concentrations of the solutions after the test are determined from their refractive indices measured by the Abbé refractometer.

4 Requirements for experimental report

(1) Calculate the K_c for water by Equation (2.53). The surface tension of water at 25 °C is 71.97×10^{-3} N/m.

(2) Tabulate the readouts and averages of the maximum pressure difference of solutions with different concentrations. Calculate the σ values and list them in the table.

(3) Plot the $\sigma - c$ curve. Fit the curve with a second-order equation.

(4) Select 10 points from the $\sigma - c$ curve. Calculate the corresponding Γ values at each point.

(5) Plot the $c/\Gamma - c$ curve. Determine the values of Γ_∞ and S.

5 Questions and Further Thoughts

(1) How does it affect surface tension if the pumping in step (5) is too fast?

(2) Can we insert the capillary tube into the solution for the measurement?

Experiment 12 Determination of Relative Molecular Weight of Polymer by Viscosity Method

Type of the experiment: Cognitive
Recommended credit hours: 4
Brief introduction: It is essential to determine the average molecular weight of polymer materials since various polymer characteristics are affected by the magnitude of the molecular weight. In this experiment, the relative molecular weight of polyvinyl pyrrolidone (PVP), a water-soluble polymer, is determined by the viscosity method. The principle and method to determine the viscosity with Ubbelohde viscometer are introduced.

1 Objectives

(1) To understand the principle of the viscosity method for determining the average molecule weight of a polymer.

(2) To learn the usage of the Ubbelohde viscometer.

(3) To determine the average molecular weight of PVP.

2 Principles

2.1 Average molecular weight of polymer

For most polymers, the molecules are in the form of long and flexible chains. Polymer molecular weight is defined as a distribution rather than a specific number because polymerization occurs in such a way as to produce different chain lengths.

There are several ways to define the average molecular weight depending on the applied statistical method.

The number-average molecular weight, M_n, is a simple arithmetic mean representing the total weight of molecules divided by the total number of molecules. A weight-average molecular weight, M_w, is based on the weight fraction of molecules within various size ranges.

Besides, the viscosity-average molecular weight, M_v, is determined by measuring the intrinsic viscosity of a polymer solution. It is the most commonly used representation of the molecular weight of polymers because the measurement uses simple equipment, and a wide range of molecular weights can be determined with high accuracy.

2.2 Definition of viscosity

Viscosity is an internal property of a fluid that offers resistance to flow. It resists the relative movement of neighboring molecules in the fluid. Hence, it can be regarded as a kind of internal friction.

Figure 2.15 shows the schematic diagram of liquid flow. When the two liquid layers with the distance of ds move at velocities, v and $v + dv$, respectively, the velocity gradient is dv/ds. When the steady flow is established, the force f' required to maintain a certain flow rate is equal to the liquid flow resistance. It is proportional to the contact area A and the velocity gradient, that is,

$$f' = \eta A \frac{dv}{ds}, \tag{2.58}$$

where the constant η is the viscosity coefficient in a unit of Pa·s. Thus, the viscous resistance per unit area is

$$f = \eta \frac{dv}{ds}. \tag{2.59}$$

This formula is called Newton's law of viscosity.

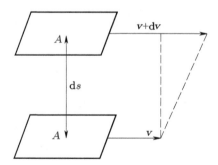

Fig. 2.15 Schematic illustration of liquid flow.

2.3 Viscosity of polymer solution

The viscosity of a dilute polymer solution mainly reflects the internal friction in the flow of the solution. The internal friction between pure solvent molecules produces the pure solvent viscosity, which is denoted as η_0. In addition, there are internal frictions between polymer molecules, and between polymer molecules and solvent molecules. The sum of these three kinds of frictions is the viscosity of the solution, which is denoted as η. Generally speaking, at a given temperature, $\eta > \eta_0$. The fraction of the increase in the viscosity due to the addition of solute is defined as the specific viscosity, η_{sp}, and calculated by

$$\eta_{sp} = \frac{\eta - \eta_0}{\eta_0}. \tag{2.60}$$

The ratio of the solution viscosity to the pure solvent viscosity is defined as the relative viscosity, η_r, and calculated by

$$\eta_r = \frac{\eta}{\eta_0}. \tag{2.61}$$

Both η_r and η_{sp} are dimensionless. η_r represents the viscous behavior of the entire solution, while η_{sp} is the net viscosity without the internal friction between the solvent molecules. Their relationship is expressed as

$$\eta_{sp} = \frac{\eta}{\eta_0} - 1 = \eta_r - 1. \tag{2.62}$$

As for a polymer solution, η_{sp} increases with the increase in solution concentration c. The reduced viscosity, η_{sp}/c, and the inherent viscosity, $\ln \eta_r/c$, are defined for the convenience of comparison.

In order to further eliminate the internal friction between the polymer molecules, the polymer solution must be infinitely diluted so that polymer molecules are far apart from each other and the mutual interference can be ignored. Then the viscosity behavior of the solution reflects the internal friction between the polymer molecules and the solvent molecules. The limit value of the viscosity is

$$[\eta] = \lim_{c \to 0} \frac{\eta_{sp}}{c}, \quad \text{or} \quad [\eta] = \lim_{c \to 0} \frac{\ln \eta_r}{c}. \tag{2.63}$$

Here, $[\eta]$ is the intrinsic viscosity that is independent of concentration.

It has been proved that $[\eta]$ is only related to the average molecule weight M_v when the polymer, the solvent, and the temperature are given. The semi-empirical relationship between $[\eta]$ and M_v is expressed by the Mark Houwink equation,

$$[\eta] = K M_v^\alpha, \tag{2.64}$$

where K is a constant, α is an empirical constant related to the molecular shape. They both depend on temperature, the nature of the polymer and solvent but are independent of the molecular weight in a certain range. Their values can be determined using other methods, such as the osmotic pressure method, the light scattering method. Then M_v can be calculated by Equation (2.64) after determining the $[\eta]$ by the viscosity method.

2.4 Measurement of viscosity

The viscosity of a liquid can generally be measured by three methods: to measure the time required for the liquid to flow through a capillary tube by a capillary viscometer; to measure the falling speed of a ball in the liquid with a ball viscometer; to measure the relative rotation between the liquid and a coaxial cylinder with a rotation viscometer. Of these methods, the most convenient way to determine the $[\eta]$ value of a polymer is to use a capillary viscometer.

Figure 2.16 shows the schematic illustration of a Ubbelohde viscometer. The liquid that flows through the capillary tube B of the viscometer under gravity obeys the Poiseuille law expressed as

$$\frac{\eta}{\rho} = \frac{\pi h g r^4 t}{8lV} - m\frac{V}{8\pi l t}, \qquad (2.65)$$

where ρ is the liquid density, l is the capillary length, r is the capillary radius, t is the flow time, h is the average height of the liquid column through the capillary, g is the acceleration of gravity, V is the volume of the liquid flowing through the capillary, m is the constant related to the geometry of the apparatus. For a capillary, m is equal to 1 when $r/l \ll 1$.

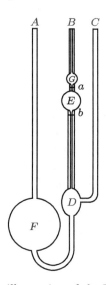

Fig. 2.16 Schematic illustration of the Ubbelohde viscometer.

For a specific viscometer, Equation (2.65) can be written as

$$\frac{\eta}{\rho} = pt - \frac{q}{t}, \qquad (2.66)$$

where $p = \frac{\pi h g r^4}{8lV}$, $q = m\frac{V}{8\pi l}$, $q < 1$ due to the geometry of the capillary. When $t > 100$ s, the term q/t in the equation can be ignored. Assuming that the solution density ρ approximately equals the solvent density ρ_0, η_r can be calculated by

$$\eta_r = \frac{\eta}{\eta_0} = \frac{t}{t_0}. \tag{2.67}$$

Thus, after measuring the flow time, t and t_0, of the solution and solvent, respectively, one gets η_r by Equation (2.67). Then, η_{sp}/c and $\ln\eta_r/c$ can be calculated. If the flow times of a series of solutions with different concentrations are measured, the curves of η_{sp}/c and $\ln\eta_r/c$ against c can be plotted as shown in Fig. 2.17. $[\eta]$ will be finally obtained by extrapolating the two straight lines to zero concentration.

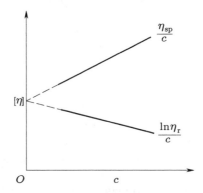

Fig. 2.17 Determination of $[\eta]$ by the extrapolating method.

3 Experimental

3.1 Preparation

(1) Chemicals: PVP, deionized water.

(2) Apparatuses: Ubbeluhde viscometer, thermostatic water bath, drying oven, electronic balance.

(3) Tools: 50mL volumetric flask, 5 mL and 10 mL pipettes, rubber suction bulb, rubber pipe, spring clip, stopwatch.

3.2 Procedures

(1) Accurately prepare 50 mL PVP water solution with a concentration of 0.02 g/mL in the volumetric flask.

(2) Adjust the temperature of the thermostatic water bath to 30 °C.

(3) Connect the upper end of tube C of the viscometer with a rubber pipe. Then place the viscometer vertically into the thermostatic water bath until bulb G is fully immersed.

(4) Pipet 10 mL water into tube A. Then wait 10 min to stabilize the viscometer and the solution to the preset temperature.

(5) Clamp the rubber pipe on tube C with a spring clip. Pump the solution at the end of the rubber tube on tube B with a rubber suction bulb to move the solution from bulb F to the middle of bulb G.

(6) Remove the rubber suction bulb and release the spring clip of tube C to let tubes B and C open to the air. The liquid in the capillary tube B flows through the capillary under gravity. Record the time precisely for the liquid flowing from line a to line b.

(7) Repeat steps (5) and (6) three times. The difference between the measured flow times should be less than 0.3 s. Average the flow times to obtain the flow time of water t_0.

(8) Remove the viscometer from the water bath and remove the rubber pipe. Use a little ethanol to rinse the viscometer and then dry it completely in the drying oven at 100 °C for 10 min.

(9) Repeat steps (3) to (7) to obtain the flow time of the 0.02 g/mL PVP solution.

(10) Pipet 2 mL, 3 mL, 4 mL, and 5 mL water in turn into tube A to dilute the PVP solution. Pump each solution from bulb F to bulb G several times to homogenize the solution. After the solution temperature is stabilized, repeat steps (5) to (7) to determine the flow times of the solutions with different PVP concentrations.

4 Requirements for experimental report

(1) Calculate and tabulate the values of η_{sp}, η_r, η_{sp}/c, $\ln \eta_r /c$ and the flow times for the solutions with different concentrations.

(2) Plot the curves of η_{sp}/c and $\ln \eta_r /c$ against c. Extrapolate the curves to zero concentration to get the value of $[\eta]$.

(3) Calculate the viscosity average molecular weight, M_v, using Equation (2.64). The parameters K and α for PVP water solution at 30 °C are 3.39×10^{-2} cm^3/g and 0.59, respectively.

5 Questions and further thoughts

(1) What is the function of tube C of the Ubbeluhde viscometer? Can we measure the viscosity without tube C?

(2) How does it affect the measurement result if the capillary radius is too large or too small?

6 Cautions

(1) The viscometer must be vertically immersed into the water bath. Avoid vibration during the measurement.

(2) Pay attention to avoid the adhesion of the solution on the inner wall of tube A when the solution is pipetted into the viscometer.

(3) The measurement should be conducted after the solution temperature is stabilized to the preset temperature.

Experiment 13   Determination of Crosslink Density of Rubbers by Swelling Method

Type of the experiment: Cognitive

Experiment 13 Determination of Crosslink Density of Rubbers by Swelling Method

Recommended credit hours: 4

Brief introduction: Solvent swelling is considered as one of the most reasonable phenomena reflecting the characteristics of the crosslink structure in polymers. In this experiment, the swelling test is carried out to measure the crosslink densities of several commonly used rubbers in the laboratory or daily life, including hair rubber band, sealing O-ring, and glue head of dropper.

The experiment is time-consuming, as it takes several days to reach an equilibrium swelling state. It is suggested to use the class time to understand the experimental principles and conduct the main experimental steps. Students can use their spare time to weigh the swollen samples.

1 Objectives

(1) To understand the principle of solvent swelling.

(2) To learn the method for determining the crosslink density by swelling measurement.

(3) To determine the crosslink densities of several commonly used rubbers.

2 Principles

2.1 Rubber

Rubber is an important type of polymer material, which is characterized by a long-range elasticity. There are many kinds of rubbers, but they all fall into two broad types: natural rubbers grown from plants and synthetic rubbers made artificially in chemical plants or laboratories.

Natural rubbers are polymers of isoprene with the chemical formula $(C_5H_8)_n$. They are known as elastomers because their polymer chains can be pulled apart and untangled easily, but they spring straight back together if they are released. Generally, natural rubbers are very squashy, pretty smelly, and not very useful.

Synthetic rubbers are synthesized using petrochemicals. One of the best-known examples is neoprene, made by reacting together with acetylene and hydrochloric acid. Another example of synthetic rubber is emulsion styrene-butadiene rubber (E-SBR), which is widely used for making vehicle tires.

2.2 Vulcanized rubber

Vulcanization is a process that increases the elasticity but decreases the plasticity of rubber. Most useful rubber products such as tires and elastic bands cannot be made without vulcanization because unvulcanized rubber is generally not very strong and can be very sticky.

Vulcanization chemically produces network junctures by inserting crosslinks between polymer chains, as shown in Fig. 2.18. A crosslink may be a group of sulfur atoms in a short chain, a single sulfur atom, a carbon to carbon bond, a polyvalent organic radical, an ionic cluster, or a polyvalent metal ion. The process is usually carried out by heating the rubber mixed with vulcanizing agents in a mold under pressure.

The network structure is defined by several parameters, i.e., the number of crosslinks, their functionality and distribution, network defects (dangling chains and loops), and entanglements. The properties of rubbers strongly depend on these features of the network structure.

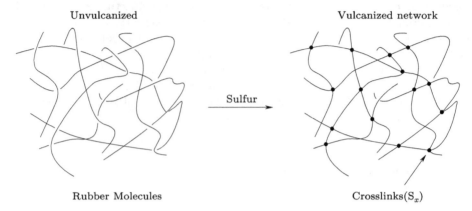

Fig. 2.18 Network formation of the vulcanized rubber.

2.3 Determination of crosslink density

Crosslink density can be determined by modulus or swelling measurement. Determining crosslink density from the stress-strain data of the compression test requires the test sample to be fabricated with precise geometries. However, many polymers are synthesized by suspension polymerization as small, irregular particles or beads. Thus, in many cases, the compression test method for determining the crosslink density may not be feasible.

In contrast, the equilibrium swelling experiment can be used for rubber samples with any geometry. A disadvantage of this method is that there is no universally accepted standard. The general practice is to immerse a sample in a solvent and record a range of parameters, such as the change in weight, volume, thickness, or hardness. Sometimes users specify their own test requirements under given conditions. This experiment will use this method to determine the crosslink densities of several commonly used rubbers.

The solvent swelling of vulcanized rubber is illustrated in Fig. 2.19. In a dry amorphous polymer crosslinked into a network, the chains are in an entangled and relaxed conformation between two network junctions. When a solvent enters the polymer, it does not dissolve it but instead swells it because the solvent molecules move the network junctions away from each other.

The driving force for swelling is the increase in entropy of dilution. During the swelling process, the chains expand.Meanwhile, the chains also want to pull back to their original positions. When these two forces become equal, the polymer refuses to accept more solvent, and a state of equilibrium is reached. Since the amount of swelling is controlled by the crosslink density and the magnitude of the interactions between the solvent and the network, measurements of the amount of swelling and the interactions between the solvent and the molecules can be used to calculate the crosslink density.

The equilibrium swelling theory states that a polymer will absorb the solvent

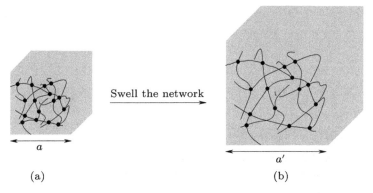

Fig. 2.19 Schematic illustrations of (a) crosslinked rubber before swelling and (b) swollen rubber after swelling.

until the solvent chemical potential in the polymer equal to that in the free solution. Assuming all contributions to the swelling pressure are independent of one another, this statement can be written in terms of osmotic swelling pressure, which is zero at equilibrium, i.e.

$$\Pi_{\text{mix}} + \Pi_{\text{elastic}} + \Pi_{\text{ion}} + \Pi_{\text{elec}} = 0, \qquad (2.68)$$

where the Π_{mix} is the swelling pressure because of the tendency of the polymer to dissolve into the solvent, Π_{elastic} is the elastic response of the network due to crosslinking, Π_{ion} is the contribution to osmotic pressure due to differences in ion concentration, and Π_{elec} is the effect of the electrostatic interactions on the polymer chains.

Several models have been established based on this equation. A commonly accepted calculation is described here. The concentration of crosslink agent in the polymer network is directly related to the molecular weight between crosslinks, M_c. The distance between two crosslink junctions can be translated from M_c, and M_c can be estimated through the swelling method.

A widely accepted expression of the crosslink density V_T is calculated by

$$V_T = \frac{1}{2M_c}. \qquad (2.69)$$

M_c can be determined according to the well-known Flory-Rehner equation and its modifications. For rubbers, it is calculated by

$$M_c = \frac{\rho_{\text{rubber}} \cdot V_{\text{solvent}} (c^{1/3} - 0.5c)}{\ln(1-c) + c + X_{\text{ps}} c^2}, \qquad (2.70)$$

where ρ_{rubber} is the density of rubber, V_{solvent} is the molar volume of solvent, X_{ps} is the polymer-solvent interaction parameter. c is the volume fraction of rubber in equilibrium swollen rubber, calculated by

$$c = \frac{w_0}{V_s \cdot \rho_{\text{rubber}}}, \qquad (2.71)$$

where w_0 is the weight of unswollen rubber. V_s is the volume of swollen rubber, which can be calculated by

$$V_s \approx \frac{w_0}{\rho_{\text{rubber}}} + \frac{(w_s - w_0)}{\rho_{\text{solvent}}}, \qquad (2.72)$$

where ρ_{solvent} is the density of solvent, and w_s is the weight of swollen rubber.

3 Experimental

3.1 Preparation

(1) Materials: three pieces of vulcanized rubber with different colors and elasticity. For instance, they can be hair rubber band (white), sealing O-ring (black), and glue head of dropper (red).

(2) Chemical: toluene.

(3) Apparatus: electronic balance.

(4) Tools: graduated cylinder, 50 mL sealed bottles.

3.2 Procedure

(1) Observe the appearance of the rubbers before swelling and take photos.

(2) Measure the weights and volumes of the samples. The volumes are measured by the drainage method.

(3) Put the three samples in the sealed bottles, respectively. Then add about 50 mL toluene to each bottle. Meanwhile, start timing.

(4) Take out the swollen rubbers every 10 min. Measure the weights after dry the specimens with drinking paper. Then put them back into the bottles. The weight should be measured at least three times, and the difference between them should be less than 0.01 g.

(5) After 1 h, increase the measurement intervals according to the rate at which the weight increases. The intervals can be 0.5 h, 1 h, 2 h, or even 10 h.

(6) Stop weight measurement when the weight does not increase, i.e., the equilibrium state is reached. It takes about 2 h for the black rubber, about 30 days for the white rubber, and 2 days for the red rubber.

(7) Observe the appearance of the rubbers after swelling and take photos.

4 Requirements for experimental report

(1) Tabulate the weights recorded during the experiment.

(2) Plot the curves of percentage increase in weight versus time.

(3) Calculate the crosslink densities of the specimens. Several known parameters X_{PS}, ρ_{toluene}, and V_{toluene} for the calculation are 0.44, 862.27 kg/m^3, and 106.69 mL/mol, respectively.

(4) Give the photos of the rubbers before and after swelling. Compare their morphologies.

5 Questions and further thoughts

(1) Please list several methods for determining the crosslink density of polymers. Compare their advantages and disadvantages.

(2) What are the effects of toluene volatility on the experiment?

6 Cautions

(1) The experiment must be carried out in a fume hood because toluene is a toxic and volatile chemical.

(2) The bottles must remain sealed when the rubbers are immersed in toluene.

Extended readings

Atkins P, de Paula J, Keeler J. Physical Chemistry. 11th Edition. Oxford University Press, 2018.

Bell C L, Peppas N A. Equilibrium and dynamic swelling of polyacrylates. Polymer Engineer and Science, 2010, 36(14): 1856-1861.

Carland C W, Nibler J W, Shoemaker D P. Experiment in Physical Chemistry. 8th Edition. McGraw-Hill Book Company, 2009.

Gao Z. Experimental Physical Chemistry. Higher Education Press, 2005.

Valentin N J L, Carretero-Gonzalez J, Mora-Barrantes I, et al. Uncertainties in the determination of cross-link density by equilibrium swelling experiments in natural rubber. Macromolecules, 2008, 41(13): 4717-4729.

Chapter 3 Fundamentals of Organic Chemistry

Experiment 14 Isolation of Alcohol from Red Wine by Simple Distillation

Type of the experiment: Cognitive
Recommended credit hours: 4
Brief introduction: Red wine is an alcoholic beverage resulting from the fermentation of grape juice by yeasts with appropriate processing and addition. The content of alcohol in the wine is generally defined as the percentage of alcohol by volume (ABV), which is somewhat higher than 10%. This experiment introduces the principle and operation of simple distillation techniques to purify and identify organic liquids. Then, alcohol will be isolated from red wine by this technique, and the ABV will be determined by an alcoholmeter.

1 Objectives

(1) To understand the principle of simple distillation.
(2) To learn the operation of simple distillation.
(3) To determine the ABV of red wine by distillation and subsequent measurement of specific density of the distillate.

2 Principles

2.1 Distillation

Distillation is a process that involves evaporating a liquid into a gas phase, condensing the gas back into the liquid, and then collecting the liquid in a clean receiver. It is a relatively inexpensive and simple method to separate two or more liquid compounds, purify liquids, and identify liquids. In the food industry, it is used to concentrate the alcohol in wines from the natural fermentation of grapes. In the chemical industry, distillation is used to separate the economically important components of fossil fuels, including natural gas, gasoline, kerosene, heating oil, and lubricants.

Distillation is a physical separation process based on the difference in boiling points of liquids. Since each component in a solution boils at a specific temperature, it is theoretically possible to separate and discretely condense the vapor of each component. The normal boiling point of a liquid occurs when the vapor pressure of the liquid equals the ambient atmospheric pressure. An idealized distillation system is governed by Raoult's Law, which assumes that a component contributes to the total vapor pressure of the mixture in proportion to its percentage in the mixture and its vapor pressure when it is pure. Besides, Dalton's Law assumes that the total pressure exerted by a gaseous mixture is equal to the sum of the partial pressures of each component in the gas mixture.

Interactions between the components of the mixture may produce unique boiling behavior. Such interactions result in a constant-boiling azeotrope that behaves like a pure compound that boils at a single temperature instead of a range. As an azeotrope, the mixture contains the given components in the same proportion as the vapor. Therefore, evaporation does not change the purity, and distillation does not affect separation. In the case of the mixture of ethanol and water in the present experiment, the phase diagram is shown in Fig. 3.1. It can be seen that ethanol and water will form an azeotrope at a concentration of 95.6 wt% ethanol, and the azeotrope boils at 78.2 °C. This temperature is lower than the boiling point of ethanol or water, which are respectively 78.5 °C and 100 °C at one atmosphere. This implies that it is impossible to distill pure ethanol from this mixture atone atmosphere.

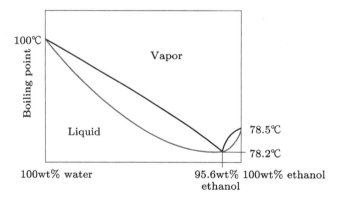

Fig. 3.1 The ethanol-water phase diagram at one atmosphere.

2.2 Distillation techniques

2.2.1 Simple distillation

Simple distillation is referred to as involving one theoretical plate, i.e, one theoretical vaporization-condensation cycle. It is used to purify liquids by removing non-volatile impurities or separating mixtures of immiscible liquids differing in boiling point by at least 25 °C, which must also present a boiling point lower than 150 °C at atmospheric pressure.

The most common setup for simple distillation is illustrated in Fig. 3.2. It has a heat source, a round-bottom flask, a distillation adapter, a thermometer, a thermometer adapter, a water-jacketed condenser, a receiving pipe, and a collector flask. The round-bottom flask should not be more than two-thirds full. The distillation adapter is a three-way adapter that allows the thermometer to be placed and the vapor to be diverted to the water-jacketed condenser. The thermometer should be placed with its bulb in the vapor stream of the distillation. The water-jacketed condenser is the simplest design of water-cooled condensers. The cooling circuit must be connected in the way that the water input is close to the receiving pipe, and the exit is close to the distillation adapter.

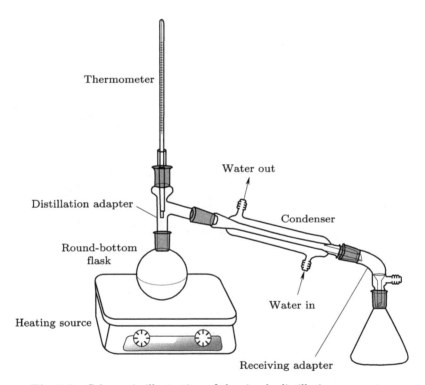

Fig. 3.2 Schematic illustration of the simple distillation apparatus.

2.2.2 Fractional distillation

Fractional distillation is based on the establishment of a large number of theoretical vaporization-condensation cycles. This technique is used to separate liquid components that differ by less than 25 °C in boiling point. The simple distillation apparatus is modified by inserting a fractionating column between the distillation flask and the distillation head, as shown in Fig. 3.3. The fractionating column provides a large surface area in which the initial distillate is redistilled and condensed again. This process continues as the vapor rises in the column until the vapor finally goes into the condenser. The vapor and the final distillate will contain a greater percentage of the liquid with lower boiling point. Thus, continuous repetition of the re-distillation process in fractional distillation gives good separation of the volatile liquid components.

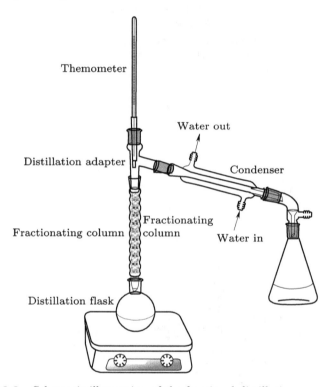

Fig. 3.3 Schematic illustration of the fractional distillation apparatus.

Compared with simple distillation, fractional distillation is more efficient, while simple distillation is generally used to purify mixtures containing only a small amount of impurities with much higher or lower boiling points.

2.3 Determination of alcohol content in red wine

The content of alcohol in wine is generally defined as the percentage of ABV. In a simple mixture of alcohol and water, this can be determined directly by measuring the density. However, wine is a complex mixture containing water, alcohol, sugar, organic acids, phenolics, and other ingredients. Therefore, the ABV of wine cannot be directly determined from the density. Instead, a simple distillation process becomes necessary to isolate the alcohol from wine. It is noted that the water and alcohol in red wine form azeotrope boiling at 78.2 °C and composition of 95.6wt% alcohol (97vol%). Therefore, in red wine distillation, any fraction containing 100% alcohol cannot be produced.

The condensed distillate collected from simple distillation contains all the alcohol from the original wine but at a lesser volume. Pure water is then added to the distillate to bring it back to the same volume as the original wine. This reconstituted distillate will have the same ABV as the original wine.

Since the specific gravity of a mixture of water and alcohol varies lawfully, the percent alcohol in the reconstituted distillate can be accurately estimated by measuring its specific gravity. The specific gravities of distilled water and pure alcohol are 1.00 and 0.79, respectively. Hence, the mixture of water and alcohol must have a specific gravity of less than 1. The specific gravity measurement can then be converted to percent alcohol.

In the experiment, the specific gravity is measured using an alcoholmeter. As shown in Fig. 3.4, it is usually made of glass, and consists of a gratuated cylindrical stem and a bulb weighted with mercury or lead shot to make it float upright in the solution.

Fig. 3.4 Photograph of an alcoholmeter.

3 Experimental

3.1 Preparation

(1) Material: red wine.

(2) Apparatuses: simple distillation apparatus (consists of a heating mantle, a 250 mL round-bottom flask, a distillation adapter, a thermometer adapter, a thermometer, a water-gacketed condenser, a receiving pipe, an Erlenmeyer flask, plastic tubes, joint clips, an iron stand with iron ring), electronic balance.

(3) Tools: alcoholmeter, 100 mL graduated cylinder, conical funnel, boiling chips.

3.2 Procedure

3.2.1 Isolation of alcohol from red wine

(1) Assemble the simple distillation apparatus as shown in Fig. 3.2.

(2) Measure 100 mL red wine using the graduated cylinder and transfer it to the 250 mL round-bottom flask.

(3) Add 50 mL water to the round-bottom flask. Clean and dry the graduated cylinder for further usage.

(4) Add some boiling chips into the round-bottom flask.

(5) Turn on the cold-water intake to keep the condenser cool.

(6) Switch on the heating mantle. Heat the flask and maintain a homogeneous boiling.

(7) Record the temperature when the first drop of distillate at the right side of the distillate adapter is obtained.

(8) During distillation, pay attention to the reading at the thermometer and

estimate the component of the distillate. After the temperature reaches about 100 °C, keep heating for about 10 min.

(9) Turn off the water intake after the distillation is finished.

(10) Collect the distillate in a graduated cylinder.

3.2.2 Determination of ABV by alcoholmeter

(1) Cool the distillate to room temperature. Transfer it into the 100 mL graduated cylinder.

(2) Add deionized water to reach 100 mL, which is the same volume as the starting sample.

(3) Determine the ABV of the wine using an alcoholmeter.

(4) Weigh the distillate and calculate the specific gravity of the distillate. Consult the ABV value in Table 3.1.

Table 3.1 Specific gravity versus ABV at 20 °C

Specific gravity	ABV/%	Specific gravity	ABV/%
0.9833	9.00	0.9792	12.00
0.9826	9.50	0.9778	13.00
0.9819	10.00	0.9765	14.00
0.9805	11.00	0.9752	15.00

4 Requirements for experimental report

(1) Describe the detailed procedures to assemble the simple distillation apparatus before the distillation experiment and dismantle it after the experiment.

(2) Give the temperature when the first drop of the distillate occurs. Estimate the concentration of alcohol in the vapor according to the phase diagram.

(3) Compare the ABV values determined by the alcoholmeter and specific gravity,respectively.

(4) Discuss the possible sources of error during the experiment.

5 Questions and further thoughts

(1) Why are boiling chips added to the round-bottom flask before heating the liquid?

(2) Where should the bulb of the thermometer be placed for distillation?

6 Cautions

(1) The volume of the liquid should not be less than 1/3 and greater than 2/3 of the volume of the flask.

(2) Do not dry the round-bottom flask during distillation.

(3) The flow direction of the cooling water should be from the bottom inlet to the top outlet of the condenser.

Experiment 15 Purification of Naphthalene by Recrystallization

Type of the experiment: Comprehensive
Recommended credit hours: 4
Brief introduction: Crude naphthalene obtained from either coal tar or petroleum distillation is primarily used to manufacture other chemicals and is also used in moth repellents. It is required to be purified for further applications. In this experiment, the crude naphthalene is purified using the recrystallization technique, which is used for purifying solids based on the variation of their different solubility with temperature. The product is isolated by hot gravity filtration and subsequent vacuum filtration. Then, the purity of naphthalene is determined by measuring the melting point.

1 Objectives

(1) To understand the principle and technique of recrystallization.

(2) To learn how to select a suitable solvent for recrystallization.

(3) To master the operation of reflux, hot gravity filtration, and vacuum filtration.

(4) To purify crude naphthalene using recrystallization technique.

(5) To determine the melting point of naphthalene using micro melting point apparatus.

2 Principles

2.1 Recrystallization

Recrystallization is one of the most often used techniques for purifying solids. The recrystallization process relies on the property that the solubility of a compound in a solvent generally increases with temperature. The general process involves the dissolution of a solid in a solvent at elevated temperature and reformation of the crystals as the solution cools, allowing for impurities to remain in the solution. Once the solid has been recrystallized, it is essential to determine the purity of the recrystallized solid. An easy way to do this is by measuring the melting point of the solid.

The size and quality of the crystals are affected by the cooling rate. A slower cooling rate results in a larger crystal size. The choice of solvent in the process of recrystallization is also critical for achieving good purity and high recovery. A suitable solvent generally meets the following properties:

(1) It does not react with the substance to be purified.

(2) It has a high solubility for the solute at high temperature and low solubility at low temperature.

(3) It has a high capability of dissolving the impurities so that they will remain in the solution when the desired compound is crystallized out.

(4) Crystals of the purified substance can be formed in the solvent.

(5) It is easy to be removed after recrystallization.

Obviously, recrystallization can be conducted in water or organic solvents. It is well known that most organic compounds are insoluble in water at room temperature but soluble at higher temperatures. Therefore, when water is used as the solvent, the suspension of the solid is prepared using the minimum amount of water in a beaker, and the mixture is heated to a boil. If some of the solid does not dissolve in the hot solution, a small amount of water is added, and the solution is boiled again until the solid dissolve. If the solutionis dark in color, a small amount of activated carbon is added to remove the colored impurities.

In most cases, the organic compounds have better dissolution in organic solvents, such as ethanol, acetone, hexane, carbon tetrachloride, benzene, petroleum ether, and ethyl acetate. Volatile organic solvents are commonly used in the recrystallization of compounds. In this situation, the heating part of the solution is performed with a reflux assembly to prevent flammable, volatile vapors from causing fires.

2.2 Filtration

Filtration is used to separate the solid from the suspension. Two main types of filtration are commonly employed in the chemistry laboratory, i.e., gravity filtration and vacuum filtration. In this experiment, hot gravity filtration will be used to filter the activated carbon and undissolved impurities, and vacuum filtration will be used to isolate the purified crystals.

2.2.1 Gravity filtration

Gravity filtration can be conducted at room temperature or high temperature. It is also named simple filtration at room temperature and hot gravity filtration at high temperatures.

The experimental setup of gravity filtration includes a conical funnel, a metal ring fixed on a stand, a piece of filter paper, and an Erlenmeyer flask to collect the liquid after filtration. Before filtration, a piece of filter paper should be prepared. Generally, a simple piece of filter paper with a stem conical funnel is employed in the simple gravity filtration. In contrast, a fluted piece of filter paper with a stemless conical funnel is used in the hot gravity filtration, as shown in Fig. 3.5.

2.2.2 Vacuum filtration

This technique uses a pressure gradient to conduct filtration and hence allows for a greater rate of filtration. It is employed mainly for separating a solid suspended in a solvent when the solid is desired.

Figure 3.6 shows the setup of the vacuum filtration, including a Büchner funnel or a Hirsch funnel, a filtration flask, and a rubber cone adapter. A vacuum trap can be inserted between the vacuum pump and filtration flask to prevent liquid in the filtration flask from passing to the vacuum pump. The setup can be modified by using vacuum-adapted funnels.

2.3 Determination of purity by measuring the melting point

The melting point is the temperature at which a solid substance melts and becomes

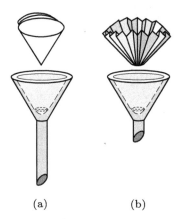

Fig. 3.5 Gravity filtration setups with (a) a stem conical funnel and (b) a stemless conical funnel.

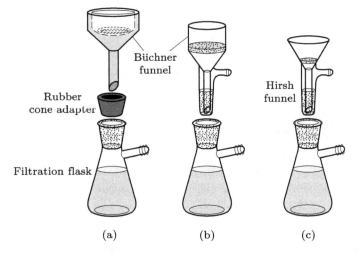

Fig. 3.6 (a) Typical setup of vacuum filtration with the Büchner funnel. The modifications of the setup use vacuum-adapted funnels, including (b) Büchner funnel and (c) Hirsch funnel.

a liquid at a certain pressure. Since this requires that the intermolecular forces that hold the solid together are overcome, the temperature at which melting occurs will depend on the structure of the molecules involved. Consequently, different compounds tend to have different melting points.

A pure, nonionic, crystalline organic compound usually has a sharp and characteristic melting point, which is usually in a range of 0.5–1.0 °C. A mixture of limited miscible impurities will depress the melting point and expand the melting point range. Consequently, the melting point of a compound is a criterion for purity as well as for identification.

The melting point of an organic solid can be determined by various methods. One method is to introduce a tiny amount into a capillary tube and attach it to the stem of a thermometer centered in a heating bath. The melting point can be determined by observing the temperatures at which melting begins and ends by heating the bath slowly. Another method is to use a micro melting point apparatus. The sample is packed between two cover glasses and then heated to melt at a low heating rate. The melting process is observed with a microscope. In this experiment, the melting temperature of purified naphthalene will be measured by a micro melting temperature apparatus.

3 Experimental

3.1 Preparation

(1) Materials: crude naphthalene.

(2) Chemical: 70% aqueous solution of ethanol.

(3) Apparatuses: reflux apparatus (consists of an electromagnetic stirrer with a heating plate, a 50 mL round-bottom flask, a reflux condenser, an iron stand with an iron ring, clamps, plastic tubes), hot gravity filtration apparatus (consists of a heating plate, an iron stand with an iron ring, a 50 mL Erlenmeyer flask, a stemless conical funnel), vacuum filtration apparatus (consists of a Büchner funnel, a filtration flask, a rubber cone adapter, a vacuum trap, a vacuum pump, an iron stand, joint clips, plastic tubes), micro melting point apparatus.

(4) Tools: 50 mL beaker, magnetic stirring bar, watch glasses, cover glasses, Pasteur pipettes.

3.2 Procedure

3.2.1 Recrystallization of naphthalene

(1) Assemble the reflux apparatus from bottom to top, as shown in Fig. 3.7(a).

(2) Transfer 2 g of crude naphthalene to the 50 mL round-bottom flask.

(3) Initially,drop about 8 mL 70% ethanol solution at the upper reflux condenser part with a Pasteur pipette. Heat the mixture to boil under stirring. Then slowly drop more ethanol solution until all the crude naphthalene is dissolved. Record the total volume of the ethanol solution used.

(4) Slightly cool the flask, add activated carbon (1–5 wt% of naphthalene) into the flask and keep stirring and refluxing for 5 min.

(5) Assemble the gravity filtration apparatus as shown in Fig. 3.7(b).

(6) Preheat the gravity filtration apparatus and a beaker with about 30 mL 70% ethanol solution.

(7) Wet the filter paper in the stemless conical funnel with a small amount of hot ethanol solution.

(8) Pour the hot solution in the round-bottom flask into the stemless conical funnel to filter the activated carbon and undissolved impurities.

(9) Collect the filtrate in the 50 mL Erlenmeyer flask, and cover it with a suitable plug.

(10) Cool the filtrate in the Erlenmeyer flask. Crystals of purified naphthalene will form in the solvent.

(11) Assemble the vacuum filtration apparatus.

(12) Wet the filter paper in the Büchner funnel with a small amount of the

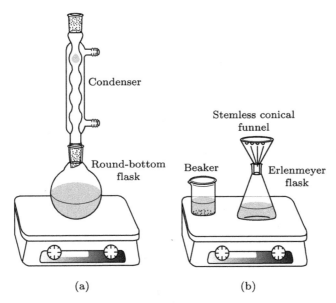

Fig. 3.7 Schematic illustrations of (a) the reflux apparatus and (b) the hot gravity filtration apparatus.

ethanol solution and make sure that the filter paper is completely stuck to the Büchner funnel.

(13) Pour all the mixture into the Büchner funnel. Then the purified crystals are isolated.

(14) Collect the purified crystals in a tared watch glass. Dry and weigh the resulting solid.

3.2.2 Determination of melting point of as-purified naphthalene

(1) Load a small number of naphthalene crystals on a piece of cover glass and then put it on the heating stage of the micro melting point apparatus.

(2) Cover the sample with another cover glass.

(3) Slowly heat the sample and then observe the sample under the microscope.

(4) Record the temperature at the first sign of melting. That is, the crystals look wet.

(5) Record the temperature when the crystals melt completely.

(6) Decrease the temperature to about 40 °C below the melting point.

(7) Repeat steps (3) to (6) three times.

4 Requirements for experimental report

(1) Describe the shape and color of the crystals.

(2) Calculate the recrystallization yield by the following equation:

$$\text{recrystallization yield} = \frac{\text{weight of solid obtained after recrystallization}}{\text{weight of solid before recrystallization}} \times 100\%. \tag{3.1}$$

(3) Tabulate the temperatures measured by the micro melting point apparatus. Calculate the average values of the higher and the lower temperatures, respectively, which determine the melting point range of naphthalene. Discuss the purity of the as-purified naphthalene by comparing the temperatures with that in the literature (80.5 °C).

5 Questions and further thoughts

(1) How much solvent is added to dissolve the crude naphthalene? How does it affect if more solvent or less solvent is added?

(2) Why is activated carbon used? What will happen if excessive activated carbon is used?

6 Cautions

(1) Do not place the activated carbon directly into a boiling liquid to avoid bumping.

(2) Make sure all the naphthalene is dissolved in the ethanol solution. Do not overheat the solution to avoid melting the naphthalene.

(3) The hot gravity filtering process should be operated as quickly as possible.

(4) When organic solvents are heated, a reflux apparatus is always required. If not handled in this way, flammable vapors can be released into the atmosphere, and nearby flames or heat sources will pose a severe risk of fire and explosion.

Experiment 16 Separation of Liquid Mixture by Liquid-Liquid Extraction

Type of the experiment: Comprehensive
Recommended credit hours: 4
Brief introduction: This experiment involves the separation of the components of a mixture consisting of three common compounds, naphthalene, p-chloroaniline, and benzoic acid. The liquid-liquid extraction technique is used for isolating organic compounds according to their acid-base properties, which change their solubilities in organic and aqueous solvents. A rotary evaporator is used to remove volatile organic solvents from the extracts.

1 Objectives

(1) To understand the principle of liquid-liquid extraction.

(2) To learn the procedure of isolating organic compounds based on their acid-base properties by changing their solubilities.

(3) To learn the operations of liquid-liquid extraction and rotary evaporator.

(4) To separate naphthalene, p-chloroaniline, and benzoic acid from their mixture.

2 Principle

2.1 Liquid-liquid extraction

Liquid-liquid extraction allows the isolation and the purification of a product resulting from a chemical reaction. The extraction can be defined as converting a substance X from a liquid phase A to another liquid phase B. Both solvents should be immiscible, thus forming two distinct phases. The concentration ratio of X sharing between phases A and B is given by the Nernst equation, which is expressed as

$$K_D = \frac{C_{B(X)}}{C_{A(X)}}, \qquad (3.2)$$

where $C_{B(X)}$ and $C_{A(X)}$ are the concentrations of X in B and A, respectively, and K_D is the partition coefficient, which depends on temperature. Given the partition coefficient for a particular system, we can calculate approximately how much compound can be extracted.

The operation of liquid-liquid extraction is typically performed between an aqueous solution and a water-immiscible organic solvent with the aid of a separatory funnel, in which an aqueous layer and an organic layer are formed. The relative position of the two layers (upper, lower) depends on their densities. Chlorinated solvents, such as CH_2Cl_2, $CHCl_3$, and CCl_4, always remain in the lower layer, as they are denser than water. Other organic solvents, such as diethyl ether, ethyl acetate, toluene, and hexane, typically have lower densities than water and therefore always remain in the upper layer. Clearly, water-miscible solvents, such as acetone, methanol, and alcohol, are not useful for this process. The separatory funnel is a pear-shaped container with a ground-glass stopper on top and a stopcock. Figure 3.8 shows the necessary steps for a successful extraction.

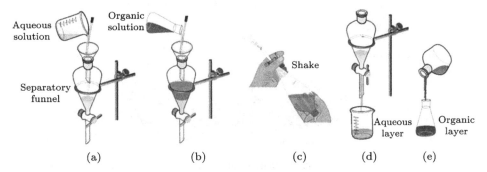

Fig. 3.8 Liquid-liquid extraction procedure with a separatory funnel: (a) pour the aqueous solution into the separatory funnel, (b) pour the organic solution, (c) shake and occasionally release the overpressure, (d) drain out the bottom layer, (e) pour out the upper layer.

When extraction occurs, the two layers are saturated with respect to the other solvent, water with the organic solvent and the organic layer with water. Therefore, water must be removed from the organic solvent to obtain the pure product. The usual procedure for drying the organic solution is to use a drying agent or desiccant, which generally is an anhydrous inorganic salt. For instance, Na_2SO_4 and $CaSO_4$ have been commonly used. The desiccant is left in contact with the solution taking up water until it becomes hydrated. Then it is separated by gravity filtration.

2.2 Rotary evaporator

A rotary evaporator is a common apparatus to remove volatile organic solvents from a reaction mixture or a process such as liquid-liquid extraction under reduced pressure. Although this operation can be conducted by simple distillation, distillation using an evaporator is more efficient.

Figure 3.9 shows the setup of a rotary evaporator. Generally, the rotary evaporator consists of an electric motor, which causes rotation of a round-bottom flask containing the solution to be distilled. The flask is partially immersed in a water bath. A condenser that circulates a liquid (water or antifreeze) is coupled to the system. It produces the condensation of the solvent, which is recovered in a collector.

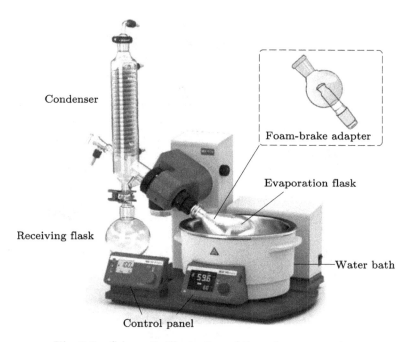

Fig. 3.9 Schematic illustration of the rotary evaporator.

During the evaporation, some solutions tend to form foams or produce splashes that move up the guide tube to contaminate the indoor coil. This problem can be avoided using a foam brake, a glass adapter between the guide tube and the round-bottom flask. It has two ground-glass joints: the female at the top connected to the guide tube and the male at the bottom connected to the flask. The body consists of a glass sphere with elbow pipes or another device to prevent any solution in the flask from ascending the guide tube.

2.3 Separate compounds of a mixture by liquid-liquid extraction

Liquid-liquid extraction used for separating compounds in a mixture can be based on their acid-base properties since the solubilities of organic acids and bases can be dramatically altered by adjusting the pH value of the solution. Most carboxylic acids are either insoluble or only slightly soluble in water. However, these organic acids can react with bases such as sodium bicarbonate, sodium carbonate, or sodium hydroxide, yielding a proton and forming the corresponding water-soluble anions. Furthermore, amines yield water-soluble ions generated by protonation, such as ammonium salts.

In this experiment, the separation of the components of a mixture consisting of three common organic compounds, naphthalene, p-chloroaniline, and benzoic acid, is taken as an example to introduce the procedure of liquid-liquid extraction. Figure 3.10 is a flowchart of the separation procedure. The compounds are dissolved in a CH_2Cl_2 solution. The addition of HCl solution can separate the

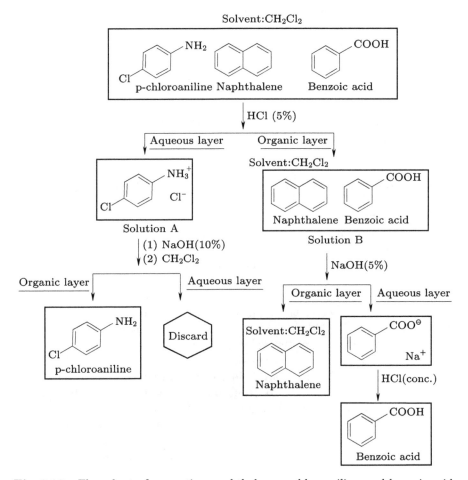

Fig. 3.10 Flow chart of separating naphthalene, p-chloroaniline, and benzoic acid.

solution into the aqueous layer (solution A), which contains water-soluble ions of p-chloroaniline, and organic layer (solution B), which contains the other two compounds. The aqueous solution A is collected and made basic by adding NaOH solution. Then the basic solution is extracted with CH_2Cl_2. The p-chloroaniline in the organic layer can then be obtained by collecting the organic layer followed by drying with anhydrous Na_2SO_4 and removing the solvent using the rotary evaporator. Solution B is extracted with NaOH solution, separating an organic layer with naphthalene and an aqueous layer with water-soluble ions of benzoic acid. Naphthalene is then obtained using the rotary evaporator. The aqueous solution extracted from solution B is acidified by HCl solution, and the benzoic acid can be obtained by filtration.

3 Experimental

3.1 Preparation

(1) Chemicals: naphthalene, p-chloroaniline, benzoic acid, 5% aqueous solution of HCl, 5% and 10% aqueous solutions of NaOH, dichloromethane (CH_2Cl_2), sodium sulfate (Na_2SO_4).

(2) Apparatuses: rotary evaporator, gravity filtration apparatus (the components have been listed in Experiment 15), vacuum filtration apparatus (the components have been listed in Experiment 15), electronic balance.

(3) Tools: separatory funnels, beakers, Erlenmeyer flasks, 100 mL round-bottom flasks, 50 mL graduated cylinder, iron stand with an iron ring, watch glass, pH test paper.

3.2 Procedure

(1) Prepare 40 mL CH_2Cl_2 solution containing 4 g naphthalene, 4 g p-chloroaniline, and 4 g benzoic acid as the starting mixture.

(2) Fix the separatory funnel on the iron stand. Lubricate the stopper and the stopcock of the separatory funnel with Vaseline and then check the separatory funnel for leaks.

(3) Transfer the CH_2Cl_2 solution into the separatory funnel.

(4) Add 15 mL HCl solution in the separatory funnel and shake it vigorously. Stand a while until the solution separates into the aqueous layer (top layer) labeled as solution A and the organic layer (bottom layer) labeled as solution B.

(5) Drain out solution B into the Erlenmeyer flask through the stopcock. Pour solution A into another beaker through the mouth of the separatory funnel.

(6) Transfer solution B into the separatory funnel and repeat steps (4) and (5) to guarantee the complete extraction of ions of p-chloroaniline.

(7) Make these aqueous extracts (solution A) basic by adding about 10 mL 10% NaOH solution. Check the pH value with pH test paper.

(8) Extract this basic solution with 10 mL CH_2Cl_2 in a separatory funnel twice. Discard the remaining aqueous layer.

(9) Collect the bottom organic layer in an Erlenmeyer flask and dry it with anhydrous Na_2SO_4 for a few minutes.

(10) Weigh a 100 mL round-bottom flask. Filter the extracts by gravity filtra-

tion into the round-bottom flask to isolate Na_2SO_4.

(11) Remove the organic solvent with the rotary evaporator and obtain the product of p-chloroaniline. Weigh the flask, and then calculate the weight and yield of the product.

(12) In a separatory funnel, extract Solution B with 15 mL 5% NaOH solution twice.

(13) Collect the extracts and dry the corresponding organic layer (bottom) with anhydrous Na_2SO_4 for a few minutes.

(14) Weigh a 100 mL round-bottom flask. Filter the extracts by gravity filtration into the round-bottom flask to isolate Na_2SO_4.

(15) Evaporate the solvent by the rotary evaporator to obtain naphthalene. Weigh the flask and calculate the weight and yield of the product.

(16) Acidify the aqueous extracts from solution B by adding a small portion (several mL) of concentrated HCl. Check the pH value with pH test paper.

(17) Cool the solution and isolate the crystalline product (benzoic acid) by vacuum filtration. Wash the benzoic acid crystals in the Büchner funnel with cold water twice. The procedure of vacuum filtration has been described in Experiment 15.

(18) Collect the benzoic acid in a watch glass. Weigh the product and calculate the yield.

4 Requirements for experimental report

(1) Please draw a detailed schematic flow chart of the separation procedure based on Fig. 3.10.

(2) List the recovery yields of the compounds.

(3) Describe the shape and color of the crystallized products.

5 Questions and further thoughts

(1) What should we pay attention to when using the separatory funnel?

(2) What are the factors that affect the recovery yields?

6 Cautions

(1) Do not discard a layer until the end of the experiment or before being confident that the layer is no longer needed.

(2) Naphthalene, benzoic acid, p-chloroaniline, and CH_2Cl_2 are harmful organic compounds. Therefore, this experiment should be conducted strictly in a fume hood. Goggles and facemask should be worn all the time during the experiment.

(3) Smear some vaseline on the joints of ground glass for lubricating while assembling the apparatuses.

Experiment 17 Extraction of Oil from Sunflower Seeds by Solid-Liquid Extraction

Type of the experiment: Comprehensive

Recommended credit hours: 4

Brief introduction: Sunflower seed oil is mainly utilized by the food industry because it can withstand high cooking temperature and has health benefits of low saturated fat and high vitamin E content. In this experiment, the oil is obtained by solid-liquid extraction. The solvent is removed by the steam distillation method. At last, the oil is separated from water using a separatory funnel.

1 Objectives

(1) To understand the principle of solid-liquid extraction.
(2) To understand the principle of steam distillation.
(3) To separate oil from sunflower seeds by solid-liquid extraction and subsequent steam distillation.

2 Principles

2.1 Sunflower seed oil

The sunflower seed oil has a light amber color with a mild flavor. It is a mixture mainly of polyunsaturated fat, monounsaturated fat, linoleic acid, and oleic acid (Fig. 3.11). The oil can withstand high cooking temperatures, making it an optimal choice for frying. It also has additional health benefits of lower saturated fat and the highest vitamin E content than any other cooking oils. Besides, sunflower seed oil is applicable as a lubricant and a cosmetic ingredient. It could potentially become a key component of vegetable oil fuel.

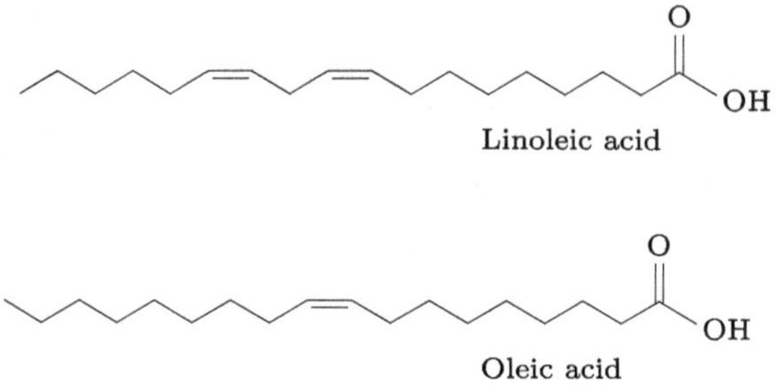

Linoleic acid

Oleic acid

Fig. 3.11 The molecular structures of linoleic acid and oleic acid in sunflower seed oil.

The vegetable oils are often produced by mechanical or chemical processes or a combination of the two. One of the most common procedures is the extraction from oil-rich plants using organic solvents such as hexane. In this experiment, the sunflower seed oil is obtained by continuous solid-liquid extraction with hexane from the sunflower seeds, in which the oil content can be as high as 40wt%. After that, the oil is isolated from the solvent by steam distillation.

2.2 Solid-liquid extraction

Solid-liquid extraction is a technique for isolating components of a solid using a liquid solvent. This process involves grinding the solid to a fine powder, mixing it with an appropriate solvent, isolating the powder by filtration, and then removing the solvent from the extracted solution.

Although a component can be extracted from a solid with a solvent by a simple extraction and subsequent filtration, this process would require many tedious extraction and filtration steps. Hence, a special apparatus called Soxhlet extractor (Fig. 3.12) was designed for isolating components from a solid in a very efficient way. This is why the solid-liquid extraction is also called Soxhlet extraction.

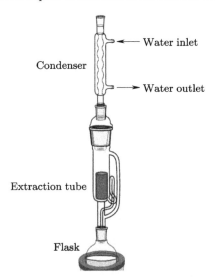

Fig. 3.12 Schematic illustration of the Soxhlet extractor.

The Soxhlet extraction apparatus mainly consists of four parts: a water-cooled condenser, a Soxhlet column extraction tube with a siphon through which the solvent falls after extraction to the flask, a Soxhlet cartridge with the solid to be extracted, and a round-bottom flask with twice the volume of the solvent.

The Soxhlet extractor can continuously extract a component from a solid material. Boiling solvent vapor goes up through the side-arm into the condenser at the top. The liquid condensate drips into the filter paper thimble, which contains the solid sample to be extracted. The extract seeps through the pores of the thimble and fills the siphon tube, where it flows back down into the round-bottom flask. Therefore, with the Soxhlet extractor, multiple extractions are performed automatically and continuously using the same solvent, which repeatedly evaporates and condenses in the Soxhlet extractor.

2.3 Steam distillation

Steam distillation is a separation process used to purify or isolate temperature-sensitive materials, like natural aromatic compounds. It is to heat and separate the components at a temperature below their decomposition point. Steam or water is added to the distillation apparatus, lowering the boiling points of the compounds. When the mixture of water and organics is distilled, the vapor is condensed. Since water and organics tend to be immiscible, the resulting liquid generally consists of two phases, water and organic distillates. Then, other techniques can be used to separate the two layers for obtaining the purified organic materials.

The advantage of steam distillation over simple distillation is that the lower boiling point reduces the decomposition of temperature-sensitive compounds. Therefore, this method is suitable for extracting plant active ingredients, which are volatile, stable, and insoluble in water.

There are two types of steam distillation apparatuses according to how the vapor is generated. One is the external source of steam, as shown in Fig. 3.13(a). In this case, the water vapor is bubbled over the solution of the mixture by an auxiliary assembly. The other type is the internal source of steam, as shown in Fig. 3.13(b). In this case, the product dissolved in an organic solvent is mixed with water in the flask. This is similar to simple distillation, except that an addition funnel is placed in the assembly with water gradually added or directly added at once.

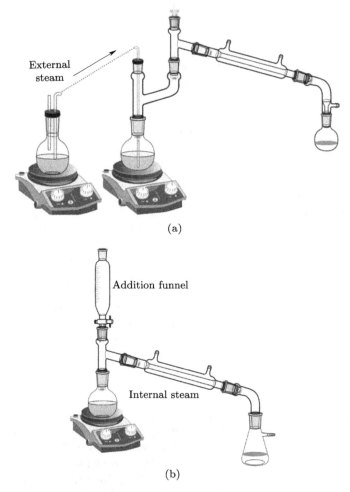

Fig. 3.13 Schematic illustrations of the steam distillation apparatuses with (a) an external steam source and (b) an internal steam source.

3 Experimental

3.1 Preparation

(1) Material: sunflower seeds.

(2) Chemicals: hexane, deionized water.

(3) Apparatuses: Soxhlet extraction apparatus (consists of a Soxhlet extractor, a Soxhlet extraction thimble, a condenser, a 250 mL round-bottom flask, an electromagnetic stirrer with a heating plate, an Al heat transfer block, an iron stand with clamps), steam distillation apparatus (consists of a 250 mL round-bottom flask, a distillation adapter, a thermometer adapter, a thermometer, a condenser, a receiving adapter, an Erlenmeyer flask, an electromagnetic stirrer with a heating plate, joint clips, an iron stand with clamps), electronic balance.

(4) Tools: mortar and pestle, stem conical funnel, magnetic stirring bar, 100 mL beakers, 10 mL and 100 mL graduated cylinders, separatory funnel.

3.2 Procedure

3.2.1 Solid-liquid extraction

(1) Weigh about 20 g sunflower seeds. Grind these seeds with mortar and pestle into fine powder. Place the seed powder into the thimble.

(2) Fill the round bottom flask with 100 mL hexane and add a stirring bar. Fill the Soxhlet extractor with 50 mL hexane and then carefully place the thimble with the seed powder in the extractor.

(3) Set up the solid-liquid extraction apparatus, i.e., the Soxhlet extraction apparatus. Switch on the stirrer and heating plate. Keep the apparatus running for more than six extraction cycles.

3.2.2 Steam distillation

(1) Prepare about 100 mL warm water in a beaker.

(2) After the Soxhlet extraction is finished, switch off the heating plate, disassemble the Soxhlet extraction apparatus and add about 100 mL warm water into the flask.

(3) Assemble the steam distillation apparatus. The flask with the extractive from the Soxhlet extraction is used.

(4) Distill the solution to remove the hexane completely. The distillate is composed of a mixture of hexane and water. The residue consists of a mixture of water and oil.

(5) Allow the round-bottom flask content to cool to room temperature. Then pour the liquid into a separatory funnel for separation.

(6) Measure the volume and weight of the separated oil in a graduated cylinder.

4 Requirements for experimental report

(1) Calculate the yield and the approximate density of the oil.

(2) Discuss the factors that influence the yield.

5 Questions and further thoughts

(1) Why is a cellulose thimble used to hold the sample in the Soxhlet extraction?

(2) How to choose a suitable solvent for Soxhlet extraction?

6 Cautions

(1) Hexane is a flammable and hazardous organic solvent. Take care to handle it during the experiment.

(2) During steam distillation, pay attention to the vapor temperature to verify the component of the distillate and the complete removal of hexane.

Experiment 18   Separation of Pigments from Spinach Leaves by Thin Layer Chromatography and Column Chromatography

Type of the experiment: Comprehensive
Recommended credit hours: 4
Brief introduction: Spinach is a common vegetable in daily life containing several kinds of pigments. In this experiment, the techniques of thin layer chromatography (TLC) and column chromatography (CC) are introduced. The components of the pigments from spinach leaves and their general polarity are identified by TLC. After that, the pigments are separated by the CC technique, including β-carotene, chlorophylls, xanthophylls, and pheophytins.

1 Objectives

(1) To understand the principles and operations of TLC and CC.
(2) To extract the pigments from spinach leaves.
(3) To separate the pigments by TLC and CC techniques.

2 Principles

2.1 Chromatographic techniques

Chromatography is a type of separation technique based on the distribution of the components in a mixture between a fixed (stationary) phase and a moving (mobile) phase. The stationary phase may be a thin layer of adsorbent on a plate or a column of adsorbent through which the mobile phase moves on. The mobile phase may be a liquid or a gas.

When a mixture is introduced into the mobile phase, its components are distributed between the stationary and mobile phases. The components travel for different times in the stationary and mobile phases, and therefore separation occurs. If a component spends most time in the mobile phase, it moves quickly, whereas if it spends most time in the stationary phase, it is much slower.

Chromatography has extraordinarily versatile variants. The most common technique used in organic chemistry is liquid-solid chromatography, including TLC and CC. In this experiment, these two techniques will be used to extract the pigments from spinach leaves.

Spinach leaves contain chlorophyll a, chlorophyll b, and β-carotene as primary pigments, as well as a small number of other pigments such as xanthophylls and pheophytins. The structures of the major components are shown in Fig. 3.14. The β-carotene, which is a hydrocarbon, is nonpolar. Xanthophylls are a family of compounds derived from carotenes with oxygen atoms. Chlorophylls contain several polar C-O and C-N bonds and also a magnesium ion chelated to the nitrogen atoms. Chlorophyll a has a methyl group in a position where chlorophyll b has an aldehyde. This makes chlorophyll b slightly more polar than chlorophyll a. Pheophytins are chlorophyll molecules without the Mg^{2+}, and two nitrogen atoms protonated instead. In this experiment, all these pigments will be identified by TLC and then separated by CC.

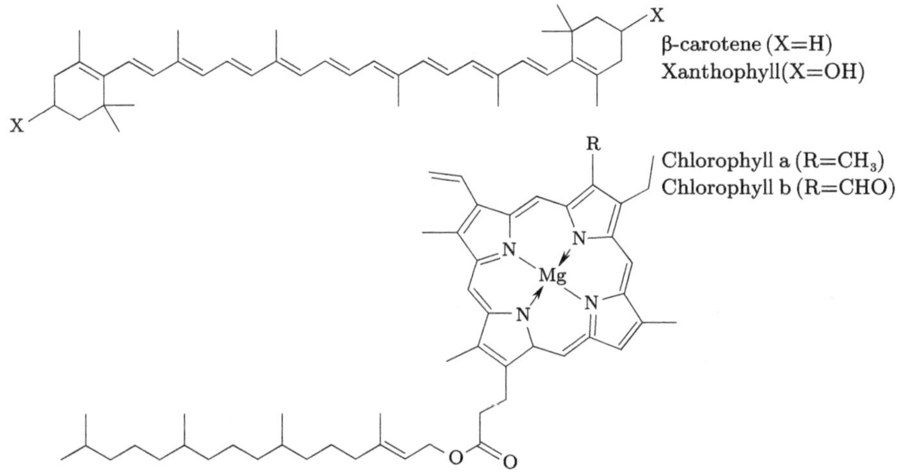

Fig. 3.14 The molecular structures of β-carotene, xanthophyll, and chlorophylls.

2.2 TLC technique

TLC is one of the most popular liquid-solid chromatography techniques. It is commonly used to identify substances or the components of a mixture and follow the process of a reaction or CC development.

TLC uses a thin, uniform layer coated on a piece of glass, metal, or rigid plastic. This layer is the stationary phase, which is commonly made of silica gel or alumina. The stationary phase often contains a substance that fluoresces in UV light too. The mobile phase is a suitable solvent or mixture of solvents.

When a TLC plate is developed, the various components in the mixture are separated. This separation is based on the distribution equilibrium between the eluent (the solvent) and the adsorbent (the stationary phase). Each compound will have a unique distribution equilibrium depending mainly upon its polarity, which is based on intermolecular forces between the compounds being separated and the adsorbent. The distribution equilibrium of a polar compound will favor the adsorbent since the adsorbent is highly polar. However, the non-polar compound will have less affinity for the polar adsorbent and will have better solubility in the mobile solvent. Consequently, polar compounds will stick to the stationary TLC plate, while non-polar compounds will be separated and move upward with the mobile phase. In other words, each compound in the mixture will ascend the plate at a different rate. Polar compounds ascend slowly, and less polar compounds ascend quickly.

Figure 3.15(a) shows the simple setup of TLC, which consists of a developing tank and a TLC plate in it. The tank is a closed container allowing the environment to be saturated with eluent vapor. The tank is filled with the eluent to a height that does not reach the area where the compounds being separated are located. Before the TLC plate is placed in the tank, a pencil line is drawn on the plate at about 1.0 cm from the bottom edge of the plate (Fig. 3.15(b)). The sample solution is transferred to a point marked on the line. After drying the solution, the plate is placed in the tank. When the eluent rises close to the top edge of the plate, take the plate out and measure the distances traveled by the eluent, L_0, and the component, L_1. In order to identify the component, the ratio R_f is often used, which is calculated by

$$R_f = \frac{L_1}{L_0}. \tag{3.3}$$

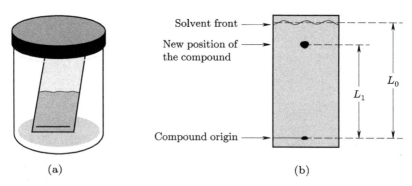

Fig. 3.15 (a) A developing tank with a TLC plate and (b) determination of R_f value on the plate.

2.3 CC technique

CC is one of the most useful methods for the separation and purification of both solids and liquids. The principle of CC is based on differential adsorption of substance by the adsorbent. The usual adsorbents are silica, alumina, calcium carbonate, calcium phosphate, magnesia, and starch. The selection of solvent is based on the nature of both the solvent and the adsorbent. The rate at which the components of a mixture are separated depends on the activity of the adsorbent and the polarity of the solvent.

Figure 3.16 shows the basic setup of CC. It uses a chromatography column as the primary device, which is filled with an adsorbent that acts as a stationary phase. The sample solution is introduced into the column, and the eluent (a solvent or solvent mixture) is added continuously from the top of the column. The eluent can run throughout the column either by gravity or by exerting gas pressure with inert gas or just air, leading to the separation of the components in the mixture. These components form separated bands running down the column and are collected successively.

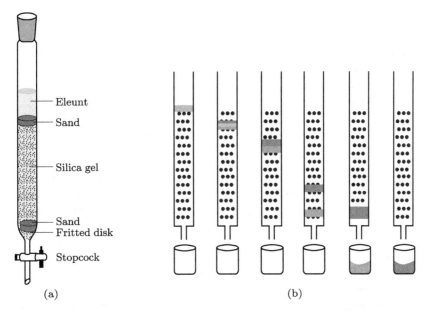

Fig. 3.16 (a) The setup of CC with silica gel as adsorbent. (b) Separation and collection of components in a mixture. (See color illustration at the back of the book)

3 Experimental

3.1 Preparation

(1) Materials: spinach leaves, sand, sand frit.

(2) Chemicals: acetone, hexane, calcium carbonate ($CaCO_3$), sodium chloride (NaCl), anhydrous sodium sulfate (Na_2SO_4), silica gel (100–200 mesh), deionized water.

(3) Apparatuses: TLC apparatus (consists of a developing tank, a watch glass, TLC plates with a size of 5 cm × 7.5 cm coated by silica gel), CC apparatus (consists of a chromatography column, an iron stand with an iron ring, clamps), gravity filtration apparatus (the components have been listed in Experiment 15).

(4) Tools: separatory funnel, mortar and pestle, 50 mL Erlenmeyer flasks, beakers, 25 mL graduated cylinder, capillaries, stem conical funnel, Pasteur pipettes, scissors, filter paper, pencil, ruler.

3.2 Procedure

3.2.1 Extraction of pigments from spinach leaves

(1) Weigh about 2 g spinach leaves and cut them into pieces.

(2) Put them in a mortar together with 22 mL acetone, 3 mL hexane, and a small amount (about one spoon) of $CaCO_3$ to prevent the degradation of photosynthetic pigments. Grind the mixture until the leaves become discolored, and the

solvent becomes deep green.

(3) Transfer the resulting liquid to a separatory funnel.

(4) Add 15 mL hexane and 15 mL 10% aqueous solution of NaCl. Shake to blend them completely. Collect the top organic layer in an Erlenmeyer flask and discard the bottom aqueous layer.

(5) Transfer the organic solution into the separatory funnel. Add 5 mL deionized water to wash the organic solution. Then collect the organic solution.

(6) Dry the organic solution with anhydrous Na_2SO_4 and remove the desiccant by gravity filtration.

(7) Divide the filtrate into two parts. Take 10 mL of the solution for CC separation and the remaining solution for TLC separation.

3.2.2 Separation and analysis of spinach pigments by TLC

(1) Prepare 500 mL eluent with hexane and acetone in a volume ratio of 7:3.

(2) Add the eluent in the developing tank to about 5 mm high and cover it with a watch glass, allowing the vapor of the organic solvent to saturate the entire tank.

(3) Draw a horizontal pencil line about 10 mm from the bottom edge of the TLC plate.

(4) Dip the sample solution using a capillary at the same spot on the pencil line several times to increase the concentration of the sample.

(5) Place the TLC plate into the tank. Keep it upright and slightly inclined so that the solvent level in the TLC plate can be seen from the outside of the tank.

(6) Cover the tankquickly to keep its atmosphere saturated with the solvent.

(7) When the solvent level reaches the position a few millimeters from the top of the TLC plate, take the plate out of the tank and quickly mark the level reached by the solvent.

(8) Dry the TLC plate. Colored spots appear along the TLC plate. Label the spots and determine their R_f values using Equation (3.3).

(9) Repeat steps (2) to (8) three times. Calculate the average R_f value for each component.

3.2.3 Separation and analysis of spinach pigments by CC

(1) Vertically fix the column on the iron stand.

(2) Put a sand frit in the bottom of the column.

(3) Transfer 10 mL eluent prepared in Section 3.2.2 into the column using a conical funnel.

(4) Weigh 20 g silica gel in a 100 mL beaker and add 80 mL eluent. Stir with a glass rod to form a slurry, which is used as the stationary phase.

(5) Fill the column with the slurry. Add sand to form a layer of 1 cm thick on the top of the stationary phase.

(6) Open the column stopcock until the liquid aligns with the top of the sand layer. Close the stopcock. Collect the solvent in an Erlenmeyer flask for reuse.

(7) Pipette 10 mL sample obtained in Section 3.2.1 into the column. Open the stopcock until the solution aligns with the top of the sand layer.

(8) Keep the column stopcock open. Add the eluent continuously and collect

it in the Erlenmeyer flask. In this process, the stopcock can be close occasionally to facilitate the operation. Never let the eluent dry in the column to avoid the formation of bubbles in the stationary phase.

(9) When a yellow-orange band of liquid, which is the carotene solution, appears at the bottom of the column, collect the liquid in another Erlenmeyer flask.

(10) Successively collect the bands in gray (pheophytins), blue-green (chlorophyll a), and green (chlorophyll b).

(11) Prepare 100 mL eluent with hexane and acetone in a volume ratio of 3:7. The higher acetone content can increase the polarity of the eluent. Add this eluent into the column, and two yellow bands (two derivatives of xanthophyll) can be separated.

(12) The solvent of the collected liquid can be further removed by the rotary evaporator.

(13) Take a photograph of the components separated by CC.

4 Requirements for experimental report

(1) Sketch the TLC plate in the report, label the spots by color and give the corresponding R_f values. Identify these components by comparing the values with those in Table 3.2. Note that some components cannot be identified because of low concentration or overlaps of spots.

(2) Give the photograph of the components separated by CC. Identify the components by comparing the color and separation sequence with those of TLC. Note that not all the components listed in Table 3.2 can be separated.

(3) Discuss the effect of the polarities of the components and the eluent on the separation sequence of TLC and CC.

(4) Discuss how these two chromatographic techniques are combined to identify and separate the components in a mixture.

Table 3.2 R_f values and colors of the pigments in spinach leaves

Pigment	Color	R_f
β-carotene	Yellow-orange	0.93
Pheophytin a	Gray	0.55
Pheophytin b	Gray	0.47–0.54
Chlorophyll a	Blue-green	0.46
Chlorophyll b	Green	0.42
Xanthophylls*	Yellow	0.41, 0.31, 0.17

* Three R_f values are given for the derivatives of xanthophyll.

5 Questions and further thoughts

(1) Explain why the pencil line on the TLC plate must not be immersed in the solvent in the developing tank?

(2) How can we uniformly fill the stationary phase in the chromatography column? How can we avoid the formation of air bubbles in the stationary phase during separation?

(3) What do the R_f values indicate about the relative attraction determined by the polarities of the components and the solvent?

6 Cautions

(1) Acetone and hexanes are skin and lung irritants. Usage of these solvents should be restricted in a fume hood to avoid exposure to harmful vapors. Protective gloves, facemasks, and goggles should be worn during operation.

(2) All organic solvents and vapors are highly flammable. No open flames should be used in the laboratory during the experiment.

Extended readings

Barich A. Recrystallization. Available at: https://chem.libretexts.org/Bookshelves/Physical_and_Theoretical_Chemistry_Textbook_Maps/Supplemental_Modules_(Physical_and_Theoretical_Chemistry)/Physical_Properties_of_Matter/Solutions_and_Mixtures/Case_Studies/RECRYSTALLIZATION. Accessed: 2 March 2021.

Isac-García J, Dobado J A, Calvo-Flores F G, et al. Experimental Organic Chemistry: Laboratory Manual. Academic Press, 2015.

Wade L G. Organic Chemistry. 8th Edition. Pearson, 2012.

Chapter 4 Fundamentals of Crystallography

Experiment 19 Ball Models of Atomic Arrangements in Pure Metal Crystals

Type of the experiment: Cognitive
Recommended credit hours: 4
Brief introduction: Four strays are artfully designed in this experiment as templates to support ping-pong balls to build up the common crystal structures of metals. The structures of closed packed planes can be understood from the distributions of the holes on the trays. Then, the face-centered cubic (FCC) structure, the hexagonal close-packed (HCP) structure, and a twin structure will be constructed by piling up ping-pong balls on the tray with the plane structure of FCC {111} or HCP {0001}. At last, the effect of crystal structure on the slip will be examined by comparing the movement of a ping-pong ball on these trays. Through this experiment, the students can intuitively understand these crystal structures and the concepts of atomic stacking, twin structure, and slip.

∎ Objectives

(1) To understand the FCC, body-centered cubic (BCC), and HCP structures.

(2) To understand the structures of close-packed planes, close-packed directions, stacking, and slip systems of each crystal.

(3) To learn the method of identification of the planes in each crystal structure.

(4) To understand the effect of crystal structure on the deformation behavior of materials.

2 Principles

2.1 Crystal structures of metals

The crystal structure is how atoms, ions, or molecules are spatially arranged in a crystalline material. For metallic materials, the most common crystal structures are BCC, FCC, and HCP structures. Their unit cells are shown in Fig. 4.1.

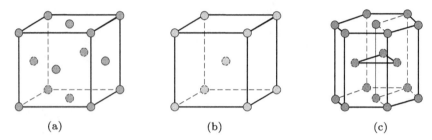

Fig. 4.1 The unit cells of (a) FCC, (b) BCC, and (c) HCP structures.

Different crystal structures lead to different mechanical properties of bulk metals. For example, FCC metals, such as Cu, Au, Ag, and Pt, are usually soft and ductile. BCC metals, such as α-Fe, Cr, W, and Mo, are less ductile but stronger. HCP metals, such as α-Ti, Mg, and Zn, are usually hard.

2.2 Close-packed planes and directions

In geometry, close-packing of equal spheres is a dense arrangement of congruent spheres in an infinite lattice. The atomic arrangements of the close-packed planes in the three metal structures are shown in Fig. 4.2, in which the atoms in the structures are considered as rigid spheres. These close-packed planes are FCC $\{111\}$, BCC $\{110\}$ and HCP $\{0001\}$, respectively. It is of interest that the atomic arrangements of the FCC $\{111\}$ plane and HCP $\{0001\}$ plane are the same. Arrows indicate the close-packed directions on the planes, which are FCC $\langle 110 \rangle$, BCC $\langle 111 \rangle$, and HCP $\langle 11\bar{2}0 \rangle$.

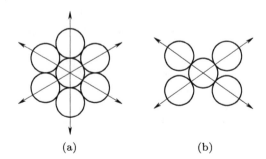

Fig. 4.2 The atomic arrangements of (a) FCC $\{111\}$ or HCP $\{0001\}$ planes, and (b) BCC $\{110\}$ planes. These planes are close-packed planes, and the arrows indicate the close-packed directions.

2.3 Stacking of close-packed planes

It has been proved that the greatest fraction of space occupied by rigid spheres that can be achieved by a lattice packing is about 0.74. Both FCC and HCP structures can achieve this highest packing density. The two different structures having the same packing density can be readily understood in terms of the stacking of atomic planes.

The crystal structures can be generated by the stacking of the close-packed planes. The difference between the FCC and HCP structures lies in the stacking sequence. As shown in Fig. 4.3, the centers of all the atoms in one close-packed plane are labeled as A. The neighboring close-packed planes can be positioned over two more possible sites B and C with respect to this reference layer. Every sequence of A, B, and C without immediate repetition of the same one is possible and gives an equally dense packing for rigid spheres of a given radius. The most regular ones are ABCABC (every third layer is the same), resulting in the FCC structure, and ABABAB (every second layer is the same), resulting in the HCP structure.

Fig. 4.3 Stacking sequences of the close-packed planes of FCC (left) and HCP (right) structures.

2.4 Slip system

Slip is a process by which plastic deformation is produced by dislocation motion. For pure metals and metallic alloys, it is the most important deformation mechanism.

Ordinarily, a dislocation traverses on a preferred plane along a specific direction. This plane is called a slip plane, and the direction is called a slip direction. The combination of a slip plane and a slip direction is a slip system. Different slip systems are presented depending on the type of crystal structure. For a particular crystal structure, slip usually occurs on close-packed planes in close-packed directions.

For the FCC crystal structure, slip occurs on the {111} planes and in the ⟨110⟩ directions. Because the FCC structure has four unique {111} planes and three independent ⟨110⟩ directions within each plane, there are 12 slip systems.

There are six {110} close-packed planes and two independent ⟨111⟩ directions on each plane for the BCC crystal structure. Thus, there are 12 slip systems. Besides, slip in BCC metals may also occur on {112} and {123} planes.

Slip in the HCP metals is much more limited than in the BCC and FCC crystal structures. Usually, the HCP crystals allow slip on the basal plane {0001} in ⟨11$\bar{2}$0⟩ directions. Obviously, there are only 3 slip systems. For arbitrary plastic deformation, additional slip or twin systems need to be activated. This typically requires a much higher resolved shear stress and results in the brittle behavior of the HCP metals.

2.5 Twinning

Another important plastic deformation mechanism is twinning. A shear force can produce atomic displacements such that atoms on one side of a plane or the twin plane are located in mirror-image positions of atoms on the other side.

Twinning, like the slip, occurs in a specific direction called a twinning direction. The displacement magnitude within the twin region is proportional to the distance from the twin plane. Compared with slip, twinning requires relatively large shear stress, and the deformation by twinning is very fast.

Among three typical structures, BCC, FCC, and HCP, twinning easily takes place in the HCP structure because there are few operable slip systems. Twinning occurs in the BCC structure at low temperature and high loading rate. In the FCC structure, there is generally no twinning deformation. The twin structures observed in the FCC structure usually result from phase transformation during an annealing heat treatment, which are called annealing twins.

3 Experimental

3.1 Preparation

(1) Models: structure models of the unit cells of FCC, BCC, and HCP structures made by splicing ping-pong balls (Fig. 4.4)

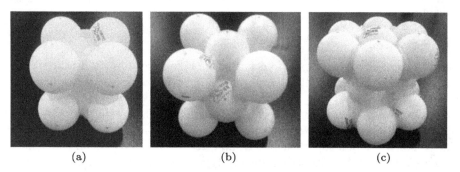

Fig. 4.4 The unit cell models of (a) BCC, (b) FCC, and (c) HCP structures.

(2) Tools: metal trays with periodic holes to fix the ping-pong balls (Fig. 4.5), ping-pong balls, personal mobile phone for taking pictures of the models built up during the experiments

3.2 Procedure

3.2.1 Identify the crystal structures

(1) Examine the unit cell models (Fig. 4.4) and identify the crystal structures.

(2) Examine the trays A, B, C, and D (Fig. 4.5). Identify the crystal planes that have the same symmetries as the holes on the trays.

(3) Take photos of the unit cells and the trays.

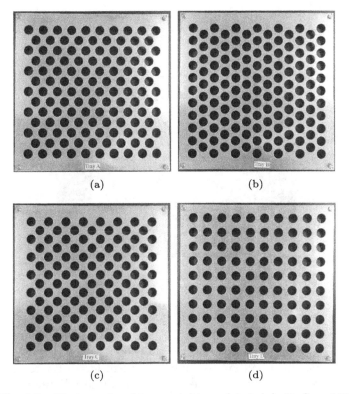

Fig. 4.5 Photographs of four metal trays labeled A, B, C, and D.

3.2.2 Build up the FCC and HCP crystal structures

(1) Build up a pyramid-like polyhedron of the FCC structure on tray B, as shown in Fig. 4.6.

(2) Verify the stacking sequence of the FCC structure.

(3) Sketch the atomic arrangement of each face of the polyhedron and identify the indices of the corresponding crystal plane.

(4) Build up a pyramid-like polyhedron of the HCP structure on tray B as shown in Fig. 4.7.

(5) Verify the stacking sequence of the HCP structure.

(6) Take photos of the structures.

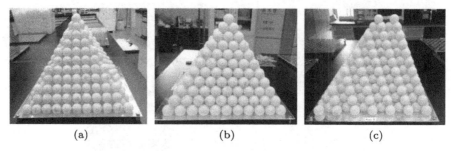

Fig. 4.6 A pyramid-like polyhedron of the FCC structure on tray B viewed from three directions.

Fig. 4.7 A pyramid-like polyhedron of the HCP structure on tray B.

3.2.3 Effect of crystal structure on slip

(1) Place a layer of ping-pong balls on trays A, B, C, and D, respectively.

(2) Put a ping-pong ball on each ping-pong ball layer. Try to move it along different directions to the first and the second nearest neighboring sites. Compare the difficulty of the movement to these sites.

(3) Sketch the atomic structures corresponding to the ping-pong ball layers on the trays. Draw directions for moving the ping-pong ball. Identify the easy directions for the movement. Determine how many easy directions can be found on each plane. Please be noted that the directions that are in the same line with opposite directions are equivalent.

(4) Put a ping-pong ball on each ping-pong ball layer. Lift the same sides of the trays slowly, as shown in Fig. 4.8. Observe the movement of the ping-pong balls on the trays. Sort the order in which the ping-pong balls start to move on the trays. Explain the result based on the crystal structure and the slip mechanism.

3.2.4 Build up a twin structure

(1) Build up a twin structure of the FCC crystal on tray B, as shown in Fig. 4.9.

Fig. 4.8 Compare the movement of a ping-pong ball on ping-pong ball layers on the trays. (See color illustration at the back of the book)

(2) Verify the stacking sequence and the twin plane.
(3) Take a photo of the twin structure.

Fig. 4.9 A twin structure on tray B. (See color illustration at the back of the book)

◪ Requirements for experimental report

(1) Give the photos taken during the experiment.
(2) Record the crystal structures, the planes, and the directions identified during the experiment.
(3) Specify the stacking sequence of the FCC, HCP, and twin structures.

◪ Questions and further thoughts

(1) List the slip systems of the BCC, FCC, and HCP structures.
(2) Discuss the effect of crystal structure on the deformation properties of materials.

Experiment 20 Construction and Analysis of Crystal Structure Models

Type of the experiment: Comprehensive
Recommended credit hours: 8
Brief introduction: In this experiment, the ball-stick structure models of several

common crystals are constructed based on understanding the crystal lattice and typical crystal structures. These crystals include carbon allotropes (diamond and graphite), ionic crystals (sodium chloride and cesium chloride), and a metal crystal (magnesium). Their crystallographic structure features will be analyzed. Then, the unit cells, the primitive cells, and the crystal directions and planes of these crystals are identified with rubber bands.

1 Objectives

(1) To understand the characteristics of 7 crystal systems and 14 Bravais lattices.
(2) To master the structures of common crystals.
(3) To understand the concepts of primitive cell and unit cell.
(4) To learn the identification method of unit cells and expressions of crystal planes and directions.

2 Principles

2.1 Crystal lattice

Crystal structures can be described by space lattice, in which every lattice point stands for one atom or atom group in the same circumstance. The atom or atom group associated with each lattice point is named motif. Consequently, a crystal structure can be defined as the collection of lattice and motif.

Because small groups of lattice points form a repetitive pattern in a lattice structure, it is convenient to subdivide a lattice into small repeat entities called unit cells. A unit cell is a parallel hexahedron. The size and shape of a unit cell can be described by three axial lengths, a, b, and c, and interaxial angles, α, β, and γ, which are called lattice constants of the unit cell. According to these lattice constants, unit cells of different types can be classified into 7 crystal systems.

The crystal systems can be further separated into 14 Bravais lattices, as listed in Table 4.1. A crystal system may include a simple lattice (P) with a lattice point at each corner of the unit cell, a body-centered lattice (I) with a lattice point at the center of the unit cell, a face-centered lattice (F) with a lattice point in the middle of each face, and a base-centered lattice (C) with a lattice point in the middle of one face. For instance, the cubic crystal system includes three Bravais lattices, simple cubic lattice, body-centered cubic (BCC) lattice, and face-centered cubic (FCC) lattice.

Table 4.1 Characteristics of the 7 crystal systems and 14 Bravais lattices

Crystal system	Bravais lattice	Lattice constants	Instances
Triclinic	P	$a \neq b \neq c$, $\alpha \neq \beta \neq \gamma \neq 90°$	K_2CrO_7
Monoclinic	P, C	$a \neq b \neq c$, $\alpha = \gamma = 90° \neq \beta$	β-S, $CaSO_4 \cdot 2H_2O$
Hexagonal	P	$a = b \neq c$, $\alpha = \beta = 90°$, $\gamma = 120°$	Zn, Cd, Mg, As
Rhombohedral	P	$a = b = c$, $\alpha = \beta = \gamma \neq 90°$	As, Sb, Bi
Orthorhombic	P, I, C, F	$a \neq b \neq c$, $\alpha = \beta = \gamma = 90°$	Fe_3C, Ga, α-S
Tetragonal	P, I	$a = b \neq c$, $\alpha = \beta = \gamma = 90°$	β-Sn, TiO_2
Cubic	P, I, F	$a = b = c$, $\alpha = \beta = \gamma = 90°$	Fe, Cr, Cu, Ag, Au

The unit cell of a Bravais lattice is defined as the one that has the highest symmetry of the space lattice and should be as small as possible. Besides, there are other ways to define a unit cell. The smallest unit cell possible for any of the lattices is the one that contains only one lattice point. It is called a primitive cell. A primitive cell, usually drawn with a lattice point at each corner. Figure 4.10 compares the unit cell and the primary cell in the FCC lattice. It can be seen that the primitive cell is much smaller than the unit cell.

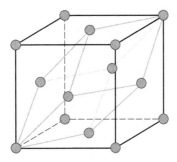

Fig. 4.10 The unit cell (outside) and primitive cell (inside) of the FCC lattice. (See color illustration at the back of the book)

2.2 Crystal direction and plane

It is often necessary to specify some particular crystal directions and crystal planes when dealing with crystalline materials. Labeling conventions have been established in which three indices are used to designate directions and planes. The basis for determining the indices is the unit cell. A coordinated system is used, in which three axes, x, y, and z, coincide with the unit cell edges.

The indices of a crystal direction are given by the coordinates u, v, and w of any point on a line parallel to the direction and passing through the origin. These coordinates are the projections along the x, y, and z axes, respectively. Because there are so many points in one line, we usually express the crystal direction with the minimum integer in square brackets. For example, since [1 1 1], [2 2 2], and [3 3 3] indicate the same direction, they are usually described by [1 1 1]. Negative indices are written with a bar over the number, such as [$\bar{1}$ $\bar{1}$ $\bar{1}$]. The equivalent directions related by symmetry that the spacing of atoms or lattice points along each direction is the same is called direction family described by $\langle u\ v\ w \rangle$. For example, the $\langle 1\ 0\ 0 \rangle$ direction family in cubic lattice has six equivalent directions [1 0 0], [0 1 0], [0 0 1], [$\bar{1}$ 0 0], [0 $\bar{1}$ 0], and [0 0 $\bar{1}$].

Crystal planes are specified by Miller indices written in round brackets $(h\ k\ l)$. If a plane is described by the Miller indices of $(h\ k\ l)$, the plane makes fractional intercepts of $1/h$, $1/k$, and $1/l$ at the unit cell edges a, b and c, respectively. A plane that lies parallel to a cell edge and never cuts it is given the index zero. When a plane intercepts at the negative side in any axis, the negative value is represented by writing a bar over the Miller indices. There are sets of equivalent lattice planes related by symmetry are called plane families described by $\{h\ k\ l\}$.

For example, the planes of a cubic lattice, (1 0 0), (0 1 0), (0 0 1) belong to the plane family {0 0 1}.

In the case of the hexagonal system, a different method for direction and plane indexing is employed. That is Miller-Bravais indices expressed in a four-axis coordinate system shown in Fig. 4.11. The unit cell of a hexagonal lattice is given by two equal and coplanar vectors a_1 and a_2 with 120° to one another and an axis c at a right angle. In addition, the axis a_3, lying on the basal plane of the hexagonal prism, is symmetrically related to a_1 and a_2.

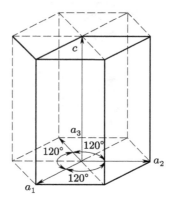

Fig. 4.11 The four-axis coordinate system of the hexagonal structure.

The directions denoted by four indices are $[u\ v\ t\ w]$. These indices are projections along the four axes. When the indices of a direction are $[U\ V\ W]$ in the three-index system, theycan be converted to the Miller-Bravais indices $[u\ v\ t\ w]$ by the following formulas:

$$u = (2U - V), \tag{4.1}$$

$$v = \frac{1}{3}(2V - U), \tag{4.2}$$

$$t = -(u+v) = -\frac{1}{3}(U+V), \tag{4.3}$$

$$w = W. \tag{4.4}$$

After calculation, it may be required to reduce u, v, t, and w to the smallest integers. For example, we can get [2 $\bar{1}$ $\bar{1}$ 0] and [1 1 $\bar{2}$ 0] from [1 0 0] and [1 1 0], respectively.

The plane denoted by four indices is $(h\ k\ i\ l)$. For the Miller-Bravais indices, the relation of h, k, i is expressed as

$$i = -(h+k). \tag{4.5}$$

With this equation, Miller indices expressed in the three-index system $(h\ k\ l)$ and Miller-Bravais indices expressed in the four-index system $(h\ k\ i\ l)$ can be readily transferred from each other.

3 Experimental

3.1 Preparation

(1) Models: structure models of 14 Bravais lattices, structure models of common crystals (diamond, graphite, sodium chloride, cesium chloride, copper, iron, and magnesium).

(2) Tools: ball-stick crystal model kit, rubber bands, personal mobile phone.

3.2 Procedure

3.2.1 Observe the structure models

(1) Observe and identify the 14 Bravais lattice models.

(2) Observe and identify the common crystals. Analyze their structure features, including lattice type, number of atoms in a unit cell, and coordination number.

3.2.2 Construct crystal structure models

(1) Construct the crystal structure models of diamond, graphite, sodium chloride, cesium chloride, and magnesium using the ball-stick crystal model kit.

(2) Show the unit cells of the crystal structure models with rubber bands.

(3) Show several directions and planes in the models with rubber bands. The crystal directions are [100], [110], [111], [121], [221], and the crystal planes are (100), (110), (111), (121), (221).

(4) Show the primitive cells of the models of diamond and sodium chloride with rubber bands.

(5) Take photos of the constructed models.

4 Requirements for experimental report

(1) Draw the unit cells of 14 Bravais lattices.

(2) Tabulate the structural features of the common crystals.

(3) Give the photos of the constructed models. Label the crystal directions and the crystal planes shown by rubber bands. For the hexagonal structure, transfer the indices of the directions and planes in the three-index system to those in the four-index system.

5 Questions and further thoughts

(1) Why are the properties of diamond and graphite so different?

(2) What are the differences among unit cell, primitive cell, and Wigner-Seitz cell?

Extended readings

Callister W D, Rethwisch D G. Fundamentals of Materials Science and Engineering. 5th Edition. John Wiley & Sons Inc., 2001.

Vainshtein B K. Fundamentals of Crystals: Symmetry, and Methods of Structural Crystallography. 2nd Edition. Springer, 1994.

Zolotoyabko E. Basic Concepts of Crystallography. Wiley-VCH, 2011.

Part III
Material Properties

Chapter 5 Mechanical Properties of Materials

Experiment 21 Tensile Test of Materials

Type of the experiment: Comprehensive
Recommended credit hours: 12
Brief introduction: This experiment introduces the principles and techniques of the tensile test. The tensile properties of some widely used metal and polymer materials are measured, and the influencing factors are analyzed. The tensile behaviors of common materials are intuitively understood. The experiment is divided into three experiments that can be conducted respectively. The first experiment introduces the principle and experimental apparatus of the tensile test and compares the tensile properties of aluminum alloy plates after cold rolling to show the effect of work hardening. The second experiment introduces the application of the extenso meter and strain gauge for precisely measuring the tensile properties of several common metals and their Poisson's ratios. The third experiment measures the tensile properties of common polymers, and the effects of molecular structure and strain rate are studied. A large range extenso meter is used to measure the relatively large elongation of polymers.

Experiment 21.1 Tensile Properties of Metals and Work Hardening

Type of the experiment: Comprehensive
Recommended credit hours: 4

1 Objectives

(1) To understand the principle of the tensile test.
(2) To learn the operation of the tensile test machine.
(3) To evaluate the tensile properties of Al alloy plates from the stress-strain curves.
(4) To learn the process of cold rolling.
(5) To understand the mechanism of work hardening and its effect on tensile properties.

2 Principles

2.1 Tensile test and stress-strain curve

A tensile test is a fundamental mechanical test where a carefully prepared sample is loaded in a very controlled manner while measuring the applied load and the elongation of the sample over some distance (Fig. 5.1). The ability to resist breaking under tensile stress is one of the most important and widely measured properties of materials used in structural applications. The results of the test are important for component design and material selection for specific applications.

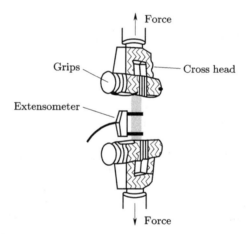

Fig. 5.1 Schematic illustration of the tensile test.

As shown in Fig. 5.2, a plot of engineering stress, σ, versus engineering strain, ε, is constructed during a tensile test, which can be obtained automatically on the software provided by the instrument manufacturer. Mathematically, the stress is calculated by

$$\sigma = \frac{F}{A_0}, \tag{5.1}$$

where F is the instantaneous force applied by the machine in the axial direction, and A_0 is the original cross-sectional area of the sample measured before the test. The strain is dimensionless, which is calculated by

$$\varepsilon = \frac{L - L_0}{L_0} = \frac{\Delta L}{L_0}, \tag{5.2}$$

where L is the instantaneous gauge length, L_0 is the initial length before the test, and ΔL is the deformation elongation at some instant.

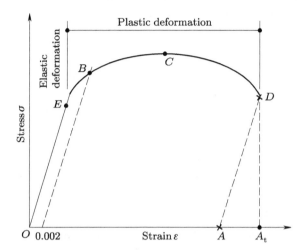

Fig. 5.2 Representative tensile stress-strain curve of metals.

In most cases, standard samples such as round rods and rectangular bars are used, while nonstandard samples can also be tested for specific purposes. In this experiment, the samples are aluminum alloy plates with a size of about 12 cm × 2 cm. They are nonstandard samples for the tensile test but are convenient for cold rolling. Moreover, to simplify the experiment, the extenso meter or strain gauge is not used. Instead, the distance between the grips after loading the sample is used as the gauge length so that the displacement of the crossheads can be simply considered as the deformation.

2.2 Tensile properties

The tensile properties can be determined from the stress-strain curve, including but not limited to elastic modulus, elastic limit, elongation, tensile strength, and yield strength. Several important properties will be tested and explained in detail.

2.2.1 Tensile elastic modulus

Tensile elastic modulus, or Young's modulus, is a property associated with elastic deformation (segment OE in Fig. 5.2). It is the ratio of stress to elastic strain, i.e.

$$E = \frac{\Delta \sigma}{\Delta \varepsilon}, \tag{5.3}$$

where E is the elastic modulus, and $\Delta \sigma$ and $\Delta \varepsilon$ are respectively the changes of tensile stress and strain in the elastic deformation range. Thus, a larger tensile modulus means that the material is rigid, and higher stress is required to produce a given amount of strain.

2.2.2 Tensile strength

Tensile strength is a measure of the force required to pull a sample to the point where it stretches. There are several typical definitions of tensile strength.

Yield strength is the stress at which permanent deformation occurs. It represents the upper limit of the load that can be safely applied to a material, which makes it an essential parameter for designing components. If the stress-strain curve has a yielding plateau, the yield strength is generally defined as the average stress associated with the lower yield point. If no yield plateau can be detected, yield strength is the stress that causes permanent deformation of 0.2% of the original dimension (the stress at point B in Fig. 5.2).

Ultimate strength is the maximum stress that a material can withstand before fracture (the stress at point C in Fig. 5.2). For plastic materials, it is the maximum stress before necking. Fracture strength is the stress that the material is ruptured (the stress at point D in Fig. 5.2).

2.2.3 Tensile elongation

Tensile elongation or ductility is the ability of a material to deform under tensile stress. It is the opposite of brittleness. The strain at fracture (segment OA_t in Fig. 5.2) and the strain after fracture (segment OA in Fig. 5.2) are often used to measure ductility. The former is measured just before the sudden decrease in force associated with fracture. The latter does not take the elastic strain into account. It is measured by fitting the two halves of the broken sample together or determined on the stress-strain curve by drawing a line (segment AD) parallel to the line of elastic deformation (segment OE).

2.3 Rolling process and work hardening

Rolling is a fabricating process in which a metal sample is passed through a pair of rollers (Fig. 5.3). Rolling is classified into two types according to the recrystallization temperature, hot rolling, and cold rolling, and they each have their specific application.

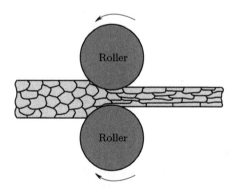

Fig. 5.3 Schematic illustration of the rolling process.

Hot rolling uses large metal pieces, such as slabs or billets, and heats them above their recrystallization temperatures. The metal pieces are then deformed between rollers to decrease the thickness. Hot rolling also reduces the average grain size of metals but maintains an equiaxial microstructure due to recrystallization.

Cold rolling is a process that passes metals through a pair of rollers at temperatures below their recrystallization temperatures. This process results in dislocation multiplication. As the dislocation density increases, the interaction between the dislocations prevents further slip, leading to the increase in the hardness and strength of metals. This phenomenon is called work hardening or strain hardening.

3 Experimental

3.1 Preparation

(1) Materials: aluminum alloy plates with a size of about 12 cm × 2 cm and a thickness of about 1.0 mm. They have been annealed at 360 °C for 2 h.

(2) Apparatuses: universal testing machine (SANS CMT4304), rolling machine (MTI corporation, MSK-2150).

(3) Tools: Vernier caliper, ruler.

3.2 Procedure

3.2.1 Cold rolling of aluminum alloy plates

(1) Cold roll three Al alloy plates, so that their thicknesses reach about 0.8 mm, 0.6 mm, and 0.4 mm, respectively. Measure the thicknesses by Vernier caliper. To avoid bending of the plates during rolling, the distance of the rollers is decreased stepwise at intervals of 0.05 mm or less.

(2) The cold-rolled plates are conditioned at room temperature for 24 h before the tensile test. In order to conduct the experiment smoothly, some plates may be cold-rolled and conditioned before the class.

3.2.2 Tensile test

(1) Take four Al plates with thicknesses of 1.0 mm, 0.8 mm, 0.6 mm, and 0.4 mm. Measure their widths and the thicknesses by Vernier caliper. Mark two parallel lines with a distance of 80 mm on each plate. The distance is used as the gauge length.

(2) Switch on the universal testing machine and the computer. Open the software.

(3) Select the test module "Tensile test of metals at room temperature" in the software and set the testing parameters accordingly. The tensile rate is 2 mm/min. Fill in the sample information, including width, thickness, and gauge length. It is worth mentioning that the rate at which a sample is pulled apart in the test will influence the results.

(4) Load a sample in the grips. The marked lines are aligned at the edges of the grips, and the displacement of the crossheads is considered as the deformation.

(5) Set the force and the displacement to zero, and then start the test. The force-displacement curve and the corresponding data are simultaneously recorded by the computer.

(6) Export the testing result as an Excel file.

(7) Repeat steps (3) to (6) for the plates with different thicknesses.

4 Requirements for experimental report

(1) Convert the force-displacement data into the stress-strain data. Plot the stress-strain curves of the Al alloy plates with different thicknesses in one figure for comparison. The data exported in the Excel files should be used. Label these curves clearly. The software Origin is recommended for the data analysis.

(2) Calculate and tabulate the tensile properties of the Al alloy plates, including elastic modulus, tensile yield strength, ultimate tensile strength, fracture strength, fracture strain, and tensile ductility.

(3) Discuss the effect of work hardening on the tensile properties by comparing the stress-strain curves and the data in the table.

5 Questions and further thoughts

(1) How is the elastic modulus obtained from a stress-strain curve?

(2) The displacement of the crossheads is simply considered as the deformation in this experiment. Which experimental data will be subject to error by such simplification?

Experiment 21.2   Precise Measurement of Tensile Properties

Type of the experiment: Comprehensive
Recommended credit hours: 4

1 Objectives

(1) To understand the principles and applications of extensometer and strain gauge for the tensile test.

(2) To measure the tensile properties of pure copper, brass, low carbon steel, and cast iron.

(3) To understand the tensile behaviors of common metal materials.

2 Principles

2.1   Extensometer for tensile test

Although tensile test machine allows the usage of the crosshead displacement to calculate the elongation of a sample, much greater accuracy is achieved by direct measurement using an extensometer. An extensometer is an apparatus that is used in conjunction with a universal testing machine. It measures elongation with high accuracy.

An extensometer consists of a sensor, an amplifier, and a recorder. The sensor is in direct contact with the sample under test. The distance between the measured points on the sample is the gauge length, and the change of the gauge length (elongation or shortening) is the deformation. As the sample deforms, the sensor transforms the deformation into a mechanical, optical, electrical, or acoustic signal, and the amplifier amplifies the signal. The recorder displays and records the amplified signal automatically.

There are many kinds of extensometers, which can be divided into mechanical extensometer, optical extensometer, and electromagnetic extensometer. These extensometers are generally as sensitive as one micron (μm). The clip-on extensometer is the most common mechanical extensometer. In practice, to protect the extensometer, it is removed before the sample rupture or before the elongation reaches the maximum measurement range of the extensometer. Further elongation after removing the extensometer can be measured from the crosshead movement or the gauge length marked on the sample.

In this experiment, the clip-on extensometer will be used for the tensile test of standard rod samples of pure copper, brass, low carbon steel, and cast iron.

2.2 Strain gauge for measurement of elastic modulus and Poisson's ratio

A strain gauge is a device used to measure the strain on an object. The most common type is the resistance strain gauge. It operates on the principle that the electrical resistance of a metal wire changes when the wire is stretched or compressed.

The gauge consists of a flexible insulating backing that supports a metallic foil pattern. It is firmly attached to the surface of the sample by a suitable adhesive, such as cyanoacrylate. The foil is deformed with the sample, causing its electrical resistance to change. This resistance change is usually measured using a Wheatstone bridge, and it is related to the strain by the quantity called the gauge factor.

For the measurement of elastic modulus using a strain gauge, an equal loading method is generally used. The tensile force F_1 on the sample is loaded step by step to minimize the measurement error. The longitudinal strain ε_l and transverse strain ε_t associated with the force can be recorded simultaneously. The elastic modulus E is equal to the slope obtained by fitting the curve of the longitudinal stress versus the longitudinal strain.

The elongation of the sample along the stretch direction by the tensile force must result in the transverse contraction. For elastic deformation, the ratio of the transverse strain, ε_t, to the longitudinal strain, ε_l, is a constant called Poisson's ratio, which is calculated by

$$\nu = -\frac{\varepsilon_t}{\varepsilon_l} \tag{5.4}$$

During applying the tensile force on the sample, a series of ε_t and ε_l values can be recorded. Then, a straight line of ε_t versus ε_l can be fitted, and the slope is the Poisson's ratio.

In this experiment, the strain gauge will be used for the tensile test of standard rectangular bar samples of pure copper, brass, low carbon steel, and cast iron. The elastic modulus and Poisson's ratio will be determined precisely.

3 Experimental

3.1 Preparation

(1) Materials: standard rod samples and standard rectangular bar samples of

pure copper, brass (H68), low carbon steel (T20), and white cast iron.

(2) Apparatus: universal testing machine (SANS, CMT4304).

(3) Tools: extensometer, strain gauges, Vernier caliper, ruler.

3.2 Procedure

3.2.1 Tensile test with an extensometer

(1) Measure the diameters of the rod samples.

(2) Switch on the universal testing machine and the computer. Open the software.

(3) Select the "Tensile test of metals at room temperature" module in the software. Set the experimental parameters. The tensile rate is 1 mm/min. The switch point of the extensometer is 0.2 mm.

(4) Load a sample in the grips. Clamp the sample with the extensometer with a gauge length of 50 mm.

(5) Set the force and displacement to zero, and then start the test. The load-displacement curve and the stress-strain curve are simultaneously recorded by the computer. Remove the extensometer when the elongation reaches the switch point.

(6) Export the testing result as an Excel file.

(7) Repeat steps (3) to (6) for all the samples.

3.2.2 Tensile test with strain gauges

(1) Measure the widths and thicknesses of the bar samples.

(2) Stick two strain gauges on both sides of each sample at the same height in transverse and longitudinal directions, respectively.

(3) Switch on the universal testing machine and the computer. Open the software and connect the machine.

(4) Select the test module "Tensile test of metals at room temperature" in the software. Set the experimental parameters.

(5) Load a sample in the grips. Using at least 6 different weights not exceeding 2000 N on the sample, take output readings from each guage for each weight.

(6) Record the corresponding transverse strain and longitudinal strain obtained from the strain gauge at each load level.

4 Requirements for experimental report

(1) Plot the stress-strain curves of the tensile test with the extensometer. The data exported in the Excel files should be used.

(2) Calculate and tabulate the tensile properties of the tensile test with the extensometer, including elastic modulus, tensile yield strength, ultimate tensile strength, and fracture strength. The calculation method has been described in Experiment 21.1.

(3) Tabulate the values of the longitudinal force F_l, longitudinal stress σ_l, longitudinal strain ε_l, and transverse strain ε_t of the samples tested with the strain gauge.

(4) Plot the $\sigma_l - \varepsilon_l$ curves. Calculate the elastic moduli by linear fitting the curves.

(5) Plot the $\varepsilon_t - \varepsilon_l$ curves. Calculate the Poisson's ratios by linear fitting the curves.

5 Questions and further thoughts

(1) In this experiment, the elastic moduli are obtained from the tensile test using the extensometer and strain gauge. Please compare the advantages and disadvantages of the two methods.

(2) To determine the Poisson's ratio, the two strain gauges are attached to both sides of a sample at the same height. How will it affect the results if they are attached to the same side of the sample?

(3) Are there any other methods that can be used to measure the elastic modulus?

Experiment 21.3 Influence of Molecular Structure and Strain Rate on the Tensile Properties of Polymers

Type of the experiment: Cognitive
Recommended credit hours: 4

1 Objectives

(1) To determine the tensile properties of common polymers with a large range extensometer.

(2) To investigate the effect of molecular structure on the tensile properties of polymers.

(3) To investigate the effect of strain rate on the tensile properties of polymers.

2 Principles

2.1 Structure of polymers

A polymer is a chemical compound whose molecules are bonded together in long repeating chains. Many common types of polymers are composed of hydrocarbons, i.e., compounds of carbon and hydrogen. These polymers are specifically made of carbon atoms bonded together into long chains. Some polymers contain only carbon and hydrogen atoms, such as polyethylene, polypropylene, polybutylene, and polystyrene. Other common manufactured polymers have backbones that include elements other than carbon. For example, nylons contain nitrogen atoms in the repeat unit backbone, and polyesters and polycarbonates contain oxygen in the backbone.

The structures of the molecular chains can be grouped into linear, branched, crosslinked, and network structures. Linear polymers are those in which the units are joined together end to end in single chains. Branched polymers have side-branch chains connected to the main ones. In crosslinked polymers, adjacent linear chains are joined one to another at various positions by covalent bonds. The chains in network polymers form three-dimensional networks.

2.2 Tensile properties of polymers

The mechanical properties of polymers generally depend on their molecular structures and molecular weights. Besides, the degree of crystallinity and the glass transition temperature, T_g, also affect the mechanical properties. Highly crystalline polymer materials with a T_g above room temperature are usually brittle. When stress is applied to the polymers, the chains stretch, leading to the elongation of polymers in both crystalline and amorphous regions. When a semi-crystalline polymer undergoes a tensile test, the amorphous chains will become aligned. This is usually evident for transparent materials, which become opaque upon turning crystalline.

The equipment for the tensile test of polymers is the same as that for metal materials as introduced in Experiment 21.1. In addition, the large range extensometer is generally used due to the relatively large elongation of polymers. Again, the stress-strain curve can be plotted as shown in Fig. 5.4. The stress-strain behavior of a polymer material depends on various parameters such as molecular characteristics, microstructure, strain rate, and temperature. The mechanical properties of materials, including elastic modulus, yield strength, ultimate tensile strength, and elongation, can be determined from the curve.

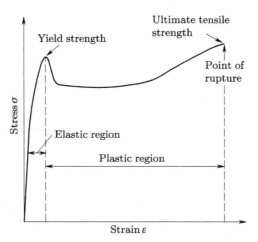

Fig. 5.4 Representative tensile stress-strain curve of polymers.

The features of the stress-strain curve of a polymer sample shown in Fig. 5.4 are greatly different from that of metal materials shown in Fig. 5.2 because of the distinct deformation mechanism of polymers. When a polymer is deformed, the stress increases with the strain at the initial elastic region. The elastic deformation takes place due to the elongation of chain molecules by bond stretching and bond rotation along the direction of the applied loading. The applied load leads to the maximum stress, which is the yield point of the polymers, indicating the start of permanent deformation.

After this yield point, necking forms, and the sample stretches out considerably in the plastic region. This is associated with the elongation of amorphous tie chains, and followed by the tilting of lamellar chains and the separation of crystalline block segments. As for semi-crystalline polymers, cold drawing occurs in this region. Cold drawing refers to the behavior of polymers that develop oriented neck upon stretching. Continuous elongation results in the orientation of lamellar chains and amorphous tie chains along the tensile axis, leading to strain hardening of materials.

Further loading will result in the ultimate rupture. At the rupture point, the corresponding stress and strain are called the ultimate strength and the elongation at break, respectively.

3 Experimental

3.1 Preparation

(1) Materials: plastic bags and ropes (polyethylene), standard bar samples of polypropylene (PP), high-density polyethylene (HDPE), polycarbonate (PC), and polystyrene (PS) prepared by injection molding.

(2) Apparatus: universal testing machine equipped with a large range extensometer (SANS CMT4304).

(3) Tools: Vernier caliper, ruler.

3.2 Procedure

3.2.1 Tensile test of plastic bag and rope by hand

(1) Pull the plastic bag and plastic rope by hand indifferent directions.

(2) Intuitively feel the difference in the tensile behaviors in different directions.

3.2.2 Tensile test of the polymer samples

(1) Measure the widths and thicknesses of the polymer samples using Vernier caliper.

(2) Switch on the universal testing machine and the computer. Open the software.

(3) Select the test module "Tensile test of plastics at room temperature" in the software and set the testing parameters.

(4) Load a sample in the grips. Clamp the sample with the large range extensometer with a gauge length of 75 mm.

(5) Set the force and displacement to zero and then start the test. The force-displacement curve and the corresponding data are simultaneously recorded by the computer.

(6) After the experiment, export the raw data as an Excel file.

(7) Repeat steps (3) to (6) for PP, PE, PC, and PS respectively at a tensile rate of 30 mm/min.

(8) Repeat steps (3) to (6) for PP at tensile rates of 50, 100, and 200 mm/min, respectively.

4 Requirements for experimental report

(1) Explain the difference in tensile behaviors of the plastic bag and rope when they are pulled along different directions.

(2) Plot the stress-strain curves of PP, PE, PC, and PS at the tensile rate of 30 mm/min in one figure using the data exported in the Excel files.

(3) Plot the stress-strain curves of PP at different tensile rates in one figure using the data exported in the Excel files.

(4) Calculate and tabulate the tensile properties of these samples from the tensile stress-strain curves, including elastic modulus, yield strength, ultimate tensile strength, and tensile ductility.

(5) Sketch the molecular structures of the polymers tested in this experiment. Explain the difference in the tensile properties in regards to their molecular structures.

(6) Discuss the influence of strain rate on the tensile properties of PP.

5 Questions and further thoughts

(1) What is the cold drawing of polymers?
(2) Discuss the influencing factors on the tensile properties of polymers.

6 Cautions

After each tensile test, take off the large range extensometer from the sample and carefully fix it on the stand.

Experiment 22   Hardness Test of Materials

Type of the experiment: Comprehensive
Recommended credit hours: 10
Brief introduction: This experiment introduces the principles and techniques of hardness test for metals and polymers, as well as the influencing factors on the hardness. The experiment is divided into two experiments that can be conducted respectively. The first experiment introduces the principles of metal heat treatment and Vickers hardness. The effects of heat treatment on the microstructure and hardness of saw blades, which is a high carbon eutectoid steel (T8 steel), are studied. The usage of saw blades as the samples can also contribute to the cognition of the application of heat treatment on engineering components. The cold-rolled aluminum alloy plates are also examined to show the effect of work hardening on the hardness. The results can be associated with those in the experiment of the tensile test (Experiment 21.1). The second experiment describes the hardness test of polymers. The Shore hardness values of several widely used polymer materials are measured.

Experiment 22.1   Heat Treatment and Hardness Test of Metals

Type of the experiment: Comprehensive
Recommended credit hours: 8

1 Objectives

(1) To understand the typical heat treatment methods.
(2) To understand the principle of Vickers hardness.
(3) To determine the hardness values of heat-treated T8 steel and cold-rolled Al alloy plates.
(4) To understand various heat treatment methods for manipulating the microstructures and mechanical properties of metals.
(5) To understand the influence of cold rolling on the hardness of metals.

2 Principles

2.1 Heat treatment

Heat treatment is the controlled heating and cooling of metallic alloys. Many microstructural features, especially phase constitution, are controlled by the heat treatment. It is often used to alter the specific physical and mechanical properties of final products without a change in their shape. It is also used to improve machinability, improve formability, and restore ductility after cold deformation.

Typical heat treatment methods usually include annealing, tempering, normalizing, and quenching. These are explained in detail as follows.

2.1.1 Annealing

Annealing involves heating metal to above its recrystallization temperature, holding at a suitable temperature, and then cooling slowly at a controlled rate in the furnace. Annealing temperatures vary with metals and alloys and with properties desired but must be within a range that prevents the apparent growth of crystals. It is mainly used to increase the ductility and reduce the hardness of a material. This change results from the reduction of dislocations in the material, leading to a near-equilibrium microstructure. Annealing is often performed after a sample has undergone a cold deformation to prevent it from brittle failure, improve machinability, or remove residual stress.

2.1.2 Normalizing

Normalizing involves heating metal to an elevated temperature, keeping it at that temperature for a period, and then air cooling. The heating temperature for normalization usually is also above the recrystallization temperature. It also results in a near-equilibrium microstructure but finer than that of annealing because of the faster cooling rate. For example, the resulting microstructure of carbon steel is a mixture of ferrite and cementite. The normalizing produces a uniform, fine-grained structure with a more uniform carbide size and distribution that could not be achieved in the preceding casting, forging, or rolling processes and, as a result, to obtain improved ductility and impact toughness.

2.1.3 Quenching

Quenching involves heating metal, keeping it at an appropriate temperature, and then rapidly cooling it in water, oil, or air. Quenching prevents undesired low-temperature phase transformation by reducing the cooling time during which these undesired reactions take place.

Quenching is most commonly used to harden steel, in which case the steel is rapidly cooled to introduce martensite. Martensite has a distorted body center tetragonal (BCT) structure with supersaturated carbon. It has high hardness and is very brittle as well. The heating temperature and time should be increased to ensure that the workpiece will be fully transformed into austenite.

2.1.4 Tempering

Tempering involves heating metal to a temperature below the critical point for a certain period, then cooling it at any rate desired. It is usually performed after quenching to reduce some of the excess hardness and to increase the toughness. The exact temperature that determines the amount of hardness removed depends on both the specific composition of the alloy and on the desired properties in the finished product. For instance, for tool steels, the hardness of which must be retained, the tempering temperature is usually at relatively low temperature from 200 to 250 °C, while springs are tempered at much higher temperatures.

In the case of carbon steels, tempering alters the size and distribution of carbides in the martensite, forming a microstructure called tempered martensite. Tempering is also performed on normalized steels and cast irons, to increase their ductility, machinability, and impact strength.

2.2 Vickers hardness

Hardness is defined as the resistance to indentation, and it is determined by measuring the permanent depth of the indentation. For a fixed load and a given indenter, the harder the material, the smaller the indentation. The hardness value is then obtained by measuring the depth or the area of the indentation.

The principle of the Vickers hardness test is shown in Fig. 5.5. It consists of indenting the test material with a diamond indenter, which is a right pyramid with a square base and an angle of 136° between opposite faces. The load selected between 1 and 1000 gf[①] is typically applied for 10 to 15 s. The two diagonals of the indentation left on the surface of the material after removing the load are measured using a microscope. The Vickers hardness is the quotient obtained by dividing the load by the area of indentation, i.e.

$$\mathrm{HV} = \frac{2F \sin \frac{136°}{2}}{d^2} = 1.854 \frac{F}{d^2}, \tag{5.5}$$

where HV is the Vickers hardness, F is the load (in kgf), and d is the mean value of the two diagonals, d_1 and d_2 (in mm).

3 Experimental

3.1 Preparation

(1) Materials: saw blades, which are T8 carbon steel with a carbon content of 0.75–0.80wt%; aluminum alloy plates with thickness of 1.0 mm, 0.8 mm, 0.6 mm, and 0.4 mm, which are obtained by cold rolling in experiment 21; Bakelite powder.

[①] 1 gf = 9.8 mN, the same below.

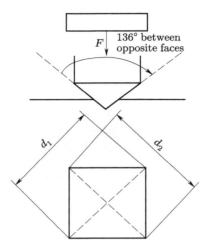

Fig. 5.5 The principle of the Vickers hardness test.

(2) Apparatuses: two muffle furnaces respectively heated to 900 °C and 400 °C before the experiment, mounting machine (Struers, CitoPress-10), Vickers microhardness tester (Shanghai optical instrument factory, HXD-1000TMC), metallographic microscope.

(3) Tools: heat treatment pliers, heat resistant gloves, Vernier caliper.

(4) Others: items for preparation of metallographic sample listed in Experiment 1.

3.2 Procedure

3.2.1 Heat treatment of saw blades

(1) Heat four saw blades in the furnace at 900 °C for 2 min to ensure that they fully transform into austenite.

(2) Conduct normalizing treatment. Take one saw blade from the furnace and cool slowly in still air. Fine pearlites are formed during the cooling process.

(3) Conduct quenching treatment. Quench the other two saw blades in water as soon as possible. The cooling rate is fast enough to prevent the formation of ferrite, pearlite or bainite, to obtain pure martensite.

(4) Conduct tempering treatment.Place one quenched blade into the furnace at 400 °C. After 5 min, cool it in still air.

(5) Conduct annealing treatment. Stop heating the furnace. Let the last saw blade stay in the furnace and cool to room temperature.

(6) Bend the four heat-treated samples by hand. Intuitively sense the difference in the toughness of the samples. The blades must be bent downward and never towards anybody.

3.2.2 Vickers microhardness test

(1) Take pieces of the heat-treated saw blades. Mount them with Bakelite powder by hot compression mounting for convenience in handling the samples.

Molding tools (two hardened steel cylinders) and a cylindrical heater placed around the mold are necessary. The general mounting procedure is as follows:

(a) A sample to be mounted is placed on the top of molding steel and inserted into the mold.

(b) The mold is filled with Bakelite powder, and the other molding steel is inserted into the open end of the mold.

(c) The prescribed pressure is exerted on the molding steels for 10 min to compress the molding material at the prescribed temperature.

(d) After cooling, the mounted sample is ejected by forcing the molding steel entirely through the mold cylinder.

(2) Grind and polish the saw blade samples. The operation has been described in Experiment 1.

(3) Measure the hardness values of the heat-treated saw blade samples. The load is 300–500 gf and the Dwell time is 10 s. For each sample, measure the hardness values at least at five positions. The general procedure for the Vickers microhardness test is as follows:

(a) Place the sample on the stage of the hardness tester, and then focus the microscope.

(b) Set the load and dwell time. Input these parameters accordingly in the measuring software.

(c) Start test.

(d) Measure the length of two diagonals, d_1 and d_2.

(e) Calculate the HV value using Equation (5.5).

(4) Measure the hardness values of the Al alloy plates with thicknesses of 1.0 mm, 0.8 mm, 0.6 mm, and 0.4 mm, respectively. The plates are placed on a fixture. The load is 25 gf, and the dwell time is 10 s.

3.2.3 Microstructure observation

(1) Etch the polished saw blade samples using an etchant of 4% nitric acid ethanol solution.

(2) Observe the samples under the metallographic microscope.

(3) Capture the images of the microstructures at optimal magnification.

4 Requirements of experimental report

(1) Tabulate the hardness values of the saw blade samples and the Al alloy plates. Calculate the average hardness values and the standard deviation.

(2) Give the metallographs of the saw blade samples. Describe the microstructure features.

(3) Discuss the effect of heat treatment on the microstructure and hardness of saw blades.

(4) Plot the hardness values of the Al alloy plates against the cold work reduction. The cold work reduction is calculated by

$$\text{cold work reduction} = \frac{t_0 - t_r}{t_0} \times 100\%, \tag{5.6}$$

where t_0 is the initial thickness (1.0 mm), and t_r is the thickness after cold rolling.

Please note that the average HV values are used, and the standard deviations should be plotted as error bars.

(5) Discuss the relationship between cold work reduction and the hardness of the Al alloy plates.

5 Questions and further thoughts

(1) Why are the samples required to be ground and polished before the microhardness test?

(2) Do you know other methods for metal hardness measurement besides Vickers microhardness?

6 Cautions

(1) Please wear heat-protective gloves during the heat treatment and mounting.

(2) Please wear acid-resistant gloves during etching.

Experiment 22.2 Shore Hardness Test of Polymers

Type of the experiment: Cognitive
Recommended credit hours: 2

1 Objectives

(1) To understand the principle of hardness test of polymers.
(2) To learn the operation of Shore durometer.
(3) To measure the Shore hardness values of common polymers.

2 Principles

2.1 Hardness of polymers

The mechanical properties of a polymer involve its behavior under stress, such as strength, stiffness, toughness, hardness, melt flow index, and ductility. In this experiment, the hardness of polymers is introduced.

The hardness of polymers is a mechanical property that provides a rapid evaluation of variations in polymer microstructure and morphology. It is most commonly measured by the Rockwell hardness test or the Shore hardness test. Both methods measure the resistance of polymers toward indentation and provide a hardness value. Note that the hardness value does not necessarily correlate well to other properties or fundamental characteristics. The Rockwell hardness is usually chosen for harder polymers, such as polycarbonate, polystyrene, and polymethacrylate. It is also used for metals. The Shore hardness is often preferred for softer polymers, such as elastomers and rubbers.

2.2 Test methods of Rockwell hardness and Shore hardness

2.2.1 Rockwell hardness

In the Rockwell hardness test, a sample is indented by a hard steel ball or a diamond cone. The Rockwell hardness scales use different loads and different indenters, as shown in Fig. 5.6. The three most common scales used for polymers are Rockwell E, M, and R. Other Rockwell hardness scales A, B, and C are commonly used for metals.

To measure the Rockwell hardness, a small preload F_0 is applied first, and the apparatus is zeroed. Then a specified larger load F_1 is applied and hold for a while. Then the load F_1 is removed while the preload is still applied. The remaining indentation is read from the scale.

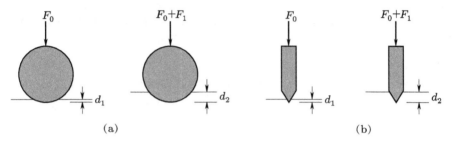

Fig. 5.6 The principle of the Rockwell hardness test with (a) spherical and (b) conical indenters.

2.2.2 Shore hardness

The Shore hardness test uses a spring-loaded steel needle with a specific shape pressing vertically under the test force into the sample surface. When the indenter foot surface is fully contacted with the sample surface, the indenter tip has a specific length L in the sample relative to the indenter plane, as shown in Fig. 5.7. The value L represents the Shore hardness. A larger L value means a lower hardness. It is worth noting that the depth of the indentation not only depends on the hardness of the material but also on its viscoelastic properties, the shape of the indenter, and the duration of the test.

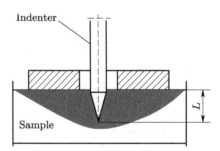

Fig. 5.7 The principle of the Shore hardness test.

Generally, there are twelve Shore hardness scales for measuring the hardness of different materials according to the ASTM D2240-00 testing standard. They are types A, B, C, D, DO, E, M, O, OO, OOO, OOO-S, and R. Each scale results in a value between 0 and 100, with a higher value indicating a harder material. The most commonly used scales are Shore A (HA) and Shore D (HD), as shown in

Fig. 5.8. The Shore A scale measures softer plastics or rubbers. The Shore D scale measures the hardness of hard rubbers, semi-rigid plastics, and hard plastics. The Shore C (HC) scale with a spherical indenter is also widely used for soft rubber, sponge, plastic foam, elastomers, and similar materials. The hardness of the three scales can be read directly from the durometers, which are calculated by the same equation

$$\text{HA, HC or HD} = 100 - \frac{L}{0.025}. \tag{5.7}$$

In this experiment, Shore durometers with scales A, C, and D are used for testing several common polymer materials.

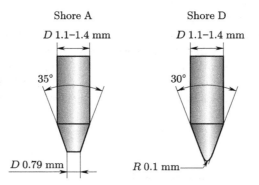

Fig. 5.8 Schematic illustration of the durometer indenters used for Shore A and D.

3 Experimental

3.1 Preparation

(1) Materials: plates of polypropylene (PP), high-density polyethylene (HDPE), polycarbonate (PC), polystyrene (PS), polyamide (PA6), rubber, and foam plastics with a thickness of about 7 mm.

(2) Apparatuses: Shore durometers with scales A, C, and D.

3.2 Procedure

(1) Select a suitable Shore durometer from Shore A, C, or D that the hardness of the selected scale should be in the range of 0 to 100.

(2) Place a sample on a hard-flat table.

(3) Press the indenter of the durometer into the sample. Ensure firmly contact the indenter foot surface with the sample.

(4) Read the hardness within one second.

(5) For each sample, measure the hardness at least at five positions.

4 Requirements for experimental report

(1) Tabulate the hardness values of the samples.

(2) Calculate the average hardness values and the standard deviations.

132　Chapter 5　Mechanical Properties of Materials

5 Questions and further thoughts

(1) Compare the similarities and differences between the two hardness testing methods for polymers.

(2) There is a duration for reading the Shore hardness value after the durometer has been firmly contacted with the sample. Analyze the effect of the duration on the hardness value.

Experiment 23　Charpy Impact Test of Polymers

Type of the experiment: Comprehensive
Recommended credit hours: 4
Brief introduction: Impact testing techniques were established to ascertain the fracture characteristics of materials, especially for some circumstances that materials fracture abruptly with minor plastic deformation. In this experiment, the principles and techniques of the impact test are introduced. The impact toughness of common polymers is measured by the Charpy impact test. The effects of molecular structure and temperature on the impact toughness of polymers are studied.

1 Objectives

(1) To understand the principle of the impact test.
(2) To learn the method of the Charpy impact test.
(3) To understand the effects of molecular structure and temperature on the impact toughness of polymers.

2 Principles

2.1　Impact toughness

Toughness is the amount of energy per unit volume that a material can absorb before rupture. It is also defined as the resistance of a material to fracture when stressed. For the static situation that the material is deformed at a very low strain rate, toughness may be measured from the area under the stress-strain curve of the tensile test. Under certain situations, material fractures abruptly with minimal plastic deformation. In this case, impact energy or impact toughness is chosen to represent the potential for fracture.

The typical impact test uses a pendulum to strike a notched sample with a defined cross-section and deform it. The energy absorbed by the sample during deformation is measured by the difference between the height from which the pendulum fell and the height to which it rose after rupturing the sample. Obviously, this energy absorption is directly related to the brittleness of the material. Brittle materials, such as ceramics or glass, tend to have lower absorption energy than ductile materials, such as copper, aluminum, and most polymers.

The Charpy and Izod notched impact tests are standardized methods used to determine the impact toughness of materials. They will be introduced in the following section. The Charpy impact test will be carried out for common polymers in this experiment.

2.2 Impact test methods

2.2.1 Charpy impact test

The Charpy impact test was invented in 1900 by Georges Augustin Albert Charpy (1865—1945). It is one of the most popular impact testing methods due to the relative ease of creating samples and obtaining results.

Figure 5.9 illustrates the apparatus of the Charpy impact test. The sample is placed horizontally on the supports. The V-shaped or U-shaped notch is placed facing away from the pendulum and helps concentrate the stress and encourage fracture. The pendulum impacts the backside of the sample. The Charpy impact test is performed according to some standards. For instance, ASTM D 6110-18isawidely accepted standard for polymers.

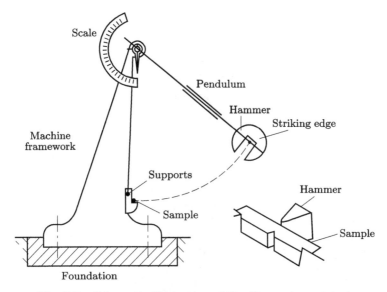

Fig. 5.9 Schematic illustration of the Charpy impact test.

2.2.2 Izod impact test

The Izod impact test was named after English engineer Edwin Gilbert Izod (1876—1946), who first described the test method in 1903. The test apparatus and sample design shown in Fig. 5.10 are very similar to those of the Charpy impact test. The primary difference from Charpy impact is the manner of sample supports. The sample is clamped vertically with the notch facing towards the pendulum. The pendulum then impacts the sample above the notch. ASTM D256-10(2018) is one of the most common standards for the Izod impact test of polymers.

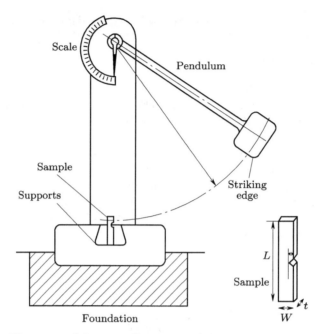

Fig. 5.10 Schematic illustration of the Izod impact test.

❸ Experimental

3.1 Preparation

(1) Materials: notched samples of polypropylene (PP), high-density polyethylene (HDPE), polycarbonate (PC), and polystyrene (PS). The sample size is 64 mm × 12.7 mm × 3.2 mm, and the notch depth is 2 mm.

(2) Apparatuses: Charpy impact testing machine (Chengde Precision Testing Machine Co. LTD., JC-5D), thermostatic water baths.

(3) Tools: beakers, tweezers.

3.2 Procedure

3.2.1 Charpy impact test of polymers at room temperature

(1) Switch on the Charpy impact testing machine.

(2) Set the experimental parameters. Select the pendulum with an impact energy of 1 J. The impact velocity is 2.9 m/s.

(3) Perform a test without a sample for the compensation of windage and friction. Press "Reset" to reset the machine. Press "Test" and raise the pendulum to the prescribed highest position. Press "Impact" and the pendulum will fall. The absorbed energy for compensation will be recorded automatically by the machine.

(4) Place the notched sample horizontally on the supports and against the anvils. Center the notch between the anvils. Raise the pendulum to the prescribed highest position and then press "Impact".

(5) Repeat step (4) for at least five samples of each type of polymer material.

(6) Consult the results and record the impact energies.

(7) Repeat steps (3) to (6) for each type of polymer material.

3.2.2 Charpy impact test of PP at various temperatures

(1) Cool some PP samples in the ice-water mixture, and heat some PP samples at 50 °C and 80 °C in the water baths.

(2) Perform the Charpy impact test for these samples. The test should be conducted as soon as possible after the samples are placed on the test machine. For the PP samples at 50 °C and 80 °C, select the pendulum with an impact energy of 5 J.

(3) Consult the results and record the impact energies.

4 Requirements for experimental report

(1) Tabulate the impact energies. Calculate the average values and standard deviations.

(2) Calculate the impact strength of each type of sample by

$$\sigma = \frac{A}{b \times d}, \tag{5.8}$$

where σ is the impact strength, A is the energy absorbed, b and d are the width and thickness of test samples at the notched cross-section.

(3) Compare the impact toughness of the polymers and figure out the fracture mechanisms of those polymers.

(4) Plot the impact energy versus the temperature for PP. Plot error bars representing the standard deviation.

(5) Discuss the influence of temperature on the impact strength of PP. Estimate the temperature over which the ductile-to-brittle transition occurs.

5 Questions and further thoughts

(1) What are the main differences between the Charpy and Izod impact tests?

(2) Analyze the possible sources of error of the Charpy impact test.

6 Cautions

The pendulum carries a risk of collision injury. Make sure there is no student in the movement direction of the pendulum before each impact test.

Experiment 24   Compression Test of Materials

Type of the experiment: Comprehensive

Recommended credit hours: 4

Brief introduction: This experiment introduces the principle and method of the compression test. The compressive properties of common materials, including aluminum alloy, low carbon steel, cast iron, and wood, are measured and compared. The effect of the length to diameter (L/D) ratio of samples is examined by comparing the compressive behaviors of the samples with different lengths.

1 Objectives

(1) To understand the principle of the compression test.
(2) To measure the compressive properties of several common materials.
(3) To understand the effect of the L/D ratio of samples on the compressive behaviors.

2 Principles

2.1 Compression test and properties

The compression test is to compress and deform a sample at various loads. It can determine the behaviors of materials under crushing loads. The compression test is often used on brittle materials to show mechanical behaviors that cannot be reflected in tensile, torsion, and bending tests.

The compression test measures the strength and plasticity of a sample during its deformation and fracture by applying axial pressure. The difference from the tensile test is that the compression test acts in the opposite direction. Therefore, the definitions of the mechanical properties and calculation formulas in the tensile test are applicable in the compression test.

During the compressive test, a plot of force versus reduction can be obtained. For both the ductile and brittle materials, the samples being compressed can withstand much larger stress than those under tension. Following yielding, failure of ductile materials will occur, and brittle materials will crush when their limits of compressive strength are reached.

For engineering application, the force-reduction plot can be converted to a stress-strain plot. As mentioned above, the definition of stress and strain is the same as those for the tensile test. For a rod sample, the compressive stress is calculated by

$$\sigma = \frac{F}{A_0}, \tag{5.9}$$

where F is the instantaneous force applied by the machine in the axial direction, and A_0 is the original cross-sectional area of the sample which is measured before the compression test. The compressive strain is dimensionless, which is calculated by

$$\varepsilon = \frac{L - L_0}{L_0} = \frac{\Delta L}{L_0}, \tag{5.10}$$

where L is the instantaneous gauge length, L_0 is the initial length before any load is applied, and ΔL is the reduction in length at some instant.

The samples for the compression test are simpler in shape than those used in the tensile test since they do not require gripping. The compression test is sometimes used for testing nonstandard samples when the samples for the tensile test are difficult to be machined.

2.2 Compressive properties

The compressive properties can be determined from the force-reduction curve. These properties include but are not limited to the elastic modulus, compressive yield strength, compressive strength. The determination of these properties is analogous to that during a tensile test.

The elastic modulus in the compressive test can be calculated with the proportional part of the force-reduction curve using the following equation

$$E_c = \frac{(F_K - F_J)L_0}{(\Delta L_K - \Delta L_J)S_0},\tag{5.11}$$

where E_c is compressive elastic modulus, F_K and F_J are the pressures of points K and J at the proportional part in Fig. 5.11, and ΔL_K and ΔL_J are the corresponding reductions. Obviously, it can also be simply determined from the slope of the stress-strain relation in the elastic range.

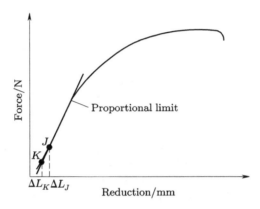

Fig. 5.11 Determination of the compressive elastic modulus.

Compressive yield strength is defined as the stress associated with the lower yield point if the force-reduction curve or stress-strain curve has a yielding plateau. For many materials, there are no yield processes during the compression test. In this case, the compressive yield strength is the stress which will cause a permanent deformation 0.2% of the original dimension.

Compressive strength is the maximum stress that a material can withstand before fracture under uniaxial pressure. For ductile materials that compressive fracture does not take place, the maximum stress is defined as that associated with a specified deformation.

2.3 Compression behaviors of ductile and brittle materials

Under uniaxial compression, ductile materials do not exhibit sudden fracture. Generally they can only be flattened and will not be damaged. There are a number of deformation modes that do not occur in the tensile test, as illustrated in Fig. 5.12. These modes are determined by the L/D ratio and the friction at the contact surfaces. A smaller L/D ratio results in a more significant effect of friction on the experimental results. Therefore, the L/D ratio can be increased to reduce the impact of friction. However, if the L/D ratio is too large, it will also lead to longitudinal instability. Besides, lubricant can be applied to the contact surface to reduce the impact of friction.

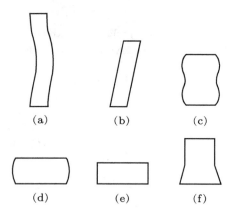

Fig. 5.12 Deformation modes in the compression test of ductile materials: (a) buckling when $L/D > 5$; (b) shearing, when $L/D > 2.5$; (c) double barreling when $L/D > 2.0$ and friction is present at the contact surfaces; (d) barreling when $L/D < 2.0$ and friction is present at the contact surfaces; (e) homogenous compression when $L/D < 2.0$ and no friction is present at the contact surfaces; (f) compressive instability due to work softening.

The brittle materials under uniaxial compression exhibit sudden fracture, and many brittle materials do not exhibit yield before failure. Some of these materials may show deviation from the linear elastic deformation during the tensile test.

Some materials show both ductile and brittle behaviors, depending on the states or compositions of the materials being tested. Steel is a typical example of dual behaviors. For instance, it is known that high carbon steel shows brittle behavior at low temperatures and is ductile at high temperatures. Besides, the carbon content in steel also changes its behavior, making it either brittle or ductile.

In this experiment, the samples for the compressive test are aluminum alloy, low carbon steel, cast iron, and wood rods with a diameter of 10 mm. Each type of sample has several lengths so that the L/D ratio will be examined. In addition, the effect of carbon content on the compression behavior can be seen by comparing the compression behaviors of low carbon steel and cast iron.

3 Experimental

3.1 Preparation

(1) Materials: rod samples of aluminum alloy, Q235 low carbon steel, cast iron, and birch wood with sizes of $\Phi 10$ mm × 10 mm, $\Phi 10$ mm × 20 mm, and $\Phi 10$ mm × 30 mm.

(2) Apparatus: universal testing machine (SANS CMT5105).

(3) Tool: Vernier caliper.

3.2 Procedure

(1) Measure the diameters and lengths of the samples three times using Vernier caliper and record the average values.

(2) Switch on the universal testing machine and the computer. Open the soft-

ware.

(3) Select the test module "Compressive test of metals at room temperature" in the software and set the testing parameters accordingly. The compressive speed is 1 mm/min.

(4) Place a sample on the center of the bottom fixture. Lower the upper fixture slowly until it is close to the test sample.

(5) Set the force and displacement to zero, and then start the test. The load-displacement curve is simultaneously recorded by the computer.

(6) Export the testing result as an Excel file.

(7) Repeat steps (3) to (6) three times for each type of sample.

4 Requirements for experimental report

(1) Plot the force-displacement curves and the stress-strain curves of the samples. The curves of the same type of samples with different lengths should be drawn in one figure for comparison.

(2) Determine and tabulate the compressive properties, including elastic modulus, compressive yield strength, compressive strength, and deformation mode for each test.

(3) Discuss the effect of the L/D ratio on the compressive behaviors of the samples.

(4) Discuss the effect of the carbon content on the compressive behaviors of low carbon steel and cast iron.

5 Questions and further thoughts

(1) Why cannot the compressive strength of low carbon steel be measured?

(2) Why does the cast iron fracture along a plane that is 45° from the top or bottom surface during compression?

Experiment 25 Three-Point Bending Test of Materials

Type of the experiment: Comprehensive
Recommended credit hours: 4
Brief introduction: This experiment introduces the principle and method of the three-point bending test. The flexural properties of several common materials, including cast iron, low carbon steel, ceramic tile, and birch wood, are measured and compared. Students can intuitively understand the flexural behaviors of different materials under bending load.

1 Objectives

(1) To understand the principle of the three-point bending test.
(2) To measure the flexural properties of several common materials.
(3) To compare the flexural behaviors of common materials.

❷ Principles

2.1 Three-point bending test

A flexural test of materials is often required for design purposes since most structural components in actual engineering applications are subjected to bending. A suitable flexural test is the most frequently employed for brittle materials because it is relatively difficult to prepare and grip brittle samples for the tensile test. In addition, the brittle materials for the tensile test require perfect alignment to avoid the presence of bending stress since they fail after a limited strain.

The flexural test is usually carried out with either the three-point or the four-point configurations. The three-point bending test is simpler, while the four-point one has the advantage of imposing a constant radius of curvature in the section of the sample between the central knife-edge supports. In this experiment, the three-point bending test will be carried out for several common materials.

The three-point bending test is to bend a rod sample having either a circular or rectangular cross-section until fracture. The span-to-depth ratio of the samples is usually specified by the relevant standards, such as ASTM C1161 for ceramic materials. The configuration is illustrated in Fig. 5.13. A sample is placed on two fixed supports, and a point force is applied in the middle. The top surface of the sample is in a state of compression, whereas the bottom surface is in tension. The stress depends on the sample thickness, the bending moment, and the moment of the cross-section inertia. For a rectangular or circular cross-section, the stress

$$\sigma = \frac{3FL_s}{2bd^2}, \quad (5.12)$$

or

$$\sigma = \frac{FL_s}{\pi R^3}, \quad (5.13)$$

where F is the applied load, L_s is the distance between support points, and other parameters are indicated in Fig. 5.13.

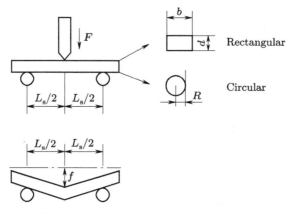

Fig. 5.13 Schematic illustration of the three-point bending test.

2.2 Flexural properties

A plot of applied force F versus deflection f is constructed during the bending test, as shown in Fig. 5.14. The flexural properties can be determined from the force-deflection curve. These properties include but are not limited to flexural modulus, flexural strength, and fracture deflection.

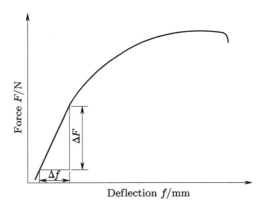

Fig. 5.14 Representative force-deflection curve of the three-point bending test.

2.2.1 Flexural Modulus

Flexural modulus is an indication of the stiffness of material under bending. It can be determined from the slope of the proportional part in the elastic bending range in the force-deflection curve. It can be calculated by

$$E_\text{b} = \frac{L_\text{s}^3}{48I}\left(\frac{\Delta F}{\Delta f}\right), \quad (5.14)$$

where E_b is the flexural modulus, ΔF and Δf are respectively the increments in the force and deflection in the elastic bending range, as shown in Fig. 5.14. I is the inertia factor, which can be calculated by

$$I = \frac{bd^3}{12} \quad (5.15)$$

for the rectangular cross-section, and

$$I = \frac{\pi R^4}{4} \quad (5.16)$$

for the circular cross-section.

2.2.2 Flexural strength and fracture deflection

Flexural strength σ_b measures the maximum stress that a material is subjected to before fracture or at a specified deflection. The corresponding deflection is the fracture deflection, which reflects the maximum deformation under bending load. The flexural strength can be calculated using the load at fracture F_b and Equations (5.12) and (5.13). Thus, for a rectangular or circular cross-section, it can be calculated respectively as

$$\sigma_b = \frac{3F_b L_s}{2bd^2} \tag{5.17}$$

or

$$\sigma_b = \frac{F_b L_s}{\pi R^3}. \tag{5.18}$$

3 Experimental

3.1 Preparation

(1) Materials: HT200 cast iron, Q235 low carbon steel, and birch wood samples with circular cross-section ($R = 5$ mm, $L = 180$ mm), and ceramic tile samples with rectangular cross-section ($b = 25$ mm, $d = 5$ mm, $L = 140$ mm). The main components in the ceramic tiles are SiO_2 and Al_2O_3.

(2) Apparatus: universal testing machine (SANS CMT5105) with three-point bending fixture.

(3) Tool: Vernier caliper.

3.3 Procedure

(1) Measure the dimensions of the samples using Vernier caliper three times and record the average values.

(2) The distance L_s between the supports is set as 160 mm for cast iron, low carbon steel and wood samples, and 120 mm for ceramic tile samples.

(3) Switch on the universal testing machine and the computer. Open the software.

(4) Select the test module "Three-point bending test at room temperature" in the software and set the testing parameters accordingly. The bending speed is 1 mm/min.

(5) Place a test sample on the two supports. Ensure that the anvil for applying the force is positioned at the middle position between the two supports. Lower the anvil slowly until it is close to the test sample.

(6) Set the force and displacement to zero, and then start the test. The force-deflection curve is simultaneously recorded by the computer.

(7) Export the testing result as an Excel file.

(8) Repeat steps (2) to (7) three times for each type of sample.

4 Requirements for experimental report

(1) Plot the force-deflection curves of the samples.

(2) Determine and tabulate the flexural properties, including flexural modulus, flexural strength, fracture deflection. Calculate the average values of each type of sample.

(3) Compare the bending behaviors of cast iron and low carbon steel. Discuss the reasons for the difference.

(4) Compare the flexural moduli obtained in this experiment with those reported in the literature. Discuss the possible reasons if there are significant discrepancies.

5 Questions and further thoughts

(1) What do you expect if the bending experiment has been carried out at an elevated temperature?

(2) Give three examples of engineering applications in which the bending properties of materials are essential.

6 Cautions

During the bending test, brittle samples will break into pieces. To avoid injury, keep a safe distance from the equipment.Do not attempt to adjust or move the sample during the test.

Experiment 26　Torsion Test of Metals

Type of the experiment: Comprehensive
Recommended credit hours: 4
Brief introduction: The principle and method of the torsion test are introduced in this experiment. The torsional properties of low carbon steel and cast iron are measured and compared. Students can intuitively understand the torsional behaviors of ductile and brittle materials. Moreover, the effect of torsional speed on the torsional properties is investigated.

1 Objectives

(1) To understand the principle of the torsion test.
(2) To measure the torsional properties of low carbon steel and cast iron.
(3) To compare the torsional behaviors of ductile and brittle materials.
(4) To understand the effect of torsion speed on the torsional properties.

2 Principles

2.1　Torsional test

A wide range of products and components, such as shafts, switches, fasteners, and automotive steering columns, are subjected to torsional forces during their engineering application. By testing these products in torsion, the actual service conditions can be simulated. Consequently, we can check the product quality, verify designs, and ensure proper manufacturing techniques.

The most notable test that demonstrates the effects of torsional force and the resulting stress is the torsion test. It measures the strength of any material against a twisting force. One end of the sample is anchored so that it cannot move or rotate, and a moment is applied to the other end so that the sample is rotated about its axis. The test consists of measuring the angle of twist at selected increments of torque, as shown in Fig. 5.15. After expressing the twist as the angular deflection per unit gauge length, one can construct a plot of torque T versus twist angle ϕ of per unit gauge length.

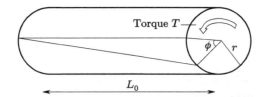

Fig. 5.15 A rod sample with circular cross-section under torsional force.

For engineering purposes, the torque-angle plot can be converted to shear stress versus shear strain plot. For a rod sample with a circular cross-section, the shear stress τ is

$$\tau = \frac{Tr}{J}, \tag{5.19}$$

and the shear strain γ is

$$\gamma = \frac{\phi r}{L_0}, \tag{5.20}$$

where r is the radius of the solid circular rod, L_0 is the length over which the relative angle of twist ϕ (in radians) is measured. J is the polar moment of inertia defined as

$$J = \frac{\pi (2r)^4}{32}. \tag{5.21}$$

2.2 Torsional properties

The torsional properties can be determined from the torque-angle curve shown in Fig. 5.16. These properties include but are not limited to shear modulus, yield shear strength, ultimate shear strength, and torsional strength.

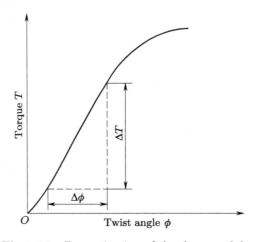

Fig. 5.16 Determination of the shear modulus.

2.2.1 Shear modulus

Shear modulus G of elasticity can be calculated using the linear slope of the torque-angle relation as

$$G = \frac{\Delta T L_0}{\Delta \phi J}. \tag{5.22}$$

Obviously, it can also be simply determined from the slope of the shear stress-shear strain relation in the elastic range.

2.2.2 Shear yield strength

Shear yield strength is generally defined as the stress associated with the lower yield point, τ_{eL}, if the torque-angle curve or stress-strain curve has a yielding plateau. If no yield plateau can be detected, a non-proportional torsion strength τ_p is considered as the yield strength, which is the stress at a specified value of non-proportional shear strain γ_p. The determination of τ_p is shown in Fig. 5.17. After the γ_p is specified, a line CA parallel to the straight line in the elastic range can be drawn, which intersects the ϕ axis at point C with a twist angle of $L_0 \gamma_p / r$. The torque T_p can be obtained from intersection point A. Then, the non-proportional torsion strength τ_p can be calculated using Equation (5.19).

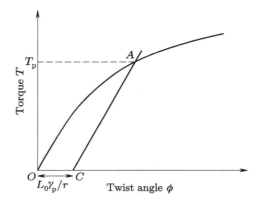

Fig. 5.17 Determination of the non-proportional torsion strength.

Generally, the non-proportional torsion strength is defined at the strain of 0.015% and 0.3%, denoted as $\tau_{p0.015}$ and $\tau_{p0.3}$. They are also called conditional torsion proportional limit and torsion yield strength, respectively.

2.2.3 Ultimate shear strength

The ultimate shear strength or torsional strength is the maximum stress that a material can withstand before fracture. It can also be obtained by determining the maximum torque from the torque-angle curve.

3 Experimental

3.1 Preparation

(1) Materials: rod samples of cast iron (Fe-3.9wt%C) and Q235 low carbon steel with a size of $\Phi 10$ mm × 200 mm.
(2) Apparatus: torsion testing machine (MTS China, CTT1202).
(3) Tools: Vernier caliper, ruler.

3.2 Procedure

(1) Measure diameters of the samples three times using Vernier caliper and record the average values.
(2) Switch on the torsion testing machine and the computer. Open the testing software.
(3) Select the test module "Metal torsional test at room temperature" in the software. Set the testing parameters accordingly. The torsion speed is 60 (°)/min. The gauge length is 100 mm.
(4) Mount a sample in the fixture of the tester.
(5) Set the torque and twist angle to zero, and then start the test. The torque-angle curve is simultaneously recorded by the computer.
(6) Export the testing results as an Excel file.
(7) Repeat steps (3) to (6) for each sample. Then, change the torsion speed to 80, 100, and 120 (°)/min, respectively, and repeat the test.

4 Requirements for experimental report

(1) Plot the torque-twist angle curves and the shear stress-shear strain curves of the samples. The curves of the same type of samples with different torsion speeds should be drawn in one chart for comparison.
(2) Calculate and tabulate the torsional properties of the samples, including shear modulus, yield strength ($\tau_{p0.015}$ and $\tau_{p0.3}$), and torsional strength.
(3) Discuss the difference between cast iron and low carbon steel in the torsional behaviors.
(4) Discuss the effect of torsion speed on the torsional properties.

5 Questions and further thoughts

(1) Why does the cast iron fracture along a plane that is 45° from the torsion axis?
(2) Is low carbon steel suitable for torsion-related applications? Why?

Extended readings

ASTM D256-10(2018), Standard Test Methods for Determining the Izod Pendulum Impact Resistance of Plastics. ASTM International, 2018.
ASTM D6110-18, Standard Test Method for Determining the Charpy Impact Resistance of Notched Specimens of Plastics. ASTM International, 2018.

ASTM E8/E8M-21, Standard Test Methods for Tension Testing of Metallic Materials. ASTM International, 2021.

Callister W D, Rethwisch D G. Fundamentals of Materials Science and Engineering. 5th Edition. John Wiley & Sons, 2001.

Dowling N E. Mechanical Behavior of Materials. 4th Edition. Pearson, 2012.

Hibbleler R C. Mechanics of Materials. 10th Edition. Pearson, 2016.

Khraishi T A, Al-Haik M S. Experiments in Materials Science and Engineering. Cognella, 2010.

Pelleg J. Mechanical Properties of Materials. Springer, 2013.

Rodriguez F. Plastic. Available at: https://www.britannica.com/science/plastic. Accessed: 2 March 2021.

Suryanarayana C. Experimental Techniques in Materials and Mechanics. CRC Press, 2011.

The Editors of Encyclopaedia Britannica. Strain gauge. Available at: https://www.britannica.com/technology/strain-gauge. Accessed: 2 March 2021.

Chapter 6 Physical Properties of Materials

Experiment 27 Determination of Ionization Rate of Plasma in Direct Current Sputtering

Type of the experiment: Comprehensive
Recommended credit hours: 4
Brief introduction: Sputtering is one of the most widely used techniques for thin film fabrication. In this experiment, the basic principle of sputtering, i.e., glow discharge, is introduced. The direct-current (DC) sputtering is used to deposit Au films on glass substrates, and the ionization rate of plasma during the glow discharge process is estimated. The effects of electric field and argon pressure in the sputtering chamber on the plasma current and ionization rate are examined. At last, the sheet resistances of the deposited Au films are measured by the four-point probe method, and the effect of ionization rate is analyzed.

■ Objectives

(1) To understand the principles of DC glow discharge and DC sputtering.
(2) To determine the ionization rate of plasma in the DC glow discharge process.
(3) To deposit Au thin films by DC sputtering technique.
(4) To measure the sheet resistances of the Au thin films using the four-point probe method.

■ Principles

2.1 DC glow discharge

A glow discharge is a gas discharge that shows a glow in a low-pressure gas. Plasma forms in the process by passing a current at 100 V to several kV through the gas, which is usually argon or another noble gas. The luminosity of the plasma is produced because the electrons gain sufficient energy to generate visible light by excitation collisions that generate photons. The glow discharge is widely used as a light source in devices such as neon lights, fluorescent lamps, and plasma-screen televisions. It is also the basis of the sputtering coating technique.

The simplest type of glow discharge is the DC glow discharge. Its simplest form consists of two electrodes in a chamber typically filled with inert gas held at low pressure. A potential of several hundred volts is applied between the two electrodes. Electrons accelerated in the electric field collide with gas molecules in the chamber. They may transfer a substantial part of their kinetic energy to neutral molecules and ionize them in inelastic collisions. The positively charged ions and electrons are respectively driven towards the cathode and the anode by the electric potential. The initial population of ions and electrons collides with other atoms and ionizes them further. As long as the potential is maintained, a population of ions and electrons will reach a dynamic equilibrium state.

2.2 Sputtering deposition technique

Sputtering is one of the most widely used thin film fabrication techniques in diverse industries for semiconductor processing and surface finishing. In the sputtering deposition, the film composition and other film properties, such as step coverage and grain structure, are readily controlled by the operating parameters. The film thickness is controlled by simply adjusting the deposition time.

The principle of sputtering deposition is based on glow discharge. Figure 6.1 shows the most basic form. It is carried out in a vacuum chamber into which an inert gas is introduced, and the most used is argon. Two items are placed into the chamber. They are a substrate on which material will be deposited and a sputtering target that provides the deposition material.

The high voltage applied across the argon results in the emission of electrons from the cathode. The emitted electrons will be accelerated by the electric field. When the energy of the accelerated electrons is higher than the ionization energy of Ar (about 15.7596 eV), inelastic collisions will add energy to Ar atoms. Thus, the Ar atoms are ionized, and the reaction formula is

$$Ar + e^- \rightarrow Ar^+ + 2e^-. \tag{6.1}$$

The electrons emitted from Ar atoms will also be accelerated by the electric field. Positively charged Ar^+ ions are accelerated toward the negatively charged target, and they strike with sufficient energy to dislodge atoms (denoted as "M" in the figure) from the target surface. The atoms escape from the target material and travel in all directions with high kinetic energy. Some of them deposit on the substrate, forming a thin film on the substrate surface.

There are two essential factors that dominate the discharge process. One is the electric field which determines the acceleration of ions and electrons. Another is argon pressure, which determines the mean free path of the particles and the average time between two collisions.

The power used to generate ions can be DC or radio frequency (RF)current. Electrical isolating materials must be sputtered by applying an RF power to pre-

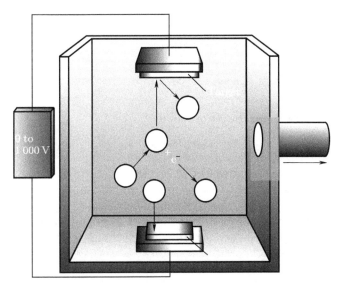

Fig. 6.1 Schematic illustration of the sputtering apparatus.

vent charging of the target. Conductive materials can be sputtered with RF or DC power. In this experiment, DC sputtering technique is used to deposit Au films on glass substrates.

2.3 Ionization rate of plasma

The ionization rate α is an important parameter affecting the sputtering efficiency. It is defined as

$$\alpha = \frac{n_i}{n + n_i}, \tag{6.2}$$

where n and n_i are densities of Ar and Ar$^+$, respectively. The sum of n and n_i is defined as

$$n_{\text{total}} = n + n_i, \tag{6.3}$$

which can be determined by

$$PV = n_{\text{total}} kT, \tag{6.4}$$

where P is the pressure in the chamber, V is the volume of the chamber, T is the temperature, and k is the Boltzmann constant.

The generation rate of electrons and ions is equal to the flux of charge per unit of time. Thus,

$$n_e + n_i = \frac{It}{e}, \tag{6.5}$$

where n_e is the density of electron, I is the current, and t is the unit time. According to the quasi neutrality of plasma,

$$n_e = n_i = \frac{It}{2e}. \tag{6.6}$$

From Equations (6.2), (6.4), and (6.6), the ionizing rate can be expressed as

$$\alpha = \frac{n_i}{n_{total}} = \frac{ItkT}{2ePV}. \quad (6.7)$$

Thus, the ionizing rate of plasma can be readily calculated after determining the relative parameters.

2.4 Sheet resistance measurement of thin film

The sheet resistance is a critical property that quantifies the ability for charges to travel along uniform thin films. Furthermore, resistivity and conductivity can be determined from the measurement of the sheet resistance.

The four-point probe method is typically used to determine the sheet resistance. As shown in Fig. 6.2, the apparatus consists of four electrical probes in a line, with equal spacing between the probes. It operates by applying a current on the outer two probes (probes 1 and 4) and measuring the resultant voltage drop between the inner two probes (probes 2 and 3). The sheet resistance can then be calculated by

$$R_{sq} = \frac{\pi}{\ln 2} \frac{V}{I}, \quad (6.8)$$

where R_{sq} is the sheet resistance ina unit of Ω/sq, V is the voltage measured between the inner probes, and I is the current applied between the outer probes.

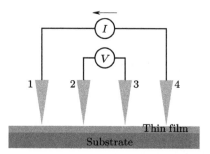

Fig. 6.2 The circuit of the four-point probe method.

It is noted that this equation is valid if the thickness of the material being tested is less than 40% of the spacing between the probes and the lateral size of the sample is sufficiently large. If this is not the case, geometric correction factors are needed to account for the size, shape, and thickness of the sample.

If the thickness of the measured material is known, the resistivity can be calculated by

$$\rho = R_{sq}t, \quad (6.9)$$

where ρ is the resistivity, and t is the thickness of the thin film.

3 Experimental

3.1 Preparation

(1) Chemical: ethanol.

(2) Substrates: glass substrates with a size of 20 mm × 20 mm.

(3) Apparatuses: DC ion sputtering system (KYKY Technology, SBC-12) with a gold target, four-point probe system (Probes Tech, RTS-8), ultrasonic cleaner.

(4) Tools: 100 mL beaker, tweezers.

3.2 Procedure

3.2.1 Determination of ionization rate of plasma

(1) Open the chamber cover of the sputtering system. Adjust the distance between the target and the stage to 3 cm. Then cover the chamber.

(2) Switch on the sputtering system and the mechanical pump.

(3) Import argon into the chamber. Adjust the pressure to about 2.0 Pa by slowly rotating the argon valve. Note that each adjustment will take several seconds to stabilize the pressure.

(4) Set the electrical voltage to 1.8 kV. Start sputtering.

(5) Record the pressure and the current.

(6) Adjust pressure to about 3.0 Pa and 4.0 Pa, respectively, and record the corresponding pressure and the current.

(7) Adjust the target-stage distance to 5 cm and 8 cm, respectively. Then, repeat steps (1) to (6).

(8) Calculate the ionization rate for different distances and pressures. The diameter and height of the chamber are 100 mm and 130 mm, respectively. The diameter of the target is 58 mm.

3.2.2 Thin film deposition

(1) Ultrasonically clean two glass substrates in ethanol for 10 min, and then dry them.

(2) Place a substrate on the stage of the sputtering chamber. Adjust the distance between the target and the stage to 5 cm.

(3) Switch on the sputtering system and the mechanical pump.

(4) Import argon into the chamber. Adjust the pressure to a value associated with a relatively high ionization rate.

(5) Set the electrical voltage to 1.8 kV and sputtering time 60 s. Start sputtering.

(6) Repeat steps (2) to (5). The pressure is set to a value associated with a relatively low ionization rate.

3.3.3 Sheet resistance measurement

(1) Switch on the four-point probe system and the computer. Open the testing software.

(2) Select the test module for thin film.

(3) Place a sample onto the probe stage.

(4) Select the appropriate current range. If the range is uncertain, the automatic measurement can be selected.

(5) Lower the probes to ensure good contact between the probes and the sample surface.

(6) Measure the sheet resistance. Then raise the probes.

(7) Move the sample to the other two positions and repeat steps (5) and (6). All these three positions should be in the vicinity of the sample center.

(8) Repeat steps (5) to (7) for the other sample.

(9) Export the measurement results as an Excel file. Although the software will automatically calculate the sheet resistances, it is suggested that students record the voltage and current and calculate the sheet resistances by themselves.

4 Requirements for experimental report

(1) Tabulate the results of electric field E, target-substrate distance D, argon pressure P, plasma current I, and ionization rate α.

(2) Plot the curves of plasma current versus E/P ratio in one chart for different target-stage distances. Discuss the relationship of the E/P ratio, the current, the ionization rate, and the target-stage distance.

(3) Calculate the sheet resistances of the two samples measured from three positions and their average values. Discuss the relationship between the ionization rate and sheet resistance.

5 Questions and further thoughts

(1) What are the advantages and disadvantages of the DC sputtering technique for thin film deposition?

(2) In this experiment, the four-point probe method is used to determine the sheet resistances of the thin films. In this case, the sample is very thin, and the area is relatively large. How can we determine the resistivity of a sample with a larger thickness or a smaller area?

Experiment 28 Coprecipitation Synthesis and Magnetic Properties of Magnetite Nanoparticles

Type of the experiment: Comprehensive
Recommended credit hours: 8
Brief introduction: Understanding the magnetic properties of nanoparticles is a central issue in magnetic materials. The magnetite, Fe_3O_4, has the highest magnetism of all the naturally occurring minerals on earth. In this experiment, Fe_3O_4 particles with a size of 2–4 nm are synthesized and characterized. The experiment is divided into two experiments that can be conducted respectively. In the first experiment, the Fe_3O_4 nanoparticles are synthesized by an economic, biocompatible coprecipitation route. Because the drying and annealing of the powder take a long time, these steps do not need to take up class time. Then, in the second experiment, the phase structure and particle size are determined by X-ray diffraction (XRD). The magnetic properties of those superparamagnetic nanoparticles are measured by vibrating sample magnetometer (VSM).

Experiment 28.1 Synthesis of Magnetite Nanoparticles by Coprecipitation Method

Type of the experiment: Cognitive
Recommended credit hours: 4

１ Objectives

(1) To understand the principle and process of the coprecipitation method.
(2) To synthesize magnetic Fe_3O_4 nanoparticles by the coprecipitation method.

２ Principles

2.1 Magnetic materials based on iron oxide nanoparticles

Magnetic nanoparticles have attracted much attention because they offer good properties in fundamental studies and technological applications. They are used as the active component of ferrofluids, recording tape, flexible disk recording media, biomedical materials, and catalysts. Furthermore, assemblies of nano-scale magnetic grains have been used as hard disk recording media, permanent magnets, and nanocrystalline soft materials.

When the particle size is equal to or less than a certain physical characteristic size, such as the light wavelength, de Broglie wavelength, the coherent length or transmission depth of the superconducting state, the periodic boundary conditions of crystals will be destroyed, the atomic density near the particle surface layer will be reduced, and the specific surface area is also significantly increased, resulting in the change of physical properties and mechanical properties. It is called the small size effect. For example, solid materials with coarse grain size have a fixed melting point, while the melting point of those with ultrafine grain size is significantly reduced.

Magnetic nanoparticles show many remarkable new phenomena, such as high field irreversibility, high saturation field, super-paramagnetism, extra anisotropy contributions, and shifted loops after field cooling. These phenomena arise from both small size effect and surface effect that dominate the magnetic behavior of individual nanoparticles.

Many crystalline materials exhibit ferromagnetism. The magnetite, Fe_3O_4, has the highest magnetism of all the naturally occurring minerals on earth. It is widely used in the form of superparamagnetic nanoparticles. In this experiment, the Fe_3O_4 nanoparticles will be synthesized, and then the magnetic properties will be characterized.

2.2 Coprecipitation method for synthesis of Fe_3O_4 nanoparticles

Considerable efforts have been made to develop synthesis methods for iron oxide nanoparticles. The commonly used methods include coprecipitation, thermal decomposition, hydrothermal synthesis, and microemulsion. Coprecipitation is one of the most widely used methods. Its preparation procedure of nano-powder is to add appropriate precipitant into the soluble salt solution, leading to the uniform precipitation or crystallization of metal ions. Then, the precipitates are generally dehydrated or decomposed.

Two kinds of coprecipitation methods are generally used to synthesize Fe_3O_4. One is the Massart hydrolysis method, in which the mixture of trivalent iron salt and divalent iron salt with a specific molar ratio is directly added into the strong alkaline aqueous solution, and the iron salt is hydrolyzed and crystallized in the strong alkaline aqueous solution to form Fe_3O_4 nanoparticles. The other is the titration hydrolysis method, in which dilute alkali solution is dropped into the mixed solution of trivalent iron salt and divalent iron salt with a specific molar ratio so that the pH value of the mixed solution increases gradually. When the pH value reaches 6–7, magnetic Fe_3O_4 nanoparticles are generated by hydrolysis.

The size, shape, structure, and magnetic properties of the nanoparticles could be affected by the preparation conditions, such as the type of Fe^{3+} or Fe^{2+} salt, the Fe^{3+}/Fe^{2+} ratio, the pH value, the reaction temperature, and the ionic strength of the media. To obtain functionalized iron oxide nanoparticles, the coprecipitation technique may be improved by adding functional materials or surface-active agents in the reaction media to reduce the aggregation and oxidation of naked iron oxide nanoparticles.

3 Experimental

3.1 Preparation

(1) Chemicals: aqueous ammonia (28wt% NH_3), iron trichloride hexahydrate ($FeCl_3·6H_2O$), ion dichloride tetrahydrate ($FeCl_2·4H_2O$), deionized water.

(2) Apparatuses: centrifuge, electronic balance, electromagnetic stirrer with a heating plate, vacuum drying oven, tube furnace.

(3) Tools: 500 mL beaker, pear-shaped separatory funnel, pH test paper.

3.2 Procedures

(1) Dissolve 54.1 g $FeCl_3·6H_2O$ and 19.9 g $FeCl_2·4H_2O$in 200 mL deionized water in a beaker. Heat the solution to 60 °C under stirring by the electromagnetic stirrer with a heating plate.

(2) Transfer about 200 mL aqueous ammonia into the separatory funnel.

(3) Drip the aqueous ammonia into the solution under continuous vigorous stirring (about 1000 r/mim) under flowing nitrogen (about 200 mL/min). The reaction will finish until the pH value reaches 9.0, and the color of the solution varies from orange-red to black gradually.

(4) After the reaction, stir the solution vigorously for 40 min to tune the size of the resultant nanoparticles.

(5) Centrifuge the solution. Discard the supernatant.

(6) To wash the powder, add 30 mL water into the tube for centrifugation. Disperse the powder ultrasonically. Then centrifuge the suspension and discard the supernatant.

(7) Repeat step (6) two or three times.

(8) Dry the powder at 60 °C for 24 h in the vacuum drying oven.

(9) Anneal the powder at 300 °C for 90 min in vacuum or flowing nitrogen using the tube furnace.

(10) Take a photo of the as-synthesized powder.

4 Requirements for experimental report

(1) Find and give the details of the crystal structure of Fe_3O_4 from the literature.
(2) Give the photograph of the as-synthesized powder.

5 Questions and further thoughts

(1) Why is the powder annealed in vacuum or flowing nitrogen?
(2) Are there any other synthetic methods of magnetic nanoparticles? Compare the advantages and disadvantages of these methods.

6 Cautions

The synthesis process must be conducted in a fume hood.

Experiment 28.2 Characterization of Structure and Magnetic Properties of Magnetite Nanoparticles

Type of the experiment: Comprehensive
Recommended credit hours: 4

1 Objectives

(1) To understand the principles of XRD and VSM.
(2) To characterize the crystal structure and particle size by XRD technique.
(3) To characterize the magnetic properties by VSM.

2 Principles

2.1 Principle of XRD

2.1.1 Generation and diffraction of X-rays

X-rays with energies ranging from 100 eV to 10 MeV are classified as electromagnetic waves. They are generated in an X-ray tube that consists of a cathode and an anode. When a high voltage with several tens of kV is applied between two electrodes, the high-speed electrons with sufficient kinetic energy are drawn out from the cathode and collide with the metallic anode. Characteristic X-ray associated with the target metal and the related electron shell is produced, which has a specific wavelength ranging from 10 to 10^{-4} nm.

Because the wavelength of X-rays is in the same order as the crystal structure features, a crystal can be used as an optical grating for X-rays. When an X-ray beam irradiates a crystal, the rays scattered by the atoms in the crystal have certain phase relations between them. These phase relations are such that destructive interference occurs in most scattering directions, but in a few directions, constructive interference takes place, leading to the formation of diffracted beams. Consequently, the crystal structure can be determined by analyzing this diffraction pattern.

The diffraction condition is simply and intuitively expressed by the Bragg equation, that is,
$$2d\sin\theta = n\lambda, \qquad (6.10)$$
where θ is the incident angle, d is the atomic plane distance, λ is the wavelength. n is called the order of reflection and is equal to the number of wavelengths in the path difference between diffracted X-rays from adjacent crystal planes. As shown in Fig. 6.3, when an X-ray beam irradiates at an incident angle θ on a crystal plane with a lattice spacing of d, a diffracted beam strengthened by superposition will be obtained in the reflected direction under the condition of Equation (6.10).

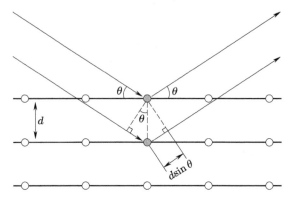

Fig. 6.3 Schematic illustration of the Bragg law.

An XRD pattern or profile, which is a plot of diffraction intensity versus 2θ angle, can be obtained by an X-ray diffractometer. The X-rays are collimated and directed on the sample. As the sample and detector are rotated, the intensity of the reflected X-rays is recorded. For the crystal planes satisfied with the Bragg equation, constructive interference occurs as peaks in the diffraction pattern.

XRD has become an important experimental method and structural analysis method in the research of materials science. The crystal structure, grain size, crystallinity, and phase constitution can be measured through XRD analysis. In this experiment, we will use XRD to qualitatively analyze the phase constitution of the synthesized material to confirm the formation of Fe_3O_4. Then the crystallite size of the sample, which is the same as the nanoparticle size, will be measured.

2.1.2 Qualitative phase analysis

The purpose of qualitative phase analysis is to determine the phase constitution of the sample. Since each phase has its specific powder diffraction profile, the primary method of qualitative phase analysis is to compare the diffraction pattern of the sample with those of known crystals. The comparison relies mainly on the positions (i.e., the 2θ angles) of the peaks in the diffraction profile and, to some extent, on the relative intensities of these peaks. Currently, powder diffraction cards, known as powder diffraction files, are widely used. Various primary data of powder diffraction profiles are listed, including the position and the intensity of each diffraction peak and the indices of the corresponding crystal plane.

Another method of qualitative phase analysis needs some calculation when the powder diffraction files are not available. A set of diffraction angles can be calculated for the crystal planes of some known crystals by combining the Bragg equation with the plane-spacing equation. One can compare the diffraction angles from the experiment with these calculated angles to determine the phase constitution.

2.1.3 Measurement of grain size

The crystallite size can cause the broadening of diffraction peaks in XRD profiles. If the instrument effects and other factors which result in peak broadening have been excluded, the crystallite size is easily calculated as a function of peak width, specified as the full width at half maximum (FWHM) peak intensity, peak position, and wavelength. The well-known Scherrer equation is generally accepted to describe this relation, which is expressed as

$$D = \frac{0.89\lambda}{\beta \cos \theta}, \quad (6.11)$$

where D is the average crystallite size, θ is the diffraction angle, and β is the corrected halfwidth of the peak after correcting for peak broadening caused by the instrument. It is calculated by

$$\beta^2 = \beta_m^2 - \beta_s^2, \quad (6.12)$$

where β_m is the measured halfwidth, and β_s is the halfwidth of a sample that is the same as the tested sample but with a larger grain size about 5–20 μm.

The crystallite size range for the application of the Scherrer equation is about 3–200 nm. The peak with a higher diffraction angle can give higher accuracy for the calculation.

2.2 VSM instrumentation

The vibrating sample magnetometer (VSM) is a sensitive and versatile instrument for studying magnetic moments in magnetic materials with the magnetic field and temperature. It is widely used to study the magnetic properties of various ferromagnetic, antiferromagnetic, paramagnetic, and diamagnetic materials. The VSM can be used to measure the primary magnetic properties, such as the hysteresis loop, magnetization curve, and the variation of magnetic properties with temperature. Then essential magnetic parameters can be determined, such as saturated strength of magnetization, left strength of magnetization, coercive force, maximum energy product, Curie temperature, and magnetic conductivity.

The VSM is mainly composed of an electromagnet system, a forced vibration system, and a signal detection system. It operates on the principle of Faraday's law of induction, which informs that a changing magnetic field will produce an electric field. This electric field can be measured and provide us information about the changing magnetic field. When a sample is placed in a uniform magnetic field, this constant magnetic field will magnetize the sample by aligning the magnetic domains or the individual magnetic spins. The magnetic dipole moment of the sample will create a stray magnetic field around the sample. As the sample is moved up and down, this stray magnetic field is changed as a function of time and can be sensed by a set of pick-up coils since the alternating magnetic field will cause an electric current in the pick-up coils according to Faraday's law of induction. Using controlling and monitoring software, the system can tell us how much the sample is magnetized and how its magnetization depends on the strength of the constant magnetic field.

3 Experimental

3.1 Preparation

(1) Materials: as-synthesized Fe_3O_4 powder in Experiment 28.1, Fe_3O_4 powder with a particle size of about 20 μm.

(2) Apparatuses: X-ray diffractometer (Rigaku, Miniflex 600), VSM (YP Magnetic Technology Development Co., Ltd., VSM-100).

3.2 Procedures

3.2.1 XRD characterization

(1) Take Fe_3O_4 powder with a particle size of 20 μm as a standard sample. Load the powder in the sample holder. Insert the sample holder into the clip.

(2) Set the testing parameters. The 2θ angle is from 10° to 80°. Cu target is used, and the X-ray wavelength of $K_{\alpha 1}$ is 1.54056 Å. Default values may be used for other parameters.

(3) Start to acquire the XRD profile. Export the profile as a ".txt" file.

(4) Repeat steps (1) to (3) for the synthesized sample.

3.2.2 Characterization of magnetic properties by VSM

(1) Switch on the VSM and preheat for 30 min.

(2) Load the synthesized powder into the fixture container.

(3) Open the VSM software. Zero the magnetic field and magnetic moment of VSM.

(4) After calibration, start the measurement and get the hysteresis loop.

4 Requirements for experimental report

(1) Find the powder diffraction card of Fe_3O_4 powder from a database or library.

(2) Plot the XRD profiles of the standard sample and the synthesized sample. Index the peaks by comparing them with the powder diffraction card. If there are extra peaks besides those of Fe_3O_4, discuss the possible reason for the formation of the extra peaks.

(3) Calculate the crystallite size of the synthesized sample by Scherrer equation. Peak (440) is recommended for the calculation.

(4) Present the hysteresis loop. Determine the saturation magnetic induction intensity B_s, remanence B_r, and coercivity H_c from the hysteresis loop.

5 Questions and further thoughts

(1) Why can the crystallite size cause the broadening of diffraction peaks in XRD profiles?

(2) Suggest one or two other techniques that can determine the crystal structure and grain size.

Experiment 29   Nanoemulsion Synthesis and Optical Properties of Iron-Gold Nanocrystals

Type of the experiment: Comprehensive
Recommended credit hours: 8
Brief introduction: Iron-gold nanocrystals have various proven applications because of the unique optical, electronic, and magnetic properties. In this experiment, the iron-gold nanocrystals are synthesized by the nanoemulsion method, which uses isotropic dispersed systems of two non-miscible liquids to form nano-sized droplets. Then the optical property of the nanoparticles is characterized by ultraviolet-visible (UV-Vis) spectroscopy.

1 Objectives

(1) To understand the principle and process of the nanoemulsion method.
(2) To prepare the Fe-Au nanoparticles by the nanoemulsion method.
(3) To understand the principle of UV-Vis spectroscopy.
(4) To characterize the optical property of the Fe-Au nanoparticles by UV-Vis spectroscopy.

2 Principles

2.1   Fe-Au nanocrystals

Nanocrystal is a term given to a cluster of atoms, which has a size of less than 100 nm. They have attracted technological interest since many of their optical, electrical, and thermodynamic properties show strong size dependence and can therefore be controlled through careful manufacturing processes.

Multicomponent hybrid nanomaterials with tunable composition and morphology have been used widely in the catalytic, optical, and magnetic research fields. The coupling between the individual components can provide enhanced functionality or new properties compared to the individual component. Their potential applications will be significantly expanded to cover more areas.

Iron-gold (Fe-Au) alloy nanocrystals show more distinctive magnetic and optical properties than the corresponding Fe and Au mono-element nanocrystals.Fehas a wide range of applications such as magnetic recording, magnetic seals, printing, magnetic resonance imaging, drug delivery, bio-detection, and cell tagging and separation because of its excellent magnetic properties and readiness to be converted to bio-friendly oxides. Au is a precious element with diverse applications in catalysis, electronics, and optoelectronics. Therefore, the amalgamation of Fe and Au into one Fe-Au alloy can offer potential functions in both magnetic and optical properties in addition to biological compatibility.

2.2 Nanoemulsion method for the synthesis of Fe-Au nanocrystals

Various methods have been reported to synthesize Fe-Au alloy nanocrystals, such as sol-thermal, hydrothermal, precipitation, and emulsion methods. In this experiment, the Fe-Au nanocrystals will be synthesized by the nanoemulsion method.

An emulsion is a thermodynamically stable system consisting of at least two immiscible liquid phases. One phase is dispersed as globules in the other liquid phase, which is the continuous phase. The dispersed phase in nanoemulsion typically comprises ultrasmall particles or droplets, with a size range of 2 nm - 200 nm, and has proper oil/water interfacial tension.

The synthesis of Fe-Au nanoparticles using the nanoemulsion method can be achieved via the reduction of gold acetate by 1, 2-hexadecanediol and the thermal decomposition of iron pentacarbonyl in the presence of the stabilizers or surfactants. The process is roughly divided into three stages,micelle formation, thermal decomposition, and alloying.

Firstly, micelles composed of many tiny colloidal particles shown in Fig. 6.4 are formed after all the reactants are stirred vigorously and heated to 80 °C. The reactants, including the sources of iron and gold as well as the reductant (1,2-hexadecanediol), are uniformly dispersed and covered by the surfactants (oleic acid and oleylamine), and thus the dispersed phase is generated. The continuous phase is the solvent (octyl ether). Secondly, when the temperature rises to 280 °C, the Fe and Au precursors are decomposed and undergo reduction to form metallic colloidal particles. Thirdly, the reduced Au and Fe atoms alloy to form the alloyed nanocrystals during the decomposition under high temperature.

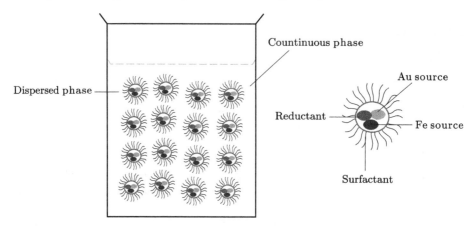

Fig. 6.4 The principle of the nanoemulsion method. (See color illustration at the back of the book)

The size, structure, and optical and magnetic properties are affected by the composition, reaction time, and reaction temperature. For instance, the size of the particles decreases with increasing the reaction time and temperature. In the visible light range, the characteristic absorption bands become broader with decreasing the Au/Fe molar ratio or increasing the reaction time and temperature.

2.3 UV-Vis spectroscopy

UV-Vis spectroscopy is the measurement of the absorption of near-ultraviolet and visible light by a sample. It can be used to analyze the composition and structure of the sample and determine the optical properties and other properties related to light absorption.

The absorption of visible and UV radiation is associated with the excitation of electrons in both atoms and molecules from lower to higher energy levels. Since the energy levels of substances are quantized, only light with the precise amount of energy that can cause transitions from one level to another will be absorbed. Therefore, the specific wavelengths that are absorbed and the intensity of the absorption give us information about the electronic structure of the sample.

UV-Vis spectrometer can be used to measure the absorbance of UV or visible light by a sample, either at a single wavelength or perform a scan over a range in the spectrum. The UV region ranges from 200 to 400 nm, and the visible region from 400 to 800 nm. A schematic diagram of a UV-Vis spectrometer is shown in Fig. 6.5, which consists of light sources, a monochromator, sample and reference cells, detectors, and recording devices. The light source emits visible and near-ultraviolet radiation over a broad range of wavelengths. A specific wavelength can be selected by using a wavelength separator such as a prism or a grating monochromator.

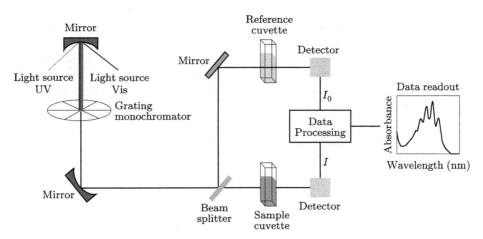

Fig. 6.5 The working principle of the UV-Vis spectrometer. (See color illustration at the back of the book)

For a sample that light can penetrate, such as liquid, it is held in an optically flat, transparent container called a cell or cuvette. Each monochromatic beam, in turn, is split into two equal intensity beams by a half-mirrored device. One beam, the sample beam, passes through the sample cell containing a solution of the compound being studied in a transparent solvent. The other beam, the reference one, passes through an identical cell containing only the solvent. For each wavelength, the intensities of lights passing through both the reference cell (I_0) and the sample cell (I) are measured by electronic detectors. The adsorption may be presented as absorbance A calculated by

$$A = \log_{10} \frac{I_0}{I}, \tag{6.13}$$

or as transmittance T calculated by

$$T = \frac{I}{I_0}. \tag{6.14}$$

According to the Beer-Lambert law, the absorbance is proportional to the concentration of the substance in the solution. It is necessary to correct the absorbance for the concentration and other operational factors if the spectra of different compounds are compared in a meaningful way. The corrected absorption value is called molar absorptivity ε calculated by

$$\varepsilon = \frac{A}{cl}, \tag{6.15}$$

where c is the sample concentration (in mol/L), l is the length of the light path through the sample (in cm).

For a sample that light cannot penetrate, it is reflected on the sample surface. As shown in Fig. 6.6, the incident light reflected symmetrically with respect to the normal line is called specular reflection, while incident light scattered in different directions is called diffuse reflection. For powders or thin films with rough surfaces, the reflection is not specular. In this case, the diffuse reflectance is generally collected using a semispherical collector known as an integrating sphere. With the integrating sphere, the measurement is performed by placing the sample in front of the incident light window and concentrating the light reflected from the sample on the detector using the integrating sphere. The obtained value becomes the relative reflectance with respect to the reflectance of the reference standard whiteboard, which is taken to be 100%.

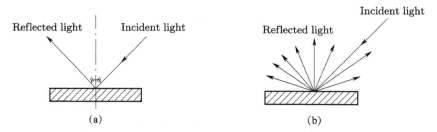

Fig. 6.6 (a) Specular reflection on a smooth surface and (b) diffuse reflection on a rough surface.

3 Experimental

3.1 Preparation

(1) Chemicals: iron acetylacetonate (Fe(acac)$_3$, 99.9%), gold acetate (Au(OOCCH$_3$)$_3$, 99.9%), 1,2-hexadecanediol (C$_{14}$H$_{29}$CH(OH)CH$_2$(OH), 90%), octyl ether, oleylamine, oleic acid, ethanol, hexane.

(2) Apparatuses: the reaction apparatus (consists of an electromagnetic stirrer with a heating plate, an oil bath, a ball condenser, a 250 mL three-necked glass flask), UV-Vis spectrometer (PerkinElmer, Lambda 25), centrifuge, ultrasonic cleaner, electronic balance.

(3) Tools: pipettes, glass bottles, quartz cuvettes.

3.2 Procedures

3.2.1 Synthesis of Fe-Au nanoparticles

(1) Assemble the reaction apparatus.Connect the flask with the condenser. Immerse the flask into the oil bath on the electromagnetic stirrer with a heating plate.

(2) Add 0.25 mmol Fe(acac)$_3$, 0.25 mmol gold (III) acetate, 2.5 mmol 1,2-hexadecanediol, 1.5 mL oleylamine, 1.5 mL oleic acid, and 10 mL octyl ether as well as a stirring bar into the flask.

(3) Under vigorous stirring, gradually increase the reaction temperature to 80 °C within about 1 h and hold the temperature for 1 h.

(4) Rapidly raise the reaction temperature to 280 °C and reflux at the temperature for 1 h.

(5) Cool down the reaction product to room temperature. Add 30 mL ethanol in the flask to precipitate the nanocrystals.

(6) Separate the product by centrifugation at 9000 r/min for 15 min. Dump the supernatant.

(7) Add 30 mL of a mixed solution of ethanol/hexane (1:2) into the tube. Disperse the nanocrystals ultrasonically. Then centrifuge the suspension.

(8) Repeat step (7) two or three times.

(9) Disperse the product in 10 mL hexane and transfer the product into a glass bottle.

3.2.2 Measurement of UV-Vis spectrum

(1) Switch on the UV-Vis spectrometer. Open the software.

(2) Fill one cuvette with about 4 mL sample suspension and fill another cuvette with 4 mL hexane. Place them in the sample holder and reference holder, respectively.

(3) Collect the UV-Vis absorption spectrum in a wavelength range from 300 to 850 nm.

(4) Export the spectrum as an excel file.

4 Requirements for experimental report

(1) Plot the UV-Vis spectrum of absorbance versus wavelength using the exported data.

(2) Compare the spectrum with that of Au in the literature. Discuss the effect of Fe on the optical property of Fe-Au nanoparticles.

5 Questions and further thoughts

(1) What is the role of each reactant used in the experiment?

(2) How can we confirm that the obtained products are Fe-Au alloy nanocrystals?

Experiment 30 Preparation of ZnO Thin Film by Sol-Gel Method and Determination of Energy Bandgap

Type of the experiment: Comprehensive
Recommended credit hours: 16
Brief introduction: ZnO has been extensively studied because of its potential applications in various fields. The principle of thin film preparation by the sol-gel method and the determination of optical energy bandgap is introduced in this experiment. Then, ZnO films with a series of thicknesses are deposited using the sol-gel method combined with spin coating. The bandgaps of the films are determined by ultraviolet-visible (UV-Vis) spectroscopy, and the effect of film thickness is analyzed.

1 Objectives

(1) To understand the principle of thin film preparation by the sol-gel method.

(2) To understand the principle of the measurement method of the optical energy bandgap.

(3) To prepare ZnO films with different thicknesses by the sol-gel method.

(4) To determine the bandgaps of the ZnO films using UV-Vis spectroscopy.

(5) To understand the effect of film thickness on the energy bandgap.

2 Principle

2.1 ZnO thin film

Metal oxide semiconductors such as SnO_2, ZnO, WO_3, and TiO_2 are wide-bandgap semiconductors. They have received considerable attention due to their optical and electrical properties. Some of them are good candidates for transparent conductive oxide films.

Among them, zinc oxide (ZnO) is an n-type semiconductor with a direct bandgap of 3.44 eV. The electrical properties of ZnO have made the compound attractive in the area of thin film transistors and light-emitting diodes. Due to its non-toxic, wide bandgap and electrical conductivity, it is thought to be ideal for solar cell devices.

2.2 Thin film preparation by sol-gel method

ZnO films have been grown by many different methods such as pulsed laser deposition (PLD), magnetron sputtering, metal-organic chemical vapor deposition

(MOCVD), spray pyrolysis, and sol-gel process. Among these methods, the sol-gel method offers a way of thin film preparation with low temperature and low cost.

The sol-gel method is a process that microparticles or molecules in a solution (sol) agglomerate and eventually link together to form a coherent network (gel) under controlled conditions. The gel can further produce nanostructured materials after drying and sintering.

Fabrication of thin films and coatings by the sol-gel method is generally accomplished by dipping, spinning, or spraying. The spin coating process is the most widely accepted in the laboratory. It is based on spinning the substrate around a perpendicular vertical axis on a stage. The process involves the following steps. Firstly, drip the sol liquid on the substrate. Then, the liquid flows radially driven by the centrifugal force during spinning up. The excessive liquid is spanned off, leaving a gradually thinner film. Finally, the volatile components of the liquid are evaporated. Compare with the formation of bulk gel, the spinning creates forced convection that increases the evaporation rate, and significantly reduces the gelation time. In general, the quality of the coating layer depends on the rheological parameters of the coating liquid and on the rotating velocity that can alter the air friction.

2.3 Determination of optical bandgap

In solid materials, as the atoms come within close proximity of one another, the electron orbitals of neighboring atoms overlap. Due to the intermixing of atoms, each distinct atomic state may split into a series of closely spaced electron states, leading to the formation of bands of energy levels, termed electron energy bands.

Semiconductors can be roughly characterized by their band structure, as illustrated in Fig. 6.7. The valence band is filled with electrons, while the conduction band is empty. However, electrons can migrate from the valence band to the conduction band if they acquire sufficient energy at least equal to the bandgap energy, which is the energy difference between the top of the valence band and the bottom of the conduction band.

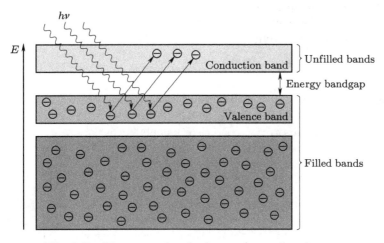

Fig. 6.7 The energy band scheme of a semiconductor.

According to the measurement method, the bandgap in semiconductors can be sorted as electronic bandgap and optical bandgap. The selection of the measurement method depends on the device that the semiconductor is used. The optical bandgap is important for applications such as solar cells. The electronic bandgap is important for devices such as light-emitting diodes and laser diodes. The measurement process of the electronic bandgap involves the removal or injection of electrons from the semiconductor valence band. The process requires additional energy because of the Coulomb interactions involved, and hence the measurement yields a slightly larger value than the optical bandgap.

In this experiment, the optical bandgap will be measured using UV-Vis spectroscopy. The measurement relies on the optical property of semiconductors, which can absorb light that has an energy equal to or greater than the bandgap energy. This absorption promotes an electron from the valence to the conduction band, leaving behind a hole in the valence band. In other words, each photon absorbed thus creates one electron-hole pair. Therefore, the wavelength-dependent absorption coefficient of the semiconductor, together with the film or wafer thickness, determines how much of the incident light is absorbed ateach wavelength.

A generic semiconductor absorption spectrum is shown in Fig. 6.8(a). The abscissa axis, energy E, is calculated from the wavelength λ by

$$E = \frac{hc}{\lambda}, \quad (6.16)$$

where h is Planck's constant (6.626×10^{-34} J·s), c is the speed of light (3.0×10^8 m/s). This absorption spectrum is characterized by a sharp increase in absorption at the bandgap energy. The onset energy of the absorption edge, which represents the minimum energy required for the electronic transition from the valence band to the conduction band, is defined as the bandgap energy, E_g. To determine the bandgap energy, two lines that best fit the slopes of the absorption edge and the background are drawn, respectively. The energy value of the intersection of the two lines is the bandgap energy.

In practice, the Tauc plot is used more widely to determine the energy bandgap through the UV-Vis spectrum. The Tauc equation is

$$\alpha h\nu = K(h\nu - E_g)^n, \quad (6.17)$$

where K is a constant, ν is the frequency of a phonon, and $h\nu$ ($= hc/\lambda$) is the photon energy. The values of n are 1/2 and 2 for the direct and indirect transitions, respectively. α is the absorption coefficient calculated by

$$\alpha = \frac{\ln T}{t} = \frac{2.303 A}{t}, \quad (6.18)$$

where T is the transmittance, A is the absorbance, t is the thickness of the sample. Then the Tauc plot of $(\alpha h\nu)^{1/n}$ versus $h\nu$ can be plotted. Extrapolation of the linear portion of the curves to zero absorption coefficient value gives the energy bandgap value, as shown in Fig. 6.8(b).

Determination of Energy Bandgap

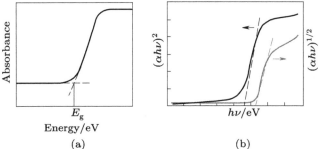

Fig. 6.8 Determination of the optical energy bandgap (a) from the absorbance-energy plot and (b) from the Tauc plot.

3 Experimental

3.1 Preparation

(1) Chemicals: zinc acetate, 2-methoxyethanol, ethanolamine, acetone, ethanol.

(2) Substrates: quartz glass substrates with a size of 2 cm × 2 cm.

(3) Apparatuses: reaction apparatus (the components have been listed in experiment 29), UV-Vis spectrometer (PerkinElmer, Lambda 850), ultrasonic cleaner, electronic balance, heating plate, muffle furnace.

(4) Tools: pipettes, quartz cuvettes.

3.2 Procedure

(1) Ultrasonically clean the quartz glass substrates for 10 min in deionized water, acetone, and ethanol, sequentially.

(2) Dissolve 6.59 g zinc acetate with 60 mL of 2-methoxyethanol in the three-necked flask.

(3) Connect the flask with the condenser. Immerse the flask into the oil bath on the electromagnetic stirrer with a heating plate.

(4) Heat the solution to 60 °C and stir for 0.5 h.

(5) Slowly drop 2 mL ethanolamine into the solution under stirring.

(6) Increase the solution temperature to 70 °C. Keep stirring for 1.5 h. Then cool down the solution to room temperature.

(7) Spin coat the above solution onto a glass substrate at the speed of 2400 r/min for 30 s.

(8) Heat the sample at 200 °C for about 5min on the heating plate.

(9) Repeat steps (7) and (8) to prepare thin film samples, on which the thin films are deposited by spin coating for 2, 4, 6, 8, 10, and 12 times, respectively.

(10) Heat the samples at 400°C for 2 h in the muffle furnace.

(11) Measure the UV-Vis spectra of the samples. The diffuse reflectance spectra are collected since the film surface is rough. The procedure has been described in Experiment 29.

4 Requirements for experimental report

(1) Plot the UV-Vis spectra of absorbance versus wavelength. These spectra should be plotted in one chart for comparison.

(2) Plot the UV-Vis spectra with energy E as the abscissa axis.

(3) Draw the Tauc plots and calculate the bandgap energies. n is $1/2$ since ZnO is a direct gap semiconductor. The thickness of each layer is about 20 nm.

(4) Plot the optical energy bandgap as a function of film thickness. Discuss the effect of film thickness on the bandgap.

5 Questions and further thoughts

(1) Why do we not use the common glass for the deposition of ZnO film?

(2) Suggest one or two other methods for measuring the bandgap.

(3) How can we adjust the bandgap width of a semiconductor? Please provide several examples.

Experiment 31   Preparation and Characterization of BaTiO$_3$ Based Piezoelectric Ceramics

Type of the experiment: Designing
Recommended credit hours: 24
Brief introduction: Piezoelectric ceramics are attractive sensor or actuator materials. In this experiment, two or three types of BaTiO$_3$ based piezoelectric ceramics are required to be designed and prepared based on a literature survey. Not only the composition but also the preparation process is required to be designed. Several examples are given in the section of the experimental procedure for reference. These designed piezoelectric ceramics will then be prepared, followed by measurement of the piezoelectric strain constant d_{33}. Through this experiment, students will understand the basic process and method of material design.

1 Objectives

(1) To understand the principle of the piezoelectric effect.

(2) To understand the common preparation method and procedure of piezoelectric ceramics.

(3) To design and prepare several BaTiO$_3$ based piezoelectric ceramics.

(4) To measure the piezoelectric strain constant d_{33} of the prepared samples.

(5) To understand the basic process of material design.

2 Principles

2.1 Piezoelectric effect

The piezoelectric effect is the ability of certain materials to generate an electric charge in response to applied mechanical stress. Due to the intrinsic characteristics of piezoelectric materials, numerous applications benefit from their usage. These involve producing and detecting sound, generation of high voltages, piezoelectric sensors, and piezoelectric motors. It is also used in scientific instruments with atomic resolution, such as scanning probe microscope.

One of the unique characteristics of the piezoelectric effect is that it is reversible. This means that the material exhibits both direct and reverse piezoelectric effects. The direct piezoelectric effect is shown in Fig. 6.9(a). When a piezoelectric material is placed under a force F, a shifting of the positive and negative charge centers in the material takes place, resulting in an external electrical field E. The compressive force and tensile force can produce an opposite electric field. Figure 6.9(b) shows the converse piezoelectric effect, by which an outer electrical field either stretches or compresses the piezoelectric material.

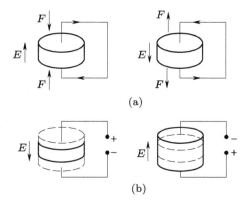

Fig. 6.9 (a) The direct and (b) converse piezoelectric effects.

The electrical response due to the direct piezoelectric effect can be expressed in tensor form as

$$P_i = d_{ij}\sigma_j (i = 1, 2, 3; j = 1, 2, \cdots, 6), \tag{6.19}$$

where P_i is polarization (C/m^2), σ_j is stress (N/m^2), and d_{ij} is the piezoelectric strain constant or piezoelectric charge constant (C/N). The tensor form of these parameters is taken in the equation because the piezoelectric material is anisotropic. The physical constants in the equation relate to both the direction of the applied mechanical or electric force and the directions perpendicular to the applied force. Direction x, y, or z is represented by the subscript 1, 2, or 3, respectively, and rotation (shear) about one of these axes is represented by the subscript 4, 5, or 6, respectively (Fig. 6.10). Therefore, the subscripts i and j of d_{ij} indicate the direction of polarization generated in the material and the direction of the applied stress, respectively.

For the converse effect, it is expressed as

$$S_j = d_{ij}^T E_i (i = 1, 2, 3; j = 1, 2, \cdots, 6), \tag{6.20}$$

where S_j is the strain, E_i is the electric field (in a unit of V/m). In this case, the subscripts i and j of d_{ij} indicate the direction of the applied field and the direction of the induced strain, respectively.

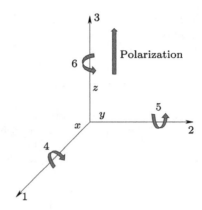

Fig. 6.10 An orthogonal system to describe the properties of piezoelectric ceramics. Axis 3 is the direction of polarization.

For example, d_{33} is the constant related with induced polarization in direction 3 (parallel to the direction in which ceramic element is polarized) per unit stress applied in direction 3 or induced strain in direction 3 per unit electric field applied in direction 3.

2.2 Piezoelectric ceramic

There are many materials, both natural and artificial, that exhibit a range of piezoelectric effects. Examples of artificial piezoelectric materials include barium titanate ($BaTiO_3$, BT) and lead zirconate titanate ($Pb(Zr, Ti)O_3$, PZT). $BaTiO_3$ was the first material to be developed as a piezoceramic. $Pb(Zr, Ti)O_3$ based piezoelectric ceramics have been commercialized in the last century for their outstanding piezoelectric and electromechanical properties. However, due to the growing environmental concern regarding toxicity in lead-containing devices, $BaTiO_3$ has come back into focus. The piezoelectric properties of BT-based ceramics are significantly enhanced by adding a small amount of $BaZrO_3$ and $CaTiO_3$. The doped BT ceramics possess great potential applications.

In this experiment, two or three types of BT-based piezoelectric ceramics with different compositions are required to be designed and prepared based on a literature survey. Since Zr-doped $BaTiO_3$ ($Ba(Zr, Ti)O_3$, BZT) is a widely studied BT-based piezoelectric ceramic, it is recommended to design piezoelectric ceramics based on BZT. Certainly, other dopant types are also acceptable.

2.3 Preparation of piezoelectric ceramic

The preparation of bulk piezoelectric ceramics is the same as general ceramic materials, which includes preparation of mixed powder or precursor, calcination, milling, forming, and sintering.

(1) Preparation of mixed powder or precursor.

The piezoelectric ceramic powder can be prepared by both the solid phase method and the liquid phase method. The solid phase method is relatively simple with a high yield. It starts from mixing and refining the raw materials using mortar or ball mill to obtain the mixed powder. The selection of mixing method and mixing time depends on the type and amounts of raw materials and the initial and final particle sizes. The liquid phase method starts from synthesis precursor by sol-gel, hydrothermal, and precipitation methods. No matter which method is used, homogeneity of the mixture is the most important at this step.

(2) Calcination.

The objectives of calcination are to remove the bound water, CO_2, and volatile matters in the mixed powder or precursor possibly introduced during the preparation process and obtain the solid solution of the final piezoelectric ceramic by diffusion and reaction. Theoretically, when calcination temperature increases, diffusion and reaction are more adequate, but the powder may have a larger particle size and severe agglomeration.

(3) Milling.

In order to refine the calcined powder and break the agglomeration, milling is required after calcination. Mortar or ball mill is generally used in the laboratory. For wet milling, liquid milling aid is added to improve the milling efficiency. This step can be omitted if the calcined powder is well dispersed and has a suitable particle size.

(4) Forming.

Forming is a step to form ceramic shapes. There are many forming techniques, such as uniaxial (die) pressing, isostatic pressing, injection molding, and extrusion. Uniaxial pressing is the powder compaction method involving uniaxial pressure applied to the powder placed in a die between two rigid punches. It is effectively used for the mass production of simple parts and is widely used in the laboratory. PVA solution is usually used as the binder in the forming process. Pressure and pressing time need to be in the appropriate range depending on the powder properties and particle size.

(5) Sintering.

During sintering, the binder will be removed at a specific lower temperature range. The particles will be bonded into a coherent, solid structure via diffusion at a specified high temperature. The sintering temperature and time have significant impacts on the crystal structure and the piezoelectric performance of ceramics. For plenty of piezoelectric ceramics, the sintering temperature is often higher than 1200 °C. Advanced sintering methods have been developed to decrease the sintering temperature and to adjust the microstructure and subsequent properties.

2.4 Characterization of piezoelectric property

The samples with specific shapes are employed to determine the piezoelectric properties, and different shapes are required for different constants. Samples with the rod, plate, and disc shapes are generally used.

The samples are painted with electrodes before characterization to collect the current and reduce the contact resistance. Ag is generally used as the electrode, which is applied by brushing the Ag paste on the major surfaces and firing at about 850 °C to ensure a good bond between the sample surfaces and the electrodes.

In the grains of the piezoelectric ceramics, there are plenty of small ferroelectric domains with random orientations of spontaneous polarization, as shown in Fig. 6.11(a). A ferroelectric polarization process is required to make the ceramic macroscopically piezoelectric as well. For this purpose, a strong electric field of several kV/mm is applied, which causes a reorientation of the spontaneous polarization. Moreover, domains with a favorable orientation to the polarity field direction grow, and those with an unfavorable orientation shrink. Simultaneously, the sample is extended under the electric field (Fig. 6.11(b)). After polarization, most of the reorientations are preserved even without applying an electric field, and the sample has a residual extension (Fig. 6.11(c)).

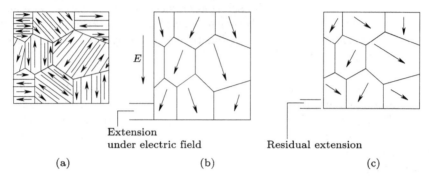

(a) (b) (c)

Fig. 6.11 Schematic illustrations of (a) unpolarized sample, (b) polarization, and (c) remnant polarization.

Polarized piezoelectric materials are characterized by several constants and relationships. These generally include piezoelectric strain constant, piezoelectric voltage coefficient, elastic compliance, frequency constant, and coupling factor. Among these constants, the piezoelectric strain constant defined in Equations (6.19) and (6.20) is the most commonly used piezoelectric constant in the literature. In this experiment, d_{33} is measured by the quasi-static d_{33} meter. The principle of the test system is to apply a low-frequency force to the sample. The electrical signals from the sample are processed and compared with a built-in reference. Then the system gives a direct reading of d_{33} in seconds.

3 Experimental

3.1 Preparation

(1) Materials and chemicals: prepare according to the experiment designed by students.

(2) Apparatuses: high voltage polarimeter (Nanjing Entai Electronic Instruments Plant, ET2673D-4), quasi-static d_{33} meter (Wuxi Shiao Technology co. LTD., YE2730A), X-ray diffractometer (Rigaku, Miniflex 600), electronic balance, ball mill, muffle furnace, mortar and pestle, press machine.

(3) Tools and others: prepare according to the experiment designed by students.

3.2 Procedure

3.2.1 General procedure

(1) Conduct a literature survey before the experiment.

(2) Design the composition and preparation process of two or three types of BT-based piezoelectric ceramics. The piezoelectric ceramics based on BZT are recommended, and other BT-based materials are also acceptable.

(3) Prepare the samples.

(4) Grind the major surfaces of the samples with abrasive papers. Paint the surfaces with Ag paste. Heat them at 850 °C for 1 h.

(5) Polarize the prepared samples at a temperature slightly below the Curie temperature. The voltage is about 3-4 kV/mm, and the time is in a range of 10–60 min.

(6) Measure the d_{33} values of the samples.

3.2.2 Example 1: $0.5Ba(Zr_{0.2}Ti_{0.8})O_3$-$0.5(Ba_{0.7}Ca_{0.3})TiO_3$ piezoelectric ceramic

(1) Prepare raw materials, including $BaCO_3$, $CaCO_3$, TiO_2, and $BaZrO_3$ powders.

(2) Weigh the raw materials according to the stoichiometric ratio.

(3) Mix the raw materials for 4 h with the ball mill using ethanol as the heat-transfer agent.

(4) Calcine the mixed powder at 1100 °C for 2 h.

(5) Add a few drops of the PVA solution (7wt%) in the powder and then mix them with a mortar and pestle.

(6) Compact the powder into discs with a size of Φ10 mm × 2 mm at a pressure of approximately 100 MPa for 2 min.

(7) Sinter the green discs at 1450 °C in the air for 2 h. The sintering process is shown in Fig. 6.12.

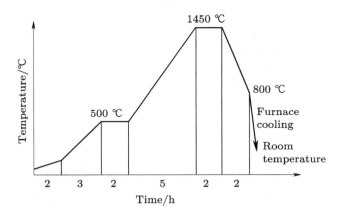

Fig. 6.12 Temperature-time plot for the sintering process.

3.2.3 Example 2: $Ca_{0.02}Ba_{0.98}(Zr_{0.02}Ti_{0.98})O_3$ piezoelectric ceramic

(1) Prepare raw materials, including $BaCO_3$, $CaCO_3$, TiO_2, $BaZrO_3$ powders.

(2) Weigh the raw materials according to the stoichiometric ratio.

(3) Mix raw materials for 6 h with a ball mill using ethanol as the heat-transfer agent.

(4) Calcine the mixed powder at 1000 °C for 2 h.

(5) Add few drops of the PVA solution (7wt%) in the powder and then mix them with a mortar and pestle.

(6) Compact the powder into discs with a size of Φ10 mm × 2 mm at a pressure of approximately 100 MPa for 2 min.

(7) Sinter the green discs at 1450 °C in the air for 2 h. The sintering process is the same as that shown in Fig. 6.12.

3.2.4 Example 3: $BaZr_{0.2}Ti_{0.8}O_3+0.5mol\%$ CuO piezoelectric ceramic

(1) Prepare raw materials including $BaCO_3$, TiO_2, ZrO_2, and CuO powders.

(2) Weigh the raw materials according to the stoichiometric ratio.

(3) Mix the raw materials for 24 h with a ball mill using ethanol as the heat-transfer agent.

(4) Calcine the mixed powder at 1200 °C for 2 h.

(5) Mix the calcined powder and CuO powder with a ball mill for 6 h using ethyl alcohol as the heat-transfer agent.

(6) Add a few drops of the PVA solution (7wt%) in the powder and then mix them with a mortar and pestle.

(7) Compact the powder into discs with a size of Φ10 mm × 2 mm at a pressure of approximately 100 MPa for 2 min.

(8) Sinter the green discs at 1250 °C in the air for 3 h. The sintering process is nearly the same as that shown in Fig. 6.12, except the sintering temperature is decreased and sintering time is increased correspondingly.

4 Requirements for experimental report

(1) Describe the design idea of the compositions of the piezoelectric ceramics prepared in this experiment.

(2) Describe the preparation process in detail.

(3) Give the d_{33} values of the piezoelectric ceramics prepared in this experiment, and compare them with those in the literature. Discuss the possible reasons if there is a significant difference.

5 Questions and further thoughts

(1) What are the advantages and disadvantages of the solid phase method and the liquid phase method for preparing ceramic powders?

(2) Describe the definitions and equations of several most frequently used parameters for piezoelectric properties.

6 Cautions

(1) After powder preparation and sintering, X-ray diffraction (XRD) should be used to confirm the phase constitution.

(2) In the experiment design, avoid using toxic or expensive chemicals and materials; avoid using apparatus that is difficult in operation.

Extended readings

Boles M, Ling D, Hyeon T, et al. The surface science of nanocrystals. Nature Materials, 2016, 15: 141-153.

Chiang I C, Chen D H. Synthesis of monodisperse FeAu nanoparticles with tunable magnetic and optical properties. Advanced Functional Materials, 2007, 17: 1311-1316.

Daoush W M. Co-precipitation and magnetic properties of magnetite nanoparticles for potential biomedical applications. Journal of Nanomedicine Research, 2017, 5(3): 00118.

Gupta A, Eral B H, Hatton T A, et al. Nanoemulsions: formation, properties and applications. Soft Materials, 2016, 12(11): 2826-2841.

Harvey D. UV/Vis and IR spectroscopy. Available at: https://chem.libretexts.org/Bookshelves/Analytical_Chemistry/Book%3A_Analytical_Chemistry_2.1_(Harvey)/10%3A_Spectroscopic_Methods/10.03%3A_UV/Vis_and_IR_Spectroscopy. Accessed: 2 March 2021.

Jagadish C, Pearton S. Zinc Oxide Bulk, Thin Films and Nanostructures. Elsevier, 2006.

Li H, Wang J, Liu H, et al. Zinc oxide films prepared by sol-gel method. Journal of Crystal Growth, 2005, 275: e943-e946.

Liu H L, Zhang W X, Hou P, et al. Facile growth of monocrystalline gold-iron nanocrystals by polymer nanoemulsion. Gold Bulletin, 2011, 44: 21-25.

Moulson A J, Herbert J M. Electroceramics: Materials, Properties, Applications, 2nd Edition. John Wiley & Sons, 2003.

Ohring M. Materials Science of Thin Films. 2nd Edition. Academic Press, 2001.

Pearton S J, Norton D P, Ip K, et al. Recent progress in processing and properties of ZnO. Progress in Materials Science, 2005, 50(3): 293-240.

Powder diffraction file (PDF) search. Available at: https://www.icdd.com/pdfsearch/. Accessed: 2 March 2021.

Rao C N R, Thomas P J, Kulkarni G U. Nanocrystals: Synthesis, Properties and Applications. Springer, 2007.

The Editors of Encyclopaedia Britannica. Piezoelectricity. Available at: https://www.britannica.com/science/piezoelectricity. Accessed: 2 March 2021.

Tian Y, Wei L, Chao X, et al. Phase transition behavior and large piezoelectricity near the morphotropic phase boundary of lead-free $(Ba_{0.85}Ca_{0.15})(Zr_{0.1}Ti_{0.9})O_3$ ceramics. Journal of the American Ceramic Society, 2012, 96(2): 496-502.

Waseda Y, Matsubara E, Shinoda K. X-Ray Diffraction Crystallography. Springer, 2011.

Wu B, Xiao D, Wu J, et al. Microstructures and piezoelectric properties of CuO-doped $(Ba_{0.98}Ca_{0.02})(Ti_{0.94}Sn_{0.06})O_3$ ceramics. Journal of Electroceramics, 2014, 33(1-2): 117-120.

Wu J H, Ko S P, Liu H L, et al. Sub 5nm magnetite nanoparticles: Synthesis, microstructure, and magnetic properties. Materials Letters, 2007, 61: 3124-3129.

Yu P Y, Cardona M. Fundamentals of Semiconductors. Springer, 2008.

Part IV
Material Preparation and Treatment

Chapter 7 Forming and Heat Treatment

Experiment 32 Microstructure Observation of Crystallization and Aluminum Alloy Ingots

Type of the experiment: Comprehensive
Recommended credit hours: 4
Brief introduction: Crystallization of metals is an essential change of state in casting. Because metal materials are untransparent and their crystallization temperature is relatively high, it is difficult to observe their crystallization process. Since the crystallization of salt from its supersaturated solution also consists of two essential steps, nucleation and crystal growth, like those of metals, the observation of salt crystallization can contribute to understanding the metal crystallization. Hence, in this experiment, the crystallization process of ammonium chloride (NH_4Cl) from its supersaturated water solution is observed. Then the structural zones in the aluminum alloy ingots are observed, and the effects of casting molds are examined by comparing the structural zones of the ingots obtained using sand and steel molds.

■ Objectives

(1) To understand the principle and process of crystallization.

(2) To observe the crystallization process of NH_4Cl from its supersaturated water solution.

(3) To observe the structure zones in aluminum alloy ingots and understand the influence of the casting process.

2 Principles

2.1 Crystallization

Crystallization is the formation of solid crystals precipitating from a solution, melt, or even from gas. It is an essential change of state in the casting of metals. Crystallization is also a chemical solid-liquid separation technique, in which a pure solid crystalline phase occurs from the liquid solution.

The crystallization process consists of two major events, nucleation and crystal growth. Nucleation is the step where the metal or solute atoms gather into a cluster at the nanoscale and become stable under operating conditions. These stable clusters constitute the nuclei. However, when the clusters are not stable, they will dissolve. Therefore, the clusters need to reach a critical size to become stable nuclei. The crystal growth is the subsequent growth of the nuclei that succeed in achieving the critical cluster size.

Under actual crystallization conditions, most crystallization begins on preexisting surfaces or interfaces, which is called heterogeneous nucleation, because it typically has a much smaller barrier than homogeneous nucleation. The morphology of the crystal growth is influenced by the temperature gradient, cooling rate, constitutional supercooling, and other factors. The dendritic crystal growth, i.e., the crystals with branches, are most often to be seen.

2.2 Crystallization of salt

When the water in a supersaturated saltwater solution, e.g., NH_4Cl solution, is vaporized, the NH_4Cl begins to crystalize. The crystallization process of a drop of the solution consists of three stages and can be seen under a microscope.

The first stage is the formation of a small ring of equiaxed crystals in the outermost layer of the droplet. This is because the outer layer of the droplet evaporates quickly, and a large number of nuclei are formed in a short time. The second stage is the formation of larger columnar crystals that grow toward the center of the droplet. This is because the droplet is evaporating more slowly, and it gradually reaches saturation from the outside. The third stage is the formation of different unaligned equiaxial crystals at the center of the droplet.

It should be noted that the crystallization of NH_4Cl is fundamentally different from that of metals. The NH_4Cl crystallizes by water evaporation to make the solution supersaturated, while metals crystallize when liquid metals are cooled with a certain degree of supercooling. In spite of this, their crystallization processes are very close.

2.3 Structural zones in alloy castings or ingots

Most engineering alloys are obtained from a starting point of being cast into a fireproof container or mold. The as-cast pieces are called castings if they are permitted to retain their shape afterward. They are called ingots if they are reshaped by deformation processing, such as rolling, extrusion, or forging.

The solidification of liquid metal also consists of three stages. As shown in Fig. 7.1, three morphological zones that correspond to these stages have been revealed by metallographic studies on the cross-section of alloy ingots.

An outer chill zone of equiaxed crystals is a very thin fine-grain surface layer. When the liquid metal is poured into the ingot mold, part of the metal liquid in contact with the cold mold wall is cooled rapidly. The product is deposited on the mold wall at a high degree of undercooling.

A columnar zone has elongated or columnar grains. The formation of this zone is because the temperature inside the ingot is higher due to the formation of the surface layer, and the crystal core is more difficult to form. Therefore, the grains of the surface layer will grow inward, forming the columnar grains perpendicular to the mold wall.

A central equiaxed zone is in the central part of the ingot. With the growth of columnar crystals, the liquid in the ingot center reaches the crystallization temperature, and many nuclei are formed. These nuclei grow in all directions, leading to the formation of the equiaxed zone. The grain size of this zone is relatively large due to the slow cooling of the central part. If the cooling rate is very fast, the equiaxed zone is hard to be seen because the columnar crystals will rapidly develop towards the center and run through the entire ingot.

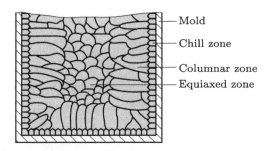

Fig. 7.1 Schematic illustration of the structural zones of mold cast ingot.

3 Experimental

3.1 Preparation

(1) Materials:cross-sectional samples from ingots of aluminum alloy 2011 (Al-Cu alloy). The ingots are obtained using sand and steel molds, respectively.

(2) Chemical: saturated NH_4Cl water solution.

(3) Apparatus: metallographic microscope.

(4) Tools: glass slide, dropper.

(5) Others: items for preparation of metallographic sample listed in Experiment 1.

3.2 Procedure

3.2.1 Observe the crystallization process of NH_4Cl

(1) Drop a droplet of saturated NH_4Cl solution on a glass slide.

(2) Observe the crystallization process during the evaporation of water under the microscope.

(3) Capture images of the typical structures.

3.2.2 Observe structural zones of Al alloy ingots

(1) Prepare the metallographic samples of the cross-sectional samples from the ingots. Details of the preparation procedure have been described in Experiment 1.

(2) Observe the microstructures of the samples under the microscope. Capture images of the typical structures.

4 Requirements for experimental report

(1) Give images of typical structures of NH_4Cl crystallization and Al alloy ingots.

(2) Describe the crystallization processes of NH_4Cl and Al alloy.

(3) Schematically draw the microstructures of the Al alloy ingots.

(4) Discuss the effects of casting molds on the structural features of the ingots.

5 Questions and further thoughts

(1) What is the essential difference between the crystallization of salt from solution and metal alloy?

(2) How can we control the structures in an ingot?

6 Cautions

(1) The droplet of saturated NH_4Cl solution for observation should be small to avoid long evaporation time.

(2) Keep the glass slide and the droplet of the solution clean to avoid the effect of inclusions on the crystallization process.

Experiment 33 Preparation of Stainless Steel Pellets by Powder Metallurgy Method

Type of the experiment: Comprehensive

Recommended credit hours: 8

Brief introduction: The powder metallurgy of stainless steel has become of great interest due to increasing demand in various applications. In this experiment, 316L stainless steel pellets are fabricated through a relatively simple process of powder metallurgy. Then, the microstructures and densities of the sintered pellets are characterized. The effects of lubricant and sintering temperature are examined.

During the experiment, the sintering of the samples takes a long time. Therefore, the experiment can be divided into two parts, preparation and characterization, so that the sintering does not occupy the class time.

1 Objectives

(1) To understand the principle and method of powder metallurgy.

(2) To prepare 316L stainless steel pellets by the powder metallurgy method.

(3) To understand the influence of sintering temperature on microstructures and densities of the sintered samples.

Experiment 33 Preparation of Stainless Steel Pellets by Powder Metallurgy Method

2 Principles

2.1 Stainless steel

The stainless steels are highly resistant to corrosion or rusting in a variety of environments, especially the ambient atmosphere. The predominant alloying element in the stainless steels is chromium, with a concentration of at least 12wt%. The addition of nickel and molybdenum may further enhance the corrosion resistance. The chromium in the steel combines with oxygen to form a thin oxide layer called the passive film. If the metal is cut or scratched and the passive film is disrupted, more oxide will quickly form and recover the exposed surface, protecting it from oxidative corrosion.

Stainless steels are divided into three classes based on the predominant phase constituent, i.e., martensitic, ferritic, or austenitic stainless steels. Austenitic steels that contain 16wt% to 26wt% Cr and up to 35wt% Ni have the highest corrosion resistance.

Type 316 is one of the most familiar austenitic stainless steel. Type 316L stainless steel has less carbon than type 316, and "L" stands for "low". Its composition is listed in Table 7.1. It is a better choice for a project that requires much welding because the lower carbon content minimizes carbide precipitation during welding, and the carbides can weaken the corrosion resistance.

Table 7.1 Composition of 316L stainless steel

Element	C	Si	Mn	P	S	Ni	Cr	Mo	Fe
Percentage/(wt%)	$\leqslant 0.03$	$\leqslant 1.00$	$\leqslant 2.00$	$\leqslant 0.04$	$\leqslant 0.03$	10.0–14.0	16.0–18.0	2.0–3.0	balance

2.3 Powder metallurgy of stainless steel

Powder metallurgy or powder sintering is a technique used for bonding powder particles together into a coherent bulk solid structure. This technique has many advantages, including cost-effectiveness, reducing machining by near-net-shape forming, minimizing material loss by decreasing the processing steps, processing materials that would otherwise be impossible to mix, processing materials with very high melting points, and achieving controlled levels of porosity. The process of powder metallurgy generally includes mixing, compacting, and sintering.

Mixing is used to regulate granularity or composition and to remove grain segregation caused by transportation. The powder can be mixed by mechanical or chemical methods. The mechanical method is to mix the powder with a ball mill and other mixing machines without chemical reaction. The chemical method is to mix the powder with a metal salt solution, or all components are mixed in a solution of some salt. Then, mixed powder with evenly distributed components is obtained after precipitation, drying, and reduction.

Compacting is the process of forming the powder into a steel block with a specific shape, size, porosity, and strength. The powder particles of stainless steel are relatively hard, so the pressing pressure is about 400–800 MPa. During the pressing process, lubricants are often added to the pre-alloyed stainless steel to improve the pressing performance. The primary role of lubricants is to reduce the friction in the powder and increase the density of the sample.

Sintering is to heat the compact body at a temperature below the melting point of its components. During the sintering, specific physical and chemical processes, such as diffusion, recrystallization, creep, and melting of powder particles, facilitate powder shrinkage and densification. The sintering results are the bonding between particles and the increase in the strength and density of the sintered body.

The powder metallurgy of stainless steel has become of great interest due to the increasing demand for powder metallurgy components of stainless steel in a variety of applications, including aerospace, automotive, chemical processing, and biomedical fields. In this experiment, the powder metallurgy is conducted using 316L stainless steel powder.

3 Experimental

3.1 Preparation

(1) Materials: 316L stainless steel powder (<100 μm) produced by water atomization.

(2) Chemical: zinc stearate.

(3) Apparatuses: press machine, atmosphere tube furnace, electronic balance, density balance, metallographic microscope.

(4) Tools: mortar and pestle, pellet press die with a diameter of 10 mm, Vernier caliper.

(5) Others: items for preparation of metallographic sample listed in Experiment 1.

3.2 Procedure

3.2.1 Powder sintering of 316L stainless steel

(1) Weigh 7.5 g 316L powder and add 0.045 g zinc stearate (0.6wt%) as a lubricant.

(2) Mix the powder with a mortar and pestle for 20 min. Separate the mixture into three copies denoted as samples A, B, and C.

(3) Weigh 2.5 g 316L powder and denoted as sample D.

(4) Put them into the pellet press die and press by the press machine. The pressure is 400 MPa, and the holding time is 2 min. Record the weights and sizes of the four compressed samples.

(5) Sinter the compressed samples A, B, and C at 1200 °C, 1250 °C, and 1300 °C respectively in the atmosphere tube furnace under argon or nitrogen gas (about 200 mL/min). Sample D is sintered only at 1250 °C. The sintering process is shown in Fig. 7.2. Record the weights and sizes of the sintered samples.

3.2.2 Characterization of the sintered samples

(1) Grind the samples to remove the oxidation layer.

(2) Measure the densities using the density balance, which works based on the Archimedes method.

(3) Grind, polish, and etch the samples. Details of the procedure have been described in Experiment 1.

(4) Observe the microstructural features of grains and pores. Take metallo-

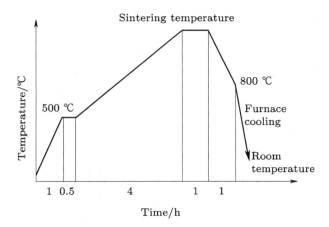

Fig. 7.2 Temperature-time plot for the sintering process.

graphs of typical regions.

4 Requirements for experimental report

(1) Give the metallographs of the samples.

(2) Tabulate the weights and sizes of the samples before and after sintering. Calculate the rates of change in weight and size.

(3) Draw a plot of density as a function of sintering temperature.

(4) Analyze the effects of sintering temperature and lubricant on the microstructures and densities of the sintered samples.

5 Questions and further thoughts

(1) During sintering of the compressed samples, the shrinking percentages in diameter and height are different. Why?

(2) In addition to the powder metallurgy process in this experiment, are there any other processes that can produce bulk solid? Please list several common processes and compare their advantages and disadvantages.

Experiment 34 Construction of Sn-Bi and Sn-Zn Binary Phase Diagrams

Type of the experiment: Cognitive
Recommended credit hours: 4
Brief introduction: Sn-Bi and Sn-Zn alloys have been applied as solder materials. Compared with the traditional Sn-Pb solder alloy, they avoid the usage of the toxic metal Pb. In this experiment, the Sn-Bi and Sn-Zn alloy systems are used as examples to introduce the construction of binary phase diagrams using the cooling curve method. The experiment can also provide an understanding of the heat release and supercooling phenomenon during cooling and phase transformation.

◼ Objectives

(1) To understand the basics of binary solid-liquid phase diagrams.
(2) To understand the techniques for constructing equilibrium phase diagrams.
(3) To understand common thermal analysis techniques.
(4) To construct the Sn-Bi or Sn-Zn phase diagrams using the cooling curve method.

◼ Principles

2.1 Binary equilibrium phase diagram

Equilibrium phase diagrams are graphics that represent the existence of phases in materials at various compositions, temperatures, and pressures. They are generally constructed by experiments in which materials are heated or cooled at a very slow rate. Thus, the phases in the materials can achieve a thermodynamic equilibrium state. They are very useful in predicting phase transformations and the resulting microstructures. In solid-liquid systems, the pressure is generally held constant, although external pressure also influences the phase structures. Therefore, the variables of composition and temperature are considered in the solid-liquid phase diagrams.

A binary phase diagram shows the phases of a binary mixture of two components formed over a range of temperatures. On the binary phase diagram, compositions run from 100% component A, through all possible mixtures, to 100% component B. Weight and atomic percentages are often used to specify the proportions of the components.

2.2 Typical binary phase diagrams

2.2.1 Binary isomorphous phase diagram

The binary isomorphous phase diagram is the most accessible type to understand. The phase diagrams and the corresponding systems are termed isomorphous because of the complete liquid (L) and solid (S) solubility of the two components, as illustrated in Fig. 7.3. A typical example is Cu-Ni alloy. The systems are comprised of completely miscible components in solid form, thus forming a continuous solid solution. According to the Hume-Rothery rules, the two components should have similar atomic radii, similar electronegativities, and the same crystal structure.

2.2.2 Binary eutectic phase diagram

Some materials are highly miscible in the liquid state but have limited mutual miscibility or solubility in the solid state. Thus, the phase diagram at low temperatures is dominated by two different solid phases. One is rich in component A, and the other is rich in component B. These binary systems are referred to as eutectic systems.

Figure 7.4(a) shows a typical binary eutectic phase diagram between two metals A and B, which have limited solubility in the solid state but unlimited solubility in the liquid state. The terminal solid solutions are denoted as α and β. The eutectic microstructure (E) in such alloys is an intimate mixture of the two phases. Systems such as Sn-Pb, Sn-Bi, and Ag-Cu alloys show this type of phase diagrams.

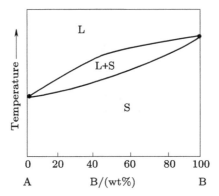

Fig. 7.3 Schematic illustration of the binary isomorphous phase diagram.

Figure 7.4(b) shows a typical eutectic phase diagram between two hypothetical metals completely miscible in the liquid state but immiscible in the solid state. Typical microstructures of hypoeutectic (A+E), eutectic(E), and hypereutectic (B+E) alloys have been included in the diagram. Systems such as Cd-Bi, Sb-Pb, and Sn-Zn show this type of phase diagram.

In this experiment, the phase diagrams of the Sn-Bi system with partial solubility (Fig. 7.5(a)) and Sn-Zn system with negligible solubility (Fig. 7.5(b)) will be constructed using the thermal analysis method.

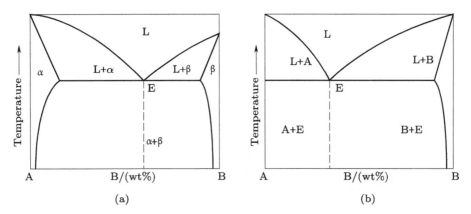

Fig. 7.4 Schematic illustrations of the binary eutectic phase diagrams with (a) partial solubility and (b) negligible solubility.

2.3 Methods for the construction of phase diagrams

2.3.1 Diffraction methods

Since crystal structures can be determined by X-ray diffraction (XRD) or electron diffraction, phase diagrams can be constructed by these diffraction methods. One is the high-temperature XRD method, which allows one to measure the changes in structure at high temperatures directly and, thus, to study the crystallographic

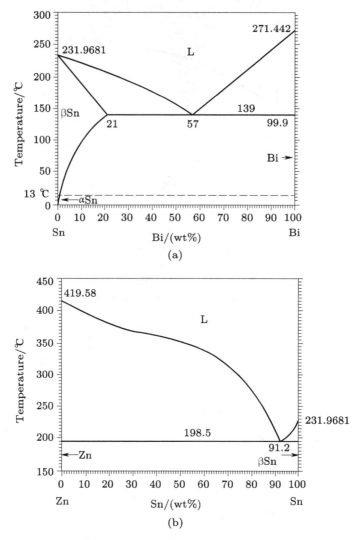

Fig. 7.5 (a) Sn-Bi and (b) Sn-Zn binary phase diagrams.

aspects of phase transformations. A similar method like electron diffraction can be used to observe the crystal structures of small precipitates of new phases at very high magnification by transmission electron microscope (TEM).

2.3.2 Thermal analysis methods

The most commonly used methods are based on thermal analysis. For example, differential scanning calorimetry (DSC) measures the heat energy expelled or absorbed by the sample as it undergoes phase transformations. Differential thermal analysis (DTA) measures changes in heating or cooling behavior as compared to a reference material. Another thermal analysis technique is dilatometry, which is used to determine the transition temperature by detecting the sudden change in volume associated with a phase transformation.

The cooling curve method probably is the most widely used method, which will be accepted in this experiment. When the samples are melted entirely and then cooled, a group of data on how temperature changes along with time can be obtained, and the cooling curves of temperature against time can be plotted. Cooling curves with a constant cooling rate provide the temperatures where phase transformations initiate and terminate.

Figure 7.6 demonstrates how the cooling curve method can be used to determine a binary eutectic phase diagram. For a pure material (curves 1 and 6) or a eutectic composition (curve 4), the cooling curve from liquid to solid shows a horizontal thermal arrest until the transformation is completed. This corresponds to the extraction of the fusion heat at the melting point. For binary materials which experience a two-phase region upon cooling from the liquid (curves 2, 3, and 5), the cooling curves show changes in slope at the beginning and end of the transformations corresponding to T_l and T_s at points l and s. When a series of samples with a number of different compositions are tested, the transformation temperatures can be identified, and the curves in the phase diagram can be determined by fitting these data.

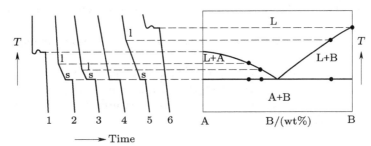

Fig. 7.6 The cooling curve method for constructing a binary phase diagram.

Ideally, the phases of the system should be determined in an equilibrium or approximately equilibrium state during the measurement. However, it takes a long time to achieve such a state. In addition, the supercooling phenomenon commonly occurs during continuous phase transformation. Figure 7.7 illustrates the supercooling phenomenon related to the crystallization of pure metal. Serious supercooling makes the arrest points deviate from the ideal phase transformation temperature, which makes temperature identification more difficult. Therefore, generally, the cooling rate must be slow enough for reasonable accuracy.

3 Experimental

3.1 Preparation

(1) Materials: Sn, Bi, and Zn particles.

(2) Apparatuses: furnace (Nanjing Sangli Electronic Equipment Factory, KWL-10), digital temperature measurement and control system (Nanjing Sangli Electronic Equipment Factory, SWKY-II), electronic balance.

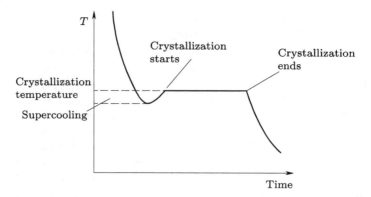

Fig. 7.7 Crystallization of pure metal with the supercooling phenomenon.

   (3) Tools: stainless steel tubes, thermocouples, 50mL beaker, spoon.

3.2 Procedure

   (1) Weigh 50.0 g mixtures of six Sn-Bi or Sn-Zn alloys with different compositions. To finish the experiment within a reasonable time and obtain most features of the phase diagram, the pure metals and the eutectic composition must be included.

   (2) Place them in stainless steel tubes, respectively. Thoroughly mix the alloys by shaking the tubes.

   (3) Insert the thermocouples in the tubes. The thermocouple tip must be inserted into the middle part of the mixture. Place the tubes in the furnace. Connect the thermocouples with the digital temperature measurement and control system, as shown in Fig. 7.8.

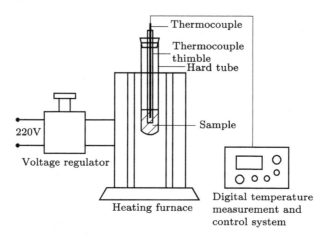

Fig. 7.8 Experimental setup for measuring cooling curves.

   (4) Switch on the furnace and the digital temperature measurement and control system. Open the software.

(5) Fill in the composition parameters of the samples in the software.

(6) Switch the digital temperature measurement and control system to the "set parameter" mode, and set temperature to destination temperature (320 °C, 460 °C for Sn-Bi, Sn-Zn alloys, respectively). Then switch to the "work" mode. The furnace starts to heat the tubes.

(7) When the temperature arrives at the desired temperature, hold the temperature for 10 min to completely melt the samples.

(8) Stop heating the samples. Switch on the cooling fan. Set the voltage value to about 5 V and control the cooling rate to about 5 °C/min. The cooling curves will be automatically recorded by the software.

(9) Identify the temperatures at the beginning and end of the transformations or the thermal arrest temperature for pure materials or the eutectic composition from the cooling curves.

(10) Export the temperature-time data of the samples as Excel files.

4 Requirements for experimental report

(1) Plot six cooling curves in one chart and separate these curves horizontally using Origin software, as shown in Fig. 7.6.

(2) Tabulate the T_l and T_s temperatures recorded from the cooling curves.

(3) Draw the binary phase diagrams by plotting temperature against composition using Origin software.

5 Questions and further thoughts

(1) Why does the thermal arrest temperature occur in the cooling curve?

(2) How can we determine the composition of an unknown Sn-Bi or Sn-Zn alloy? Please list several methods.

(3) Compare the crystallization temperature of pure metals and the eutectic temperature with those in literature. Discuss possible sources of error.

Experiment 35 Microstructures and Diffusion of Iron-Carbon Alloys

Type of the experiment: Comprehensive
Recommended credit hours: 12
Brief introduction: Iron-carbon alloys are possibly the most important binary alloy system since steels and cast irons, which are primary structural materials for engineering applications, are essentially iron-carbon alloys. In this experiment, the microstructures and diffusion of iron-carbon alloys are studied. The experiment is divided into two experiments that can be conducted respectively. The first experiment examines the equilibrium and non-equilibrium microstructures of a series of typical Fe-C alloys based on the understanding of the Fe-C phase diagram and time-temperature-transformation (TTT) diagram. The second experiment introduces the carburization treatment of low carbon steel. The microstructure and the depth of the carburized layer of T20 carbon steel will be observed to further understand the influence of carbon content and diffusion on the microstructure.

Experiment 35.1 Observation of Equilibrium and Non-Equilibrium Microstructures of Iron-Carbon Alloys

Type of the experiment: Cognitive
Recommended credit hours: 4

1 Objectives

(1) To grasp the features of the iron-carbon equilibrium diagram based on the observation of the equilibrium microstructures of Fe-C alloys.

(2) To grasp the features of the TTT diagram based on the observation of the non-equilibrium microstructures of Fe-C alloys.

(3) To understand the formation of the equilibrium and the non-equilibrium microstructures of Fe-C alloys.

(4) To understand the influence of heat treatment on microstructures and properties of Fe-C alloys.

2 Principles

2.1 Iron-carbon equilibrium phase diagram

Iron-based alloys are the most widely used metallic materials. Their simplest form is iron-carbon (Fe-C) alloy. A study of the constitution and structure of all steels and cast irons must start with the Fe-C equilibrium diagram shown in Fig. 7.9 since many essential features of this system influence the behavior of even the most complex alloy steels.

The composition in Fig. 7.9 extends to 6.70wt% of carbon corresponding to the

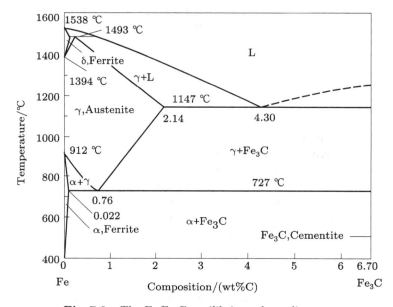

Fig. 7.9 The Fe-Fe$_3$C equilibrium phase diagram.

intermediate compound of iron carbide (Fe_3C). The Fe_3C represented by a vertical line on the phase diagram divides the Fe-C system into two parts: an iron-rich part shown in Fig. 7.9 and a part not shown in the figure for the composition range from 6.70wt% to 100wt% carbon. In most cases, the Fe-Fe_3C system is considered since all steels, and most cast irons have carbon contents less than 6.70wt% in practice.

In the classification scheme of ferrous alloys based on carbon content, there are three types, iron, steel, and cast iron. Pure iron contains less than 0.022wt% carbon, while commercially pure iron generally contains less than 0.008wt% carbon. Steels are iron-carbon alloys with carbon in a range of 0.022wt% to 2.14wt%. In practice, steel generally contains less than 1.0wt% carbon. Cast irons are classified as ferrous alloys that contain carbon between 2.14wt% and 6.70wt%. Commercial cast irons generally contain less than 4.5wt% carbon.

2.2 Equilibrium microstructures of Fe-C alloys

In the Fe-C system, carbon is an interstitial solute in iron and forms solid solutions of α, δ, and γ phases. Other phases are iron carbide and free carbon in the form of graphite. According to their microstructure features, they are also called ferrite, austenite, and cementite, respectively.

The α and δ phases are both called ferrite, in which only a limited amount of carbon is soluble. In the αferrite, the maximum solubility is 0.022wt% at 727 °C. It shows bright equiaxed grains under the microscope. The δ ferrite exists at high temperatures, while its microstructural feature is the same as α ferrite. Its maximum solubility is 0.09wt% at 1493 °C.

Austenite (γ phase) is stable in a temperature range of 727 to 1493 °C. The maximum solubility of carbon in austenite is 2.14wt% at 1147 °C. The austenite also shows bright equiaxed grains under the microscope. Moreover, annealing twins are frequently observed in the grains.

Cementite (Fe_3C) forms in a wide temperature range. It coexists with the γ phase between 727 °C and 1147 °C and also forms when the solubility limit of carbon in α ferrite is exceeded below 727 °C. The cementite shows various morphology depending on the carbon content and formation condition. They can be thick lath form in cast iron, network along the austenite grain boundaries, or lamellar mixture with ferrite.

These microstructures or phases can further combine to form other microstructures through eutectoid and eutectic reactions, respectively. During slow cooling of Fe-C alloy, pearlite forms by a eutectoid reaction at 0.76wt% C and 727 °C, which is composed of alternating layers of α-ferrite and cementite. It is named pearlite because it has the appearance of the mother of pearl when viewed under the microscope at low magnification. Ledeburite is formed by a eutectic reaction at 4.30wt% C and 1147 °C, which is a mixture of austenite and cementite. When the temperature is further decreased to below 727 °C, the pearlite is formed by the eutectoid reaction of the austenite in the ledeburite.

2.3 TTTdiagram

Although the Fe-C phase diagram clarifies the equilibrium microstructures, time is not considered as a variable.Hence, the effects of the cooling rate on the microstructures are not revealed. These effects on the microstructures can be shown by TTT diagrams or isothermal transformation diagrams. For each steel composition, a different TTT diagram is obtained.

Figure 7.10 shows theTTT diagram of T8 steel (0.8wt% C). Austenite is stable above the eutectoid temperature, 727 °C. When the steel is cooled to a temperature below this eutectoid temperature, austenite is transformed into its transformation product according to the temperature and time required to complete the transformation. TTT diagram indicates transformation product by the C-shaped curves shown in Fig. 7.10. The left one is the transformation-beginning curve, while the right one is the transformation-ending curve. The region to the left of the beginning curve corresponds to austenite. The region to the right of the end curve corresponds to the complete transformation of austenite. The interval between these two curves indicates the partial transformation of austenite.

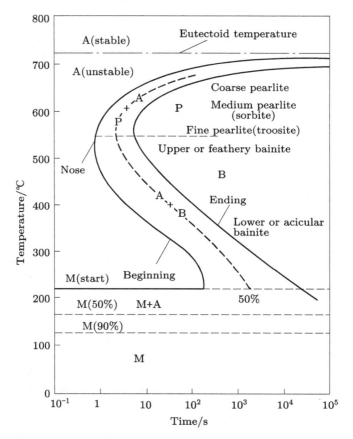

Fig. 7.10 TTT diagram of T8 steel. A, P, B, and M represent austenite, pearlite, bainite, and martensite, respectively.

2.4 Non-equilibrium microstructures of Fe-C alloys

Steels can be heat treated to produce a great variety of microstructures and properties. The heat treatment methods have been described in detail in Experiment 22.1, which are usually classified as annealing, tempering, normalizing, and quenching. Annealing is generally considered to produce the microstructure close to the equilibrium microstructure, while other methods result in non-equilibrium phase transformation during heating and cooling. For isothermal transformation, the microstructures can be predicted and analyzed based on the TTT diagrams. Several typical microstructures of Fe-C alloys are listed as follows.

(1) Coarse pearlite.

At temperatures just below the eutectoid temperature, austenite decomposes into pearlite. This microstructure is called coarse pearlite because relatively thick layers of both the α ferrite and Fe_3C are produced.

(2) Medium and fine pearlite.

With decreasing temperature, the carbon diffusion rate decreases, and the layers in pearlite become progressively thinner. Medium and fine pearlite will form at about 600 °C and 550 °C, which are also called sorbite and troostite, respectively.

(3) Bainite.

When the temperature is lower than the "nose" temperature, the microstructure called bainite forms, which also consists of ferrite and cementite. Bainite consists of the needle- or plate-like ferrite, depending on the temperature of the transformation.

(4) Spheroidite.

If a steel alloy with either pearlite or bainite microstructures is heated to a temperature below the eutectoid temperature for a sufficiently long period, the Fe_3C phase appears as sphere-like particles embedded in a continuous α phase matrix. This microstructure is called spheroidite.

(5) Martensite.

Martensite is formed when austenite is quenched to a relatively low temperature. The martensitic reaction begins when the martensite start temperature (M_s) is reached and completes when the lower transformation temperature M_f is reached. It is a non-equilibrium α phase with a supersaturated carbon content that results from a diffusionless transformation of austenite. The grains take on a plate-like or needle-like appearance.

3 Experimental

3.1 Preparation

(1) Samples: metallographic samples of Fe-C alloys. Detailed composition, heat treatment, and microstructures are listed in Table 7.2.

(2) Apparatus: metallographic microscope.

(3) Others: standard metallographs of the samples.

Table 7.2 Metallographic samples of Fe-C alloys

No.	Sample	Heat treatment	Microstructure
1	Industrial pure iron	Annealing	Ferrite
2	No. 20 steel	Annealing	Ferrite + pearlite
3	No. 45 steel	Annealing	Ferrite + pearlite
4	No. 65 steel	Annealing	Ferrite + pearlite
5	T8 steel	Annealing	Pearlite
6	T12 steel	Annealing	Pearlite + cementite
7	T12 steel	Annealing	Pearlite + cementite
8	Hypoeutectic cast iron	Cast	Ledeburite + pearlite + cementite
9	Eutectic cast iron	Cast	ledeburite
10	Hypereutectic cast iron	Cast	Cementite + ledeburite
11	T8 steel	Normalizing	Sorbite
12	T8 steel	Isothermal quenching	Troostite
13	T8 steel	Isothermal quenching	Upper bainite + martensite + retained austenite
14	T8 steel	Isothermal quenching	Lower bainite + martensite + retained austenite
15	No. 20 steel	Quench	Lath martensite
16	T8 steel	Quench	Martensite + retained austenite
17	No. 45 steel	normalizing	Ferrite + sorbite
18	No. 45 steel	Oil quenching	Martensite + troostite
19	No. 45 steel	860 °C water quenching	Medium carbon martensite
20	No. 45 steel	860 °C water quenching + low-temperature tempering	Tempered medium carbon martensite
21	No. 45 steel	860 °C water quenching + medium-temperature tempering	Tempered troostite
22	No. 45 steel	860 °C water quenching + high-temperature tempering	Tempered sorbite
23	No. 45 steel	780 °C water quenching	Ferrite + martensite
24	No. 45 steel	1100 °C water quenching	Rough martensite
25	T12 steel	Spheroidizing annealing	Globular martensite
26	T12 steel	780 °C water quenching + low-temperature tempering	Tempered martensite + granular cementite
27	T12 steel	1100 °C water quenching + low-temperature tempering	Tempered martensite + retained austenite

3.2 Procedure

(1) Observe the samples using the metallographic microscope. Take metallographs at magnifications of 200× and 500×.

(2) Compare the metallographs with the standard metallographs to confirm the observations.

(3) Sketch and label the microstructures.

(4) Analyze the formation of the microstructures.

4 Requirements for experimental report

(1) Select one annealed sample. Estimate the phase contents using the corresponding metallographs and compare them with those calculated from the Fe-C

phase diagram.

(2) Compare the microstructures of the samples with the same composition or heat treatment method. Discuss the effects of the composition and heat treatment methods on the microstructures.

5 Questions and further thoughts

(1) Please discuss the correlations between microstructures and properties. Several samples can be taken as examples.

(2) Some microstructure features are very fine and cannot be recognized under the optical microscope. Could you suggest several other techniques to observe the microstructures? What detailed features do you expect to be observed?

6 Cautions

(1) Do not touch the surface of the provided samples.
(2) While sketching the microstructures, please focus on the typical features.

Experiment 35.2 Surface Carburizing Heat Treatment of Low Carbon Steel

Type of the experiment: Comprehensive
Recommended credit hours: 8

1 Objectives

(1) To understand the mechanisms of diffusion in solid metals.
(2) To grasp the method and process of carburizing treatment.
(3) To observe the microstructure of the carburized layer and determine its thickness.
(4) To understand the factors that influence the thickness of the carburized layer.

2 Principles

2.1 Solid state diffusion

Many reactions and processes that are important in the treatment of materials rely on the transfer of mass either within a specific solid or between a solid and a liquid, a gas, or another solid. This is necessarily accomplished by diffusion, which is the phenomenon of material transport by atomic motion. The phase transformation during a heat treatment almost always involves diffusion.

From an atomic perspective, diffusion is a stepwise migration of atoms from one lattice site to another lattice site. For an atom to make such a migration, the presence of defects in solids is required. Point defects, such as vacancies and interstitial ions, are responsible for lattice diffusion. Diffusion also takes place along dislocations, grain boundaries, inner and outer surfaces. In addition, each diffused atom must have sufficient energy to break bonds with its neighboring atoms.

Diffusion is influenced by temperature, partial pressure, activities of the constituents of the materials, and microstructures. Diffusion along grain boundaries and dislocations generally needs smaller activation energy than diffusion in lattices, and as a result, becomes increasingly important at a lower temperature.

2.2 Carburization of steel

Carburization is a heat treatment process in which iron or steel absorbs carbon when the metal is heated in the presence of carbon-rich material, such as charcoal or carbon monoxide. The carbon content varies from the surface to the interior, and the metal surface becomes harder due to the higher carbon concentration. The affected depth depends on the amount of time and temperature. Longer carburizing time and higher temperature typically increase the depth of carbon diffusion. When the iron or steel is cooled rapidly by quenching, the higher carbon content on the outer surface becomes hard via the transformation from austenite to martensite, while the interior remains soft and tough as a ferritic or pearlite microstructure.

The carburization of steel is a typical application of Fick's second law. The diffusion process is a nonsteady state, i.e., the diffusion flux and the concentration gradient at some position in a solid vary with time. When a steel sample is placed in a carburizing medium, it can be considered as a semi-infinite solid in which the surface concentration is held constant. In this case, the boundary conditions are simply stated as

$$\text{for } t = 0, C = C_0 \text{ at } 0 \leqslant x \leqslant \infty; \tag{7.1}$$
$$\text{for } t > 0, C = C_s \text{ at } x = 0 \text{ and } C = C_0 \text{ at } x = \infty, \tag{7.2}$$

where t is the diffusion time, C_0 is the initial carbon concentration in the sample, x is the distance from the surface into the sample, C_s is the constant carbon concentration on the surface. The solution of Ficker's second law in terms of these boundary conditions is

$$\frac{C_s - C_x}{C_s - C_0} = \text{erf}\left(\frac{x}{2\sqrt{Dt}}\right), \tag{7.3}$$

where C_x is the carbon concentration at position x. The expression, $\text{erf}(x/(2\sqrt{Dt}))$, is the Gaussian error function.

Suppose that the depth of carburized layer is defined as the depth of the sample that achieved some specific concentration of carbon. Then the term at the left side of Equation (7.1) becomes a constant. Consequently, the term at the right side is also a constant, and hence the relationship between the depth of carburized layer and carburizing time is

$$x = K\sqrt{Dt}, \tag{7.4}$$

where K is a constant. This equation shows that the depth and diffusion time of carburized layer follows the parabola law.

2.3 Carburizing process

The general process of carburization is to place the steel sample into a carbon-rich environment and then increase the temperature so that the steel is austenitized for a certain amount of time and the carbon atoms penetrate the surface layer. According to the state of carburizing medium, carburizing can be classified into solid carburizing, liquid carburizing, and gas carburizing.

(1) Solid carburizing.

Samples are packed in a solid material with a high carbon content such as bone char, charcoal, or barium carbonate. During heating, carbon monoxide is produced, which is a reducing agent. Then, the reduction occurs on the steel surface with the release of carbon diffused into the surface. The resulting surface carbon concentration depends on the process environment. The case depth is approximately 0.1 to 2.0 mm. Usually, this method is used for soft steels with a carbon content of 0.2wt% or less, which cannot be hardened by simple heating and quenching. In this experiment, we will use this method to carburize T20 carbon steel.

(2) Gas carburizing.

Gas carburizing is one of the most successful and popular carburizing techniques used when large quantities of workpieces are required. This process contains gases capable of forming atomic carbon at high temperatures, such as carbon monoxide, natural gas, a mixture of methane, ethane, and propane. It takes 5 to 10 h for the general gas carburizing process, and the carbon-bearing environment is replenished continuously to maintain a high carbon content.

(3) Liquid carburizing.

In this process, the steel samples are immersed in a liquefied carbon-rich bath of molten salts. The molten salt contains a mixture of sodium carbonate, sodium chloride, and silicon carbide. The reaction of the mixture produces carbon into the metal. This method takes 35 to 55 min to carburize the metal surface of 0.2 to 0.3 mm.

2.4 Microstructure and case depth

If a low carbon steel sample is cooling slowly after carburization, the microstructure of the carburized layer has the equilibrium microstructures corresponding to the Fe-C phase diagram. The carburized layer has a hypereutectoid region, a eutectoid region, and a hypoeutectoid region sequentially from the surface to the interior.

In order to obtain a microstructure and corresponding performance to meet specific applications, the carburized workpieces are generally heat treated by quenching and low-temperature tempering to obtain the martensite structure. Besides the case depth, the thickness of the martensite needle, the carbide distribution, the content of residual austenite, and the ferrite content in the interior region are required to be examined in the quality inspection of carburized workpieces.

An accurate method to measure the case depth is essential for the quality control of the process and to confirm the specifications. Various methods for measuring the case depth have been adopted based on the difference in either chemical composition or mechanical properties. A proper choice of method to be used is essential because different methods may lead to different values of case depth.

Both the metallographic method and microhardness method are widely accepted. The metallographic method determines the case depth by observing the microstructure of the cross-sectional sample. For carbon steel and low carbon alloy steel, the depth is determined from the surface to half of the transition layer. The microhardness method measures the case depth through a hardness traverse at known intervals from the surface to the core of the sample after preparing the cross-sectional sample. This method is accurate and is preferred for measuring the effective case depth, i.e., depth up to 50 HRC or approved equivalent. In this experiment, the metallographic method will be used.

3 Experimental

3.1 Preparation

(1) Materials: T20 carbon steel samples with a size of 10 mm × 10 mm × 20 mm.

(2) Chemicals: graphite powder, barium carbonate ($BaCO_3$), calcium carbonate ($CaCO_3$).

(3) Apparatuses: atmosphere tube furnace, metallographic microscope.

(4) Tools: crucibles.

(5) Others: items for preparation of metallographic sample listed in Experiment 1.

3.2 Procedure

3.2.1 Carburization of T20 steel

(1) Prepare two carburizers. One is the mixture of graphite (83%–88%), $BaCO_3$ (10%–15%), $CaCO_3$ (2%). The other only use graphite. The total weight of each carburizer is 20 g.

(2) Grind two steel samples with abrasive papers to remove surface machining marks, rust, and other contaminants.

(3) Tightly mount the samples by the carburizers in two crucibles, respectively.

(4) Carburizing in the atmosphere tube furnace with the protection of flowing argon or nitrogen (about 200 mL/min). The carburizing heat treatment process is shown in Fig. 7.11.

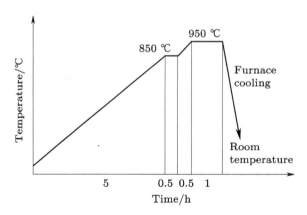

Fig. 7.11 Temperature-time plot for the carburizing heat treatment process.

3.2.2 Measurement of the case depth

(1) Select one face of each carburized sample for observation. Thermal mount the samples.

(2) Prepare the cross-sectional metallographic samples through the procedure of grinding, polishing, and etching. Details of the procedure have been described in Experiment 1.

(3) Observe the carburized layers of the two samples under the metallographic microscope and take metallographs. Identify the microstructures from the surface to the interior. Measure the case depths at more than 5 positions of each sample.

4 Requirements for experimental report

(1) Provide the metallographs of the carburized layers. Describe the regions and microstructures of the carburized layers.

(2) Tabulate the measured values, average values, and standard deviations of the case depths of the samples.

(3) Explain the effects of $BaCO_3$ and $CaCO_3$ by comparing the case depths of the two samples.

5 Questions and further thoughts

(1) Please describe the differences in the preparation of the carburized sample and the normal sample for microstructure observation.

(2) From the microstructure of the carburized layer, a faster diffusion along the grain boundaries can be seen. Why?

Experiment 36 Heat Treatment of Aluminum Alloy for Precipitation Hardening

Type of the experiment: Comprehensive
Recommended credit hours: 8
Brief introduction: Solution and aging heat treatment is often required to meet the application requirements of aluminum alloys. In this experiment, alloy 6061, one of the most widely used aluminum alloys, is heat-treated. The heat treatment method of the aluminum alloys is introduced. The effects of aging time on the microstructure and harness are examined.

During the experiment, the heat treatment of the samples takes a long time. Therefore, the experiment can be divided into two parts, heat treatment and characterization, so that the heat treatment process does not occupy the class time.

1 Objectives

(1) To understand the principle and method of solution and aging heat treatment.

(2) To understand the mechanism of precipitation hardening.

(3) To carry out the solution and aging treatment on 6061 aluminum alloy.

2 Principles

2.1 Aluminum alloy

An aluminum alloy is an alloy in which aluminum is the predominant metal, and other elements are added to enhance its properties, primarily to increase its strength. These alloying elements include iron, silicon, copper, magnesium, manganese, and zinc at a level that together may account for as much as 15wt%.

Aluminum and its alloys are characterized by a relatively low density, high electrical and thermal conductivities, and a resistance to corrosion in common environments. Many of these alloys are easily deformed because of their high ductility.

Generally, aluminum alloys are categorized as either cast or wrought according to the processing method. The deformed aluminum alloy is further divided into non-heat-treatable aluminum alloy and heat-treatable aluminum alloy. The mechanical properties of the non-heat-treatable type cannot be improved through heat treatment but only through cold deformation. In contrast, the mechanical properties, physical properties, and corrosion resistance of heat-treatable type can be improved by heat treatment, such as quenching and aging.

Aluminum alloys are assigned a four-digit number, in which the first digit identifies a general class or series, characterized by its main alloying elements. There has been considerable industrial interest in 6XXX series alloys (Al-Mg-Si alloys) because they are the first choice for structural applications. The 6XXX series are versatile, heat treatable, highly formable, weldable, and have moderately high strength coupled with excellent corrosion resistance. These materials can be heat treated to produce precipitation. Extrusion products from alloy 6061 are the most widely used alloy in this series and are often used in truck and marine frames. It is also used in this experiment.

2.2 Heat treatment for precipitation hardening

The strength and hardness of metal alloys may be enhanced by the precipitation of uniformly dispersed particles of a second phase within the original phase matrix. The presence of the precipitates and the resultant strain fields in the matrix surrounding the precipitates provide higher strength by obstructing the movement of dislocations.

The precipitation hardening must be accomplished by phase transformations that are induced by appropriate heat treatment. The heat treatment generally involves two stages shown in Fig. 7.12.

The first is a solution heat treatment in which all solute atoms are dissolved to form a single-phase solid solution by soaking the alloy at a temperature sufficiently high and for a time long enough to achieve a nearly homogeneous solid solution. This procedure is followed by rapid cooling or quenching to form a single-phase solid solution supersaturated with the solute elements.

After solution treatment and quenching, hardening is achieved by either natural aging, which is a treatment at room temperature, or artificial aging, which is a treatment at elevated temperature. In some alloys, sufficient precipitation occurs in a few days at room temperature to yield stable products with adequate

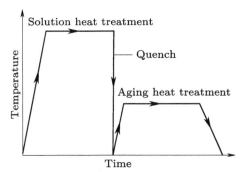

Fig. 7.12 Schematic temperature-time plot of both solution and precipitation heat treatments for precipitation hardening.

properties for many applications. Other alloys with slow precipitation reaction at room temperature are always precipitation heat-treated before being used. After an appropriate aging time, the alloy is cooled to room temperature. Generally, the cooling rate is not an important consideration.

Aluminum alloys exhibit relatively significant enhancement in mechanical properties by a heat treatment that produces considerable precipitation. In the case of the 6XXX series alloys, the alloying elements, silicon, and magnesium are approximately in the proportions required for the formulation of magnesium silicide (Mg_2Si) for precipitation hardening. In the case of the 6061 alloy, the heat treatment is first performed at above 500 °C to obtain the supersaturated solid solution. Artificial aging is obtained by heating to about 200 °C for various amounts of time.

3 Experimental

3.1 Preparation

(1) Materials: 6061 aluminum alloy samples with a size of 20 mm × 20 mm × 6 mm.

(2) Apparatuses: muffle furnace, Vickers microhardness tester (Shanghai optical instrument factory, HXD-1000TMC), metallographic microscope.

(3) Tools: heat treatment pliers, heat resistant gloves.

(4) Others: items for preparation of metallographic sample listed in Experiment 1.

3.2 procedure

(1) Grind 5 samples to remove the machining marks and other possible contaminations.

(2) Conduct solution heat treatment at 530 °C for 1 h.

(3) Conduct artificially aging treatment at 180 °C for 4 h, 8 h, 12 h, 16 h, and 20 h, respectively.

(4) Grind, polish, and etch the samples. Observe the microstructure and take metallographs. Details of the procedure have been described in Experiment 1.

(5) Measure the hardness of the samples. The loading is 25–100 gf and the dwell time is 10 s. For each sample, measure the hardness values at least at five

positions. The method of the microhardness test has been described in detail in Experiment 22.1.

4 Requirements for experimental report

(1) Tabulate the hardness values. Calculate the average value and the standard deviation of each sample.

(2) Plot the hardness as a function of aging time. Analyze the relationship of the hardness, aging time, and microstructure.

5 Questions and further thoughts

(1) What is overaging? What are the effects of overaging on the microstructures and mechanical properties of aluminum alloys?

(2) How to determine the heating temperature for solution treatment and aging treatment? What are the effects of higher or lower temperatures?

Experiment 37 Forming Process of Polymers and Application in 3D Printing

Type of the experiment: Comprehensive and designing
Recommended credit hours: 20
Brief introduction: The forming process of thermoplastics generally is realized by softening or melting the polymers at elevated temperatures and then forming the products by various methods. This experiment introduces the traditional forming processes, extrusion and injection molding, and an advanced forming process, 3D printing. The effects of the forming process on the tensile properties are examined.

The experiment is divided into three experiments that can be conducted respectively. The first experiment introduces the concept of melt flow index (MFI), which is regularly used in the plastics industry to control the quality of thermoplastic products. The MFI values of several common polymers, polypropylene (PP), high-density polyethylene (HDPE), and polylactic acid (PLA), are measured. Then, a blend of PLA, PP, and HDPE is designed in the second experiment. 3D printing filament and tensile samples are prepared by extrusion and injection molding, respectively. The third experiment introduces the software for 3D modeling and general operation of the 3D printer. The tensile samples are prepared using 3D printing. The tensile properties of the samples from injection molding and 3D printing are measured and compared.

Experiment 37.1 Determination of Melt Flow Index of Thermoplastics

Type of the experiment: Cognitive
Recommended credit hour: 4

Experiment 37 Forming Process of Polymers and Application in 3D Printing

❶ Objectives

(1) To understand the basic features of thermoplastics and thermosets.
(2) To understand the principle of MFI.
(3) To determine the MFI values of several common polymers.
(4) To understand the effect of molecular weight on the melt flow of thermoplastics.

❷ Principles

2.1 Thermoplastics and thermosets

Polymers are materials consisting of long chains or branches that are made up of many repeated monomer units covalently bonded together. These chains or branches are bonded together by Van der Waals forces, hydrogen bonds, and covalent crosslinks. The polymers can be categorized into thermoplastics and thermosets according to their thermal processing behavior.

Thermoplastics are polymers that become soft when they are heated and become rigid when cooled. This process is reversible and may be repeated. This quality allows thermoplastics to be remolded and recycled. Typical thermoplastics include polymethyl methacrylate (PMMA), acrylonitrile butadiene styrene (ABS), polyamide (nylon), polycarbonate (PC), PP, and PE.

In contrast to thermoplastics, thermosetting polymers become permanently hard when they are heated. They show relatively high mechanical strength compared with thermoplastics. However, they provide poor elasticity or elongation. Since thermosets cannot be reheated and melted to be shaped, they are ideal materials for high-temperature applications such as electronics and appliances. Since their shape is permanent, they are not recyclable as a source for newly made plastic. Common thermosetting polymers include polyester resin, vinyl ester resin, epoxy, phenolic, and urethane.

In Experiment 21.3, it has been introduced that the structures of the molecular chains can be grouped into linear, branched, crosslinked, and network structures. Most linear and branched polymers with flexible chains are thermoplastics. In the case of thermosetting polymers, covalent crosslinks are formed between adjacent molecular chains.Moreover, the heavily crosslinked structure is directly responsible for the high strength and low ductility.

2.2 Determination and application of MFI

The MFI is a measure of the ease of flow of melted thermoplastics. It is defined as the mass of polymer flowing in ten minutes through a capillary with a specific diameter and length under prescribed pressure and temperature. It is expressed in a unit of g/10 min. MFI is an indirect measure of molecular weight.Ahigh melt flow rate means a low molecular weight.

The MFI is inversely proportional to the viscosity of the melt under the conditions of the test. It is often used in the plastic industry to measure the viscosity indirectly. Consequently, it is regularly used in the geosynthetics industry to control the quality of thermoplastic products. Plastics engineers usually correlate the MFI value with the polymer grade to choose polymer for different processes.

They should select a polymer material with an MFI high enough that the molten polymer can be easily formed into the designed product but low enough that the mechanical strength of the final product is sufficient for its application.

The MFI is generally characterized by the melt flow indexer or extrusion plastometer. The method is described in the standards ASTM D1238 and ISO 1133. Figure 7.13 shows a general configuration of the melt flow indexer. The thermoplastic material is extruded through a die using a piston actuated by a specified weight, usually 2.16 kg or 5 kg. The die has an opening of typically around 2 mm diameter. In this experiment, the MFI values of PP, HDPE, and PLA will be measured.

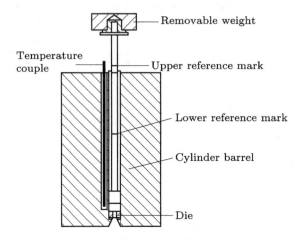

Fig. 7.13 Schematic illustration of the melt flow indexer.

3 Experimental

3.1 Preparation

(1) Materials: PP, HDPE, PLA.

(1) Apparatuses: melt flow indexer (Chengde Precision Testing Machine Co. LTD., XRL-400C), electronic balance.

(1) Tools: 50 mL beaker, tweezers, heat resistant gloves.

3.2 Procedure

(1) Inspect the melt flow indexer for cleanliness. All surfaces of the cylinder barrel, die, and piston shall be free of any residue from previous measurements.

(2) Insert the die and the piston into the cylinder barrel. Preheat the barrel to the prescribed temperature (190 °C for HDPE and PLA, 230 °C for PP).

(3) Remove the piston from the barrel. Add 6–7 g sample into the cylinder barrel. They are appropriately packed inside the barrel to avoid the formation of bubbles.

(4) Insert the piston into the cylinder barrel. The sample is preheated for about 5 min to melt completely.

(5) When the barrel temperature reaches the setting value, a specified weight of 2.16 kg is applied to the piston. Then start the measurement. The weight exerts a force on the molten polymer, and it immediately starts flowing through the die. The machine cuts off the filament automatically every 10 s. The measurement stops when the piston has moved down for about 30 mm. Discard the extrudates that contain visible bubbles or contaminations.

(6) Weigh each short filament of the extrudate.

(7) Clean up the cylinder barrel carefully.

(8) Repeat steps (1) to (7) for HDPE, PP, and PLA, respectively.

4 Requirements for experimental report

(1) Determine the MFI values of the samples. Since the MFI is expressed in a unit of g/10 min, it can be calculated by

$$\text{MFI} = \frac{600 \times w}{t}, \tag{7.5}$$

where w is the average mass (g) of polymer flowing through the die during time t(s).

(2) Tabulate the mass of the filaments of the extrudates, the MFI values, and the molecular weights of the raw materials. Discuss the effect of molecular weight on the MFI values.

5 Questions and further thoughts

(1) If there are bubbles in the extrudates, what is the effect on the MFI value?

(2) Discuss the factors that influence the determination of MFI values of thermoplastics.

6 Cautions

(1) This experiment involves high-temperature operations. Please follow the instructions of the teachers.

(2) Please wear heat-protective gloves while operating the melt flow indexer.

Experiment 37.2 Extrusion and Injection Molding of Polymers

Type of the experiment: Designing
Recommended credit hours: 4

1 Objectives

(1) To understand the principles of screw extrusion and injection molding.

(2) To design a blend of PLA, PP, and HDPE with enhanced ductility comparing with PLA.

(3) To prepare 3D printing filament of the blend by extrusion.

(4) To prepare tensile samples of the blend by injection molding.

2 Principles

2.1 Forming process of thermoplastics

Practical polymer materials or products with desired shapes are obtained by various forming processes. The choice of forming process mainly depends on the type of polymers (thermoplastics or thermosets), the starting form (powder, granule, solution, and dispersion), and the shape and size of the product.

Because the thermoplastics soften at high temperatures, they can be readily reshaped. There are various forming processes of thermoplastics, including extrusion, injection molding, calendaring, blow molding, thermoforming, compression molding, and rotational molding. Different processing methods and technological conditions should be chosen for different application requirements.

Among these methods, extrusion and injection molding are the most important polymer processing operations for thermoplastics in industrial manufacture. The two processes both involve the following steps: mixing the raw material evenly, heating and melting the materials, forming the molten polymer into the required shape and dimensions, solidifying, and cooling. The details of the methods will be described in the following sections.

In order to arrive at specific properties for applications, the polymers are almost always combined with other ingredients, or additives, which are mixed in during processing and fabrication. These additives include plasticizers, colorants, reinforcements, and stabilizers. Besides, the properties of a polymer can be modified by blending with other types of polymers.

In this experiment, extrusion and injection molding will be used for the PLA polymer. Particularly, the PLA filament obtained by extrusion will be further used for 3D printing (experiment 37.3). In order to enhance the ductility of PLA, it is blended with PP and HDPE. The contents of PP, HDPE, and the colorant, are required to be designed based on a literature survey.

2.2 Screw extrusion

In extrusion, a melted polymer is forced through a die with an orifice, and a continuous shape is formed with a constant cross-section similar to that of the orifice. The screw extruder is a commonly used apparatus in this process. It comprises one or two screws rotating in a heated barrel, as shown in Fig. 7.14. The twin-screw extruders (TSEs) offer greater control over residence time distribution and mixing than single-screw extruders (SSEs) and have superior heat and mass transfer capabilities. Therefore, TSEs have been more widely used in mixing, blending, compounding thermoplastic polymers with additives, devolatilization, and reactive extrusion.

In the screw extrusion process, polymers in the form of pellets, powders, or flakes flow from a hopper to the gap between a rotating screw and a heated barrel. The depth of the conveying channel in the screw is contoured from large to small in the flow direction, resulting in the change of the pressure and density from the particulate solid to the extruded molten polymer. The structure change in this process is illustrated in Fig. 7.15. In the first zone of the extruder, the solid polymer particles are compacted together in the screw channel by the rotation of

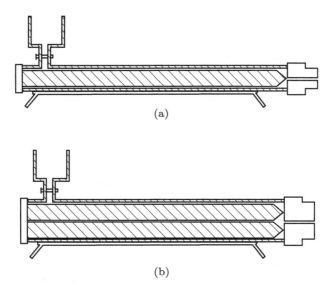

Fig. 7.14 Cross-sections of (a) single and (b) twin-screw extruder barrels.

the screw to form a solid bed of material. At the next extruder section, the melting zone, molten polymer forms in the gap between the solid bed and the barrel wall. The molten polymer is subjected to intense shearing in the gap. In the last zone of the extruder, the melt flow is stabilized in the shallow screw channels, and finally, the polymer passes through the die at the end of the machine.

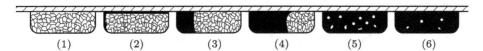

Fig. 7.15 Structure change of the polymer material in the screw channel.

2.3 Injection molding

Injection molding is a process that injects molten materials into molds. In a reciprocating screw injection molding machine shown in Fig. 7.16, material in the form of particles flows under gravity from the hopper onto a turning screw (Fig. 7.16(a)). The mechanical energy supplied by the screw together with auxiliary heaters converts the polymer material into a molten state that fills the barrel (Fig. 7.16(b)). Then the screw injects molten polymer into the cooled mold (Fig. 7.16(c)). Once the polymer has solidified in the mold, the mold is unclamped and opened, and the product is pushed from the mold by automatic ejector pins (Fig. 7.16(d)). For small products, cycles can be as rapid as several injections per minute.

3 Experimental

3.1 Preparation

(1) Materials: PLA, PP, HDPE, color masterbatch.

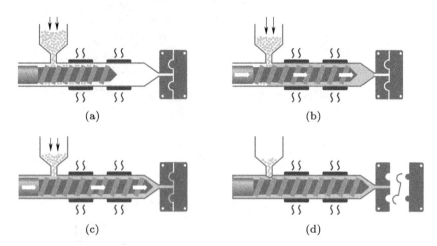

Fig. 7.16 Injection molding process with a single screw extruder: (a) add polymer material from the hopper; (b) the molten polymer fills the barrel; (c) inject the molten polymer into the mold; (d) open the mold and eject the product. (See color illustration at the back of the book)

(2) Apparatuses: single screw extruder (HAPRO EP-20), rotary thread rolling machine, pelletizer, injection molding machine (HAAKE MiniJet-Pro), electronic balance.

(3) Tools: scrolls, Vernier caliper.

3.2 Procedure

3.2.1 Extrusion of designed polymer material

(1) Conduct a literature survey before the experiment.

(2) Design a blend of PLA, PP, and HDPE, and color masterbatch. The PLA is more than 60wt%, and the color masterbatch is less than 0.5wt%.

(3) Weigh the polymer materials and simply mix the materials in a beaker or a sealed bag. The total mass of the blend is 300 g.

(4) Switch on the extruder, and set the temperature to 190 °C. Start rotating the screw after the temperature is stable.

(5) Add the blend via the hopper of the extruder.

(6) After the filament is extruded, insert the filament into the rotary thread rolling machine. Make the diameter of the filament to be 1.75 mm by adjusting the extrusion speed. The filament will be used for 3D printing in the following Experiment 37.3.

(7) Roll up the filament with a scroll.

3.2.2 Injection molding

(1) Weigh about 60 g of the filament and cut it into granules using the pelletizer.

(2) Switch on the injection molding machine. Place the mold into the mold cavity. The sample prepared by the mold is shown in Fig. 7.17. They are tensile bars that will be used for the tensile test in the following Experiment 37.3.

(3) Set the temperature of the material heating barrel to 190 °C. Set the mold temperature to 80 °C. The injection pressure is 40 MPa, the holding pressure is 30 MPa, and the holding time is 12 s.

(4) Add the appropriate number of granules (about 10 g) to the heating barrel and heat them for about 5 min to ensure complete melting. Then start the injection. Pay attention to whether the bottom of the heating barrel is aligned with the mold before injection.

(5) After injection, take out the mold. Open the mold and take out the sample when the temperature is below 50 °C.

(6) Repeat steps (4) and (5) to produce five tensile samples.

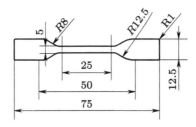

Fig. 7.17 The shape and size of the tensile bars. The thickness is 4 mm.

4 Requirements for experimental report

(1) Describe the design idea of the composition of the blend prepared in this experiment.

(2) Summarize the main points that need attention in the extrusion and injection molding processes.

5 Questions and further thoughts

(1) Analyze the possible causes for fracture or peeling of the inject molded samples.

(2) Describe several other common forming processes used to make plastic products.

6 Cautions

(1) The provided parameters of extrusion and injection molding are only a reference. They should be adjusted according to the composition of the blend.

(2) If fracture or peeling occurs after extrusion or injection molding, adjust the parameters of extrusion or injection molding, or redesign the composition of the blend.

Experiment 37.3   3D Modeling, Printing, and Tensile Properties

Type of the experiment: Comprehensive
Recommended credit hour: 12

1 Objectives

(1) To master the primary usage of the Solidworks and Flashprint software.
(2) To master the operation of the 3D printer.
(3) To measure the tensile properties of the 3D printed and injection molded samples.
(4) To understand the effect of the filling ratio on the tensile properties of the 3D printed samples.
(5) To understand the effect of forming process on the tensile properties of polymers.

2 Principles

2.1 3D modeling using Solidworks

The principles of 3D printing and several well-developed techniques have been introduced in Experiment 2. 3D printing is a forming process of making 3D solid objects using an additive process. Compared with traditional forming processes for plastic products, such as extrusion and injection molding, 3D printing is easier and cheaper. Expensive molds or tools are not required.

All the 3D printing techniques start from 3D modeling, which is the process of using software to create a mathematical representation of a 3D object or shape called a 3D model. We can use a 3D scanner, haptic device, or 3D modeling software to create a 3D model. Solidworks is one of the most usedcomputer-aided design (CAD) programs by professional 3D designers. It is a mechanical design application that lets designers create structural models quickly and precisely. It can also be used across many fields and industries, such as engineering, industrial design, sculpture, and architecture.

Although Solidworks is comparably easier to navigate and operate, users of Solidworks will need to properly acquaint themselves with all its features and interface before being able to edit and create projects smoothly. Solidworks uses a component-based approach in model creation. The Solidworks projects usually start with a 2D sketch, which is one profile of the model. A plane (top, front, right) is selected to draw the sketch. The sketch must be closed for extruding it to 3D. After sketching, users can add more solid features to their design, and these features can be moved around and arranged as accurately as possible by using the snap tools.

2.2 Slicing 3D model using Flashprint

In order to convert a digital 3D model into printing instructions for a 3D printer, the model should be cut into horizontal layers. A slicer is a program that cuts the 3D model based on the print settings and calculates how much material will be used and how long it will take to print the object. All the information is bundled up into a G-code file. This file type is a huge list of G-code commands that a3D printer will process one by one to achieve a complete object.

In this experiment, the 3D slicer program Flashprint will be used. The software not only can be used to observe the model from multiple angles but also includes movement, rotation, scaling, cutting, and other functions. The print settings havea significant impact on print quality. The basic settings include layer height, shell thickness, filling ratio, print speed, supports, initial layer thickness, print bed temperature, and print temperature. These parameters are explained as follows.

(1) Layer height specifies the height of each filament layer in the object. The object made with thinner layers will create more details with a smoother surface, but it takes more time since there will be more layers that make up the object.

(2) Shell thickness defines the thickness of the outer walls. Increasing this value will create thicker walls and improve the strength of the object.

(3) Filling ratio refers to filling the space inside the outer shell of an object. The higher the percentage of infill, the stronger and heavier the object, and the more time and filament it takes to print.

(4) Print speed refers to the speed at which the extruder travels while it lays down filament. Of course, everyone wants to print their object as quickly as possible, but a fast print speed can cause complications and muddy prints.

(5) Supports are structures that help to holdup 3D objects that do not have enough base material when they are being printed.

(6) The initial layer thickness is the very first layer on the print bed. A thicker initial layer can make a sturdier base for the print.

(7) Print bed temperature is crucial for3D printing and always needs to be set to a specific range. The best range for PLA is between 55 °C and 70 °C.

(8) Print temperature is the extrusion temperature at the nozzle. The general range for PLA is around 190 to 220°C. If the nozzle is too hot, the PLA filament can become extra soft and flimsy. This can cause the printed object to be messy and droopy.

2.3 General operation of 3D printer

In this experiment, fused deposition modeling (FDM) 3D printer will be used to print tensile bar samples, as shown in Fig. 7.17. These tensile bars will be used for the tensile test. The filament of the blend design in Experiment 37.2 will be used. The principle of the FDM technique has been introduced in experiment 2. The general operation of the printer is summarized here.

After uploading the G-code file to the printer, the print bed of the printer is leveled using the screws under the print bed. A leveled print bed is crucial for extruding material evenly across the entire build surface. The gap between the nozzle and the print bed is also adjusted to balance the filament flow and bed adhesion. After that, the nozzle is heated. The 3D printing filament is inserted after the nozzle temperature reaches the setting value. Then, the printing process is atomically conducted by the printer. Depending on the size of the object and the materials used, this process could take hours or even days to complete.

3 Experimental

3.1 Preparation

(1) Materials: 3D printing filament prepared by extrusion in Experiment 37.2, tensile samples prepared by injection molding in Experiment 37.2

(2) Apparatuses: 3D printer (FlashForge Corporation, Dreamer dual extruder),

computer installed with Solidworks and Flashprint software, universal testing machine (SANS CMT4304).

(3) Tools: extensometer, Vernier caliper.

3.2 Procedure

3.2.1 3D modeling and printing

(1) Produce the model of the tensile bar shown in Fig. 7.17 using Solidworks software. Save the model as a ".stl" file.

(2) Use the Flashprint software to slice the model. Set the printing parameters. The temperatures of the nozzle and the print bed are 205 °C and 50 °C, respectively. The thickness of each layer is 0.18 mm. The numbers of the shell layer, the top layer, and the sublayer are 4, 4, and 3, respectively. The filling ratios are 10%, 20%, 30%, 40%, respectively.

(3) Save the files of these filling ratios as ".gx" files on an SD card.

(4) Switch on the 3D printer. Plug the SD card in the 3D printer.

(5) Level the print bed, and adjust the gap between the nozzle and the print bed.

(6) Print the tensile bars with different filling ratios. Print 5 samples for each filling ratio.

3.2.2 Tensile test of the 3D printed and injection-molded samples

(1) Measure the widths and thicknesses of the samples using Vernier caliper.

(2) Switch on the universal testing machine and the computer. Open the software.

(3) Select the test module "Tensile test of plastics at room temperature" in the software and set the testing parameters.

(4) Load a sample in the grips. Clamp the sample with the extensometer with a gauge length of 25 mm.

(5) Set the force and displacement to zero and then start the test. Force-displacement curve and data are simultaneously recorded by the computer.

(6) After the experiment, export the raw data as an Excel file.

(7) Repeat steps (3) to (6) for the samples from 3D printing and injection molding obtained in Experiment 37.2.

4 Requirements for experimental report

(1) Plot a stress-strain curve of the 3D printed samples and a stress-strain curve of the injection-molded samples for each filling ratio using the data exported in the excel files.

(2) Calculate and tabulate the tensile properties of these samples from the tensile stress-strain curves, including elastic modulus E, yield strength σ_y, ultimate tensile strength σ_{UTS}, and tensile ductility. Details of these tensile properties have been described in Experiment 21. Calculate their average values and standard deviations.

(3) Plot a bar graph of the average values of the tensile strength and the elastic modulus of the samples. The standard deviations should be plotted as

error bars. Discuss the effects of filling ratio on the tensile properties of the 3D printed samples. Compare the tensile properties of the 3D printed samples and the injection-molded samples.

5 Questions and further thoughts

(1) If the distance between the nozzle and the print bed is too large or too small, what will happen during the printing process?

(2) Please list several widely used 3D modeling software and compare their advantages and disadvantages.

Extended readings

Callister W D, Rethwisch D G. Fundamentals of Materials Science and Engineering. 5th Edition. John Wiley & Sons, 2001.

Kurgan N, Sun Y, Cicek B, et al. Production of 316L stainless steel implant materials by powder metallurgy and investigation of their wear properties. Chinese Science Bulletin, 2012, 57(15): 1873-1878.

Mohamed A M A, Samuel F H. A review on the heat treatment of Al-Si-Cu/Mg casting alloys. In: Czerwinski F, editor, Heat Treatment: Conventional and Novel Applications. IntechOpen, 2012.

Rodriguez F. The polymers. Available at: https://www.britannica.com/science/plastic/Injection-molding. Accessed: 2 March 2021.

Schneider M J, Chatterjee M S. Introduction to surface hardening of steels. In: Dossett J L, Totten G E, editor, ASM Handbook, Volume 4A: Steel Heat Treating Fundamentals and Processes. ASM International, 2013.

Voort G F V, Lucas G M. Microstructural characterization of carburized steels. Heat Treating Progress, 2009, 9(5): 37-42.

Chapter 8 Organic and Polymeric Materials

Experiment 38 Preparation of Copolymer with LCST Phase Transition by Atom Transfer Radical Polymerization

Type of the experiment: Cognitive
Recommended credit hours: 4
Brief introduction: The phase transition with low critical solution temperature (LCST) is that the thermal responsive copolymers become insoluble from the solution when the temperature is higher than a specific temperature. In this experiment, the LCST phase transition is studied. P(MEO$_2$MA-co-OEGMA) copolymer is prepared by atom transfer radical polymerization (ATRP) of 2-(2'-methoxyethoxy) ethyl methacrylate (MEO$_2$MA) and oligo (ethylene glycol) methacrylate (OEGMA). Then, the LCST phenomenon of the product is observed.

1 Objectives

(1) To understand the principle of ATRP.
(2) To prepare a thermal responsive copolymer with LCST phase transition by ATRP.

2 Principles

2.1 Thermal responsive copolymer materials

The environmentally responsive polymer material is a new type of intelligent polymer material, which shows significant changes in the physical or chemical properties by small changes in the external environment. Different kinds of environmentally responsive polymer materials can be constructed according to different environmental stimulation signals, such as temperature, pH, light, and electromagnet.

Temperature is one of the most common stimuli used in environmentally responsive polymer systems that have attractive application prospects in biomedicine and nanotechnology.

A unique property of a temperature-responsive polymer is the presence of a critical solution temperature, that is, the temperature at which a discontinuous phase transition occurs in the polymer solution. There are two types of critical solution temperatures in temperature-responsive polymer systems: LCST and the upper critical solution temperature (UCST). When the temperature is higher than a specific temperature, the polymer becomes insoluble from the solution, and this temperature is called LCST. Otherwise, it is called UCST.

For LCST, the mutual solubility of the two liquids increases when the temperature decreases. The composition of the two liquid phases is close to each other gradually. When the temperature is reduced to a specific temperature, the two liquid phases have the same composition, and the interface between the two liquid phases disappears. Below this temperature, there is only one uniform liquid phase.

The LCST phenomenon is closely related to the temperature of the hydrophobic interaction of the polymer in the aqueous solution. The change of temperature affects the hydrophobic action of hydrophobic groups and the hydrogen bond action between macromolecular chains, resulting in the change of polymer structure. The corresponding surface phenomenon is whether the solvent becomes turbid.

2.2 ATRP of P(MEO$_2$MA-co-OEGMA) copolymer

The controlled radical polymerization techniques, such as ATRP, nitroxide-mediated polymerization (NMP), have been more and more considered as straightforward alternatives for preparing well-defined building blocks for life science. In particular, ATRP has been proven to be a very versatile pathway for preparing poly (ethylene glycol)-based amphiphiles.

ATRP uses simple organic halides as initiators and transition metal complexes as halogen carriers. A reversible dynamic balance is established through redox reaction between the active and dormant species, thus controlling the polymerization reaction. There are four important variable components of ATRP. They are the monomer, initiator, catalyst, and solvent.

The mechanism of ATRP is illustrated as shown in Fig. 8.1. Firstly, the redox reaction between initiator alkyl halide (R-X) and transition metal complexes (M_t^n) turns into primary free radical R•, and the reaction between primary free radical R• and monomer M produces monomer free radical R-M•, i.e., active species. Secondly, both R-M_n• and R-M• are active species, which can not only continue to initiate monomer free radical polymerization but also capture halogen atoms from dormant species R-M_n-X or R-M-X, and become dormant species by themselves, thus establishing a reversible balance between dormant species and active species. Thirdly, it can be seen that the basic principle of ATRP is actually to make the concentration of free radicals in the system extremely low through an alternative "activation-deactivation" reversible reaction, which forces the irreversible termination reaction to be reduced to the lowest level and, hence, to realize controllable free radical polymerization.

P(MEO$_2$MA-co-OEGMA) copolymer is a thermo-sensitive copolymer that has

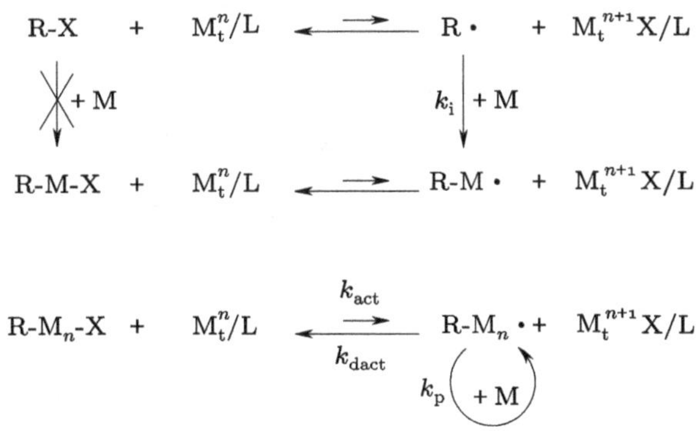

Experiment 38 Preparation of Copolymer with LCST Phase Transition

$$R\text{-}X + M_t^n/L \rightleftharpoons R\cdot + M_t^{n+1}X/L$$

$$\downarrow + M \qquad\qquad k_i \downarrow + M$$

$$R\text{-}M\text{-}X + M_t^n/L \rightleftharpoons R\text{-}M\cdot + M_t^{n+1}X/L$$

$$R\text{-}M_n\text{-}X + M_t^n/L \underset{k_{dact}}{\overset{k_{act}}{\rightleftharpoons}} R\text{-}M_n\cdot + M_t^{n+1}X/L$$

$$k_p (+M)$$

Fig. 8.1 Mechanism of metal complex-mediated ATRP. L and k denote ligand and reaction rate constant, respectively.

good biocompatibility. Therefore, the copolymer and its hydrogel have a wide application in the biomedical field, such as controlled drug delivery. MEO$_2$MA is soluble in water at room temperature. Hence MEO$_2$MA would also be a particularly tempting monomer for preparing water-soluble segments by ATRP. OEGMA is a water-soluble reagent, which is used for structural readjustment for such copolymers. POEGMA macromolecular brushes usually exhibit a much higher LCST in water due to their longer PEG side chains. Therefore, it is expected that copolymers of MEO$_2$MA and OEGMA possess higher LCST. In this experiment, P(MEO$_2$MA-co-OEGMA) copolymer will be prepared by ATRP of MEO$_2$MA and OEGMA in ethanol using CuBr/2,2'-bipyridyl (Bipy) as the catalyst and methyl 2-bromopropionate (MBP) as an initiator. The reaction is illustrated in Fig. 8.2.

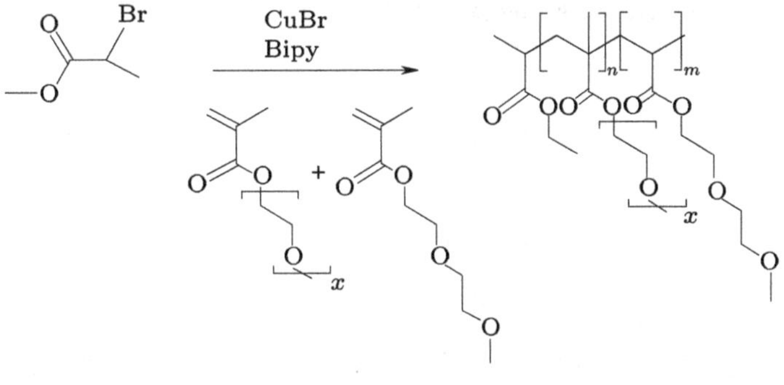

Fig. 8.2 Reaction mechanism of P(MEO$_2$MA-co-OEGMA) by ATRP.

3 Experimental

3.1 Preparation

(1) Chemicals: MEO$_2$MA, OEGMA, MBP, Bipy, CuBr, glacial acetic, ethanol.
(2) Apparatuses: electronic balance, oil bath, centrifuge, heating plate.
(3) Tools: Schlenk tube sealed with a septum, micro-syringes.

3.2 Procedure

3.2.1 Synthesis P(MEO₂MA-co-OEGMA) copolymer

(1) Weigh 26 mg CuBr and 56.6 mg Bipy. Put them into the Schlenk tube sealed with a septum. The tube is purged with dry N_2 for a few minutes.

(2) Add a degassed mixture of 3.06 g MEO$_2$MA, 862 mg OEGMA, and 4.8 mL ethanol into the tube through the septum with a degassed syringe.

(3) Add 30.2 mg MBP with a micro-syringe.

(4) Heat the mixture at 60 °C in an oil bath for 3 h.

(5) After three hours, the experiment is stopped by opening the flask and exposing the catalyst to air. The mixture is deposited by cooling in an ice bath.

(6) Separate the mixture using the centrifuge at 1000 r/min for 5 min. The final product appears as a clear oil.

3.2.2 Observation of the LCST phenomenon

(1) Heat the product on the heating plate from room temperature to 50 °C. Observe the variation of the color of the product.

(2) Take photographs of the product in different colors.

4 Requirements for experimental report

(1) Give the photographs of the product in different colors. Describe the phenomenon during the heating.

(2) Explain why the color of the as-prepared product changes.

5 Questions and further thoughts

(1) Please list several typical thermal responsive copolymers with LCST transition and their applications.

(2) What are the main applications of ATRP?

Experiment 39 Synthesis of Aspirin by Esterification Reaction

Type of the experiment: Comprehensive
Recommended credit hours: 8
Brief introduction: Aspirin or acetylsalicylic acid is one of the most widely consumed drugs in the world, which is used to treat pain and reduce fever or inflammation. In this experiment, aspirin is synthesized through the reaction of salicylic acid and acetic anhydride catalyzed by sulfuric acid. Then the crude product is purified by recrystallization using saturated $NaHCO_3$ solution for removing byproducts. The purity of the as-synthesized aspirin will be determined qualitatively by a chemical method, phenolic test, and a physical method, melting point test. The product is further identified by determining the functional group using Fourier transform infrared (FTIR) spectroscopy.

Experiment 39 Synthesis of Aspirin by Esterification Reaction

❶ Objectives

(1) To synthesize aspirin by an esterification reaction from salicylic acid and acetic anhydride.
(2) To master the methods of determining the purity of aspirin.
(3) To understand the principle of FTIR spectroscopy.
(4) To characterize the chemical structure of the as-synthesized aspirin by FTIR spectroscopy.

❷ Principles

2.1 Aspirin and its synthesis method

Acetylsalicylic acid, also known by the trade name aspirin, was firstly synthesized by the German chemist Felix Hofmann at the end of the 19th century. It is an acetyl derivative of salicylic acid that is a white, crystalline, weak acidic substance with a melting point of 137 °C. Aspirin is a nonsteroidal anti-inflammatory drug effective in treating fever, pain, and inflammation in the body. It also prevents blood clots.

The synthesis of aspirin consists of an esterification reaction catalyzed by acid (H_2SO_4 or H_3PO_4), where salicylic acid treated with acetic anhydride gives acetylsalicylic acid (Fig. 8.3). In this reaction, a hydroxyl group is converted to an ester, with acetic acid as the byproduct. The action of H_2SO_4 or H_3PO_4 is to break the intramolecular hydrogen bonding between carboxyl (COOH) and hydroxyl (OH) in salicylic acid to accelerate the reaction.

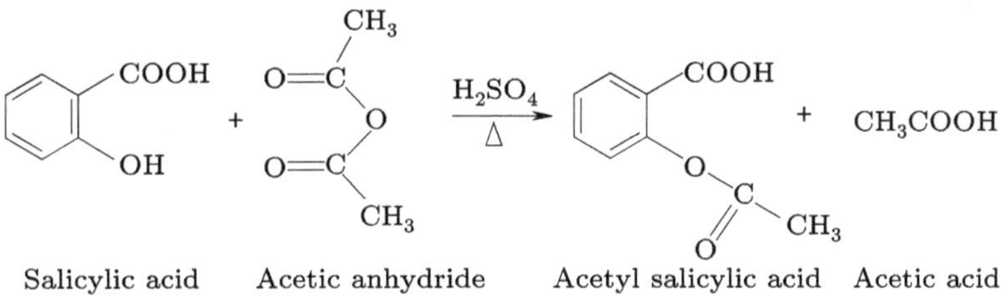

Fig. 8.3 Main chemical reaction between salicylic acid and acetic anhydride.

The side reactions shown in Fig. 8.4 occur and form byproducts such as dimer and polyester polymer accordingly. The crude product may contain a small amount of unreacted salicylic acid and polyester polymers which can be removed by filtration and recrystallization techniques (see Experiment 15).

2.2 Qualitative determination of the purity of the as-synthesized aspirin

The purity of the as-synthesized aspirin can be determined qualitatively by both chemical and physical methods, i.e., the phenolic test and melting point test. In the phenolic test, the sample containing the phenolic group is reacted with ferric chloride solution to form a purple complex. The reaction is shown in Fig. 8.5. Therefore, if the unreacted salicylic acid is not completely removed from the crude product, the purple color will be seen from the reaction.

The purity can also be confirmed by measuring the melting point of the product

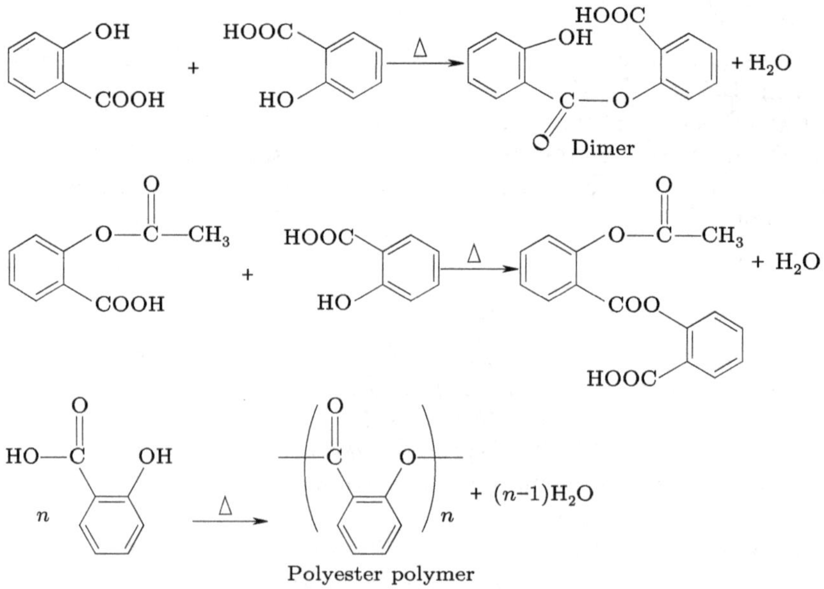

Fig. 8.4 Side reactions of aspirin synthesis.

since a mixture of limited miscible impurities will produce a depression of the melting point and an increase in the melting point range.

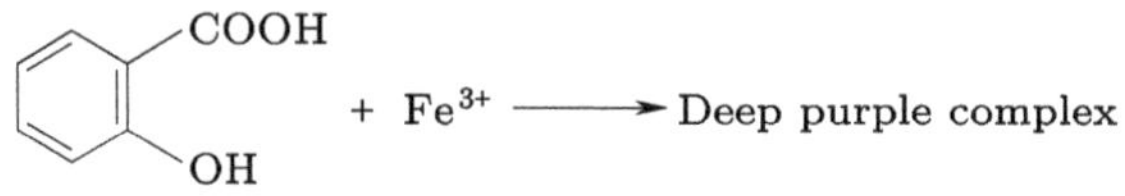

Fig. 8.5 Reaction of the phenolic test with $FeCl_3$ solution.

2.3 FTIR spectroscopy

Fourier transform infrared (FTIR) spectroscopy is a technique used to obtain an infrared spectrum of transmission and absorption of a solid, liquid, or gas. It provides information on the chemical composition of the sample. It can identify unknown materials and determine the quality and composition of a sample.

The absorption of infrared (IR) radiation is associated with the interactions between materials and electromagnetic fields in the IR region. The total internal energy of a molecule in a first approximation can be resolved into the sum of rotational, vibrational, and electronic energy levels. In the IR spectral region, the electromagnetic waves mainly couple with the molecular vibrations. In other words, a molecule can be excited to a higher vibrational state by absorbing IR radiation. The absorbing probability of a particular IR frequency depends on the actual interaction between this frequency and the molecule. A frequency will be strongly absorbed if its photon energy coincides with the vibrational energy levels of the molecule. Consequently, the specific molecular groups prevailing in the sample will be determined through spectrum data in the automated software. Therefore, IR spectroscopy can be applied for qualitative analysis of different kinds of materials. Besides, the amount of each content is indicated by the height and width of the peaks in the spectrum.

The FTIR spectrometer is an instrument that collects and digitizes the interferogram, performs the Fourier transformation, and displays the spectrum. Unlike general dispersive instruments, FTIR spectrometers collect all wavelengths simultaneously.

An FTIR spectrometer is typically based on the Michelson interferometer. Figure 8.6 shows the structure and working principle of a basic Michelson interferometer. It consists of a broad-band light source that emits light covering the mid-IR range, a beam splitter made of KBr or CsI, two front surface coated mirrors, one movable and one fixed, and a detector.

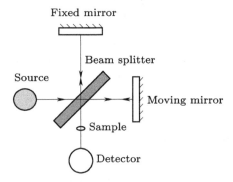

Fig. 8.6 The working principle of the Michelson interferometer.

Light from the light source is directed to the beam splitter and separates into two beams. One beam is transmitted through the beam splitter to the fixed mirror, and the other is reflected off the beam splitter to the moving mirror. The fixed and moving mirrors reflect the radiation back to the beam splitter. Again, half of the reflected radiation is transmitted, and half is reflected at the beam splitter. Therefore, the detector can receive two beams: one from the fixed mirror and the other from the moving mirror. Since the two beams reaching the detector have an optical path difference, they may interfere constructively or destructively for a particular frequency by positioning the moving mirror. If the moving mirror is scanned over a range, a sinusoidal signal against optical path difference will be detected for that frequency. The resulting signal of the interference that exits the interferometer is called an interferogram.

Because the source emits a range of frequencies, the detector output is the sum of all the interferograms. Therefore, an FTIR spectrometer simultaneously collects high spectral resolution data over a wide spectral range. The measured interferogram signal cannot be interpreted directly. Instead, Fourier transformation is required to convert the raw data into the actual spectrum.

Because molecules of organic and polymeric materials generally have a large number of bonds, and each bond has several IR-active vibrational modes, the IR spectra are complex and have many overlapping absorption bands. The benefit of the complex spectra is that the spectrum for each compound is unique. This makes IR spectra very useful for identifying compounds by direct comparison with spectra from authentic samples.

3 Experimental

3.1 Preparation

(1) Chemicals: salicylic acid, acetic anhydride, $FeCl_3$ solution (0.006 mol/L), sulfuric acid (H_2SO_4), saturated $NaHCO_3$ solution, ethanol (95%), hydrochloric acid (HCl), KBr (IR grade).

(2) Apparatuses: reflux apparatus (consists of an electromagnetic stirrer with a heating plate, a water bath, a 100 mL round bottom flask, a ball condenser, an iron stand with an iron ring, clamps, plastic tubes), vacuum filtration apparatus (the components have been listed in Experiment 15), micro melting point apparatus, FTIR spectrometer (PerkinElmer Frontier).

(3) Tools: beakers, watch glass, test tubes, Pasteur pipette, pH test paper, 10 mL and 25 mL graduated cylinders, tools for FTIR sample preparation (includes a press machine and a pellet press die with a diameter of 13mm, an infrared heat lamp, a mortar and pestle set).

3.2 Procedure

3.2.1 Synthesis of aspirin

(1) Assemble the reflux apparatus, which is the same as that described in Experiment 15, except a 100 mL round-bottom flask and a water bath are used.

(2) Weigh 3 g salicylic acid and transfer it to the 100 mL round-bottom flask. Pipette 10 mL of acetic anhydride followed by 5 drops of concentrated sulfuric acid under gently stirring.

(3) Heat the flask in the water bath at 85–90 °C for 15 min. Keep stirring during the reaction.

(4) After the reaction is completed, take the flask out of the water bath.

(5) Add dropwise about 25 mL ice-cold water under stirring.

(6) Cool the flask in an ice-cold water bath until the crystallization is completed.

(7) Collect the crude product by vacuum filtration. Wash the product three times in the Büchner funnel with ice-cold water. The procedure of vacuum filtration has been described in Experiment 15.

(8) Place a tiny amount of crude product (several mg) into a test tube for the following phenolic purity test. Transfer the other crude product to a 100mL beaker for purification.

3.2.2 Purification by recrystallization

(1) Transfer the crude product to a 100 mL beaker and slowly add saturated $NaHCO_3$ solution about 25–30 mL under stirring to dissolve the product until no more CO_2 generates and the product is completely dissolved.

(2) Conduct vacuum filtration to remove undissolved byproducts. Rinse the

byproducts in the Büchner funnel with 5 mL ice-cold water to completely separate the solution from the undissolved byproducts.

(3) Prepare an HCl solution of 4 mL concentrated HCl in 10 mL H_2O.

(4) Slowly add the HCl solution into the filtrate under stirring until the pH value reaches about 2.

(5) Cool the mixture in an ice-water bath until the crystallization is completed.

(6) Collect the solid product by vacuum filtration. Wash the product three times in the Büchner funnel with ice-cold water.

(7) Transfer the product to a pre-tared watch glass, and weigh the product after drying.

3.2.3 Determine the purity

(1) Dissolve a tiny amount of salicylic acid, the crude product, and purified product in 2 mL ethanol in three test tubes, respectively.

(2) Add 2 drops of $FeCl_3$ solution to each test tube and compare the color of the solutions.

(3) Determine the melting point of the purified product using the melting point apparatus. The procedure has been described in Experiment 15.

3.2.4 Measurement of FTIR spectra

(1) Dry the samples of salicylic acid, as-synthesized acetylsalicylic acid, and KBr powder completely with the infrared heat lamp.

(2) Grind 0.5 g KBr powder with a mortar and pestle for preparation of pure pellet for background correction.

(3) Mix the salicylic acid sample (about 3 mg) and KBr powder (about 0.5 g) thoroughly with a mortar and pestle. The weight ratio of the sample to KBr is ranging from 1:100 to 1:300.

(4) Press about 70 mg pure KBr powder and 70 mg mixture into pellets, respectively, under a pressure of 8 MPa for 1–2 min.

(5) Switch on the FTIR spectrometer. Open the software.

(6) Set the testing parameters. The scan range is 4000–450 cm^{-1}, scan resolution is 4 cm^{-1}, and scan number is 4.

(7) Load the pure KBr pellet in the beam path and collect an empty background signal.

(8) Load the sample pellet in the holder and collect the FTIR spectrum.

(9) Perform baseline correction if necessary.

(10) Export the spectrum as a ".asc" file.

(11) Repeat steps (3) to (4) and (8) to (10) for as-synthesized acetylsalicylic acid.

4 Requirements for experimental report

(1) Calculate the percent yield of the experiment, which is the percentage ratio of the weight of the purified product to the theoretical yield. Discuss the factors that affect the percent yield.

(2) Describe the phenomena during the phenolic test and melting point test, and analyze the purity of the crude and purified products. The melting point of

the acetylsalicylic acid in the literature is 139.0 °C.

(3) Plot the FTIR spectra of salicylic acid and as-synthesized acetylsalicylic acid using the exported data in one chart.

(4) Indicate the important functional groups shown in the FTIR spectra by comparing the peak positions with those reported in the literature.

(5) Analyze the purity by comparing the FTIR spectra with the results of the phenolic test and melting point test.

5 Questions and further thoughts

(1) Why must the apparatuses for synthesizing acetylsalicylic acid be anhydrous?

(2) Why do we need to add cold water into the reaction flask during the synthesis of acetylsalicylic acid?

(3) Why is it necessary to dry the samples and the KBr crystals before the measurement of FTIR spectra?

6 Cautions

(1) Carefully control the reaction temperature to avoid side reactions.

(2) Acetic anhydride is a dangerous chemical. The vapors will irritate the eyes and nose. The solution can cause burns on the skin. Therefore, the synthesis should be performed strictly in a fume hood.

(3) Sulfuric acid is a strong acid that will burn your skin or put holes in your clothes. If you spill any (even a drop), clean it up immediately using lots of water.

(4) Hydrochloric acid is a strong corrosive acid that may cause damage and severe burns to the human body. Use only in well-ventilated areas. Avoid contact with skin, eyes, and clothes. Do not breathe in the vapor or spray mist.

(5) Gloves and goggles should be worn at all times during the experiment.

Experiment 40 Synthesis of Polystyrene Microspheres by Dispersion Polymerization

Type of the experiment: Comprehensive
Recommended credit hours: 12
Brief introduction: This experiment introduces the dispersion polymerization method, which has been widely used to synthesize nano- and micro-sized materials. The polystyrene (PS) microspheres are synthesized in the experiment using polyvinyl pyrrolidone (PVP) as the stabilizer, azobisisobutyronitrile (AIBN) as the initiator, and ethanol as the reaction medium. The effect of the reaction time on the particle size of the microspheres is examined in this experiment. The reaction product is further confirmed by Fourier transform infrared (FTIR) spectroscopy.

1 Objectives

(1) To understand the basic principle of dispersion polymerization.

(2) To synthesis the mono-dispersed PS microspheres by the dispersion polymerization method.

(3) To study the effect of the reaction time on the particle size of the as-synthesized PS microspheres.

2 Principles

2.1 PS microsphere

PS is the most employed aromatic thermoplastic polymer, which has a wide range of applications. Specifically, PS microspheres are widely used in many fields such as biomedical and clinical diagnosis, high-performance liquid chromatography (HPLC) fillers, catalyst carriers, coatings and ink additives, information storage materials, and colloidal crystals. Their high sphericity and the availability of colored and fluorescent microspheres make them highly desirable for flow visualization and fluid flow analysis, microscopy techniques, process troubleshooting, and numerous research applications. The lower melting temperature enables polystyrene microspheres to create porous structures in ceramics and other materials. They are also used in biomedical applications due to their ability to facilitate procedures, such as cell sorting and immunoprecipitation.

Several synthesis methods of PS microspheres have been reported, including dispersion polymerization, emulsion polymerization, and suspension polymerization. Dispersion polymerization is an attractive method for producing 1–10 μm monodisperse polymer particles with a narrow particle size distribution in a single batch process. The PS spheres with a size of 0.06–0.5 μm are generally obtained by conventional emulsion polymerization, and those with a size of 50–1000 μm are obtained by suspension polymerization. Although the suggested methods have different synthetic procedures, they have almost the same principle of polymerization of styrene monomer under the existence of initiator.

2.2 Dispersion polymerization

The method of dispersion polymerization involves the polymerization of monomers dissolved in an organic liquid to produce insoluble polymer dispersed in the liquid. The process starts from a homogeneous solution of all the starting materials, consisting of monomer, initiator, and steric stabilizer. All these reaction materials are dissolved in the reaction medium at the beginning stage of the reaction. Then insoluble spherical polymer particles stabilized by steric stabilizer molecules are formed and dispersed in the reaction medium.

The polymerization should be carried out in suitable solvents for the monomers but poor solvents for the formed polymers. In general, dispersion polymerization of less polar conventional monomers, such as styrene in this experiment, is performed in polar protic media, such as alcohols.

One of the criteria for choosing the initiator and stabilizer is solubility in the polymerization medium. Since alcohols such as methanol and ethanol are most frequently adopted as the medium, alcohol-soluble initiators have been naturally chosen. In dispersion polymerization, the most frequently used initiators are azo-compounds such as AIBN, peroxide-compounds such as benzoyl peroxide (BPO).

PVP and hydroxypropyl cellulose (HPC) are the most commonly used stabilizers.

In this experiment, the size of polystyrene microspheres about 1–2 μm are synthesized by dispersion polymerization with PVP as the dispersant or stabilizer, AIBN as the initiator, and ethanol as the reaction medium.

3 Experimental

3.1 Preparation

(1) Chemicals: styrene, PVP (K30, MW∼40000), AIBN, absolute ethanol, KBr (IR grade).

(2) Apparatuses: reaction apparatus (the components have been listed in Experiment 29), centrifuge, ultrasonic cleaner, optical microscope, FTIR spectrometer (PerkinElmer Frontier).

(3) Tools: constant pressure funnel, 25 mL and 100 mL graduated cylinders, 50 mL centrifuge tubes, glass slides, tools for FTIR sample preparation listed in Experiment 39.

3.2 Procedure

3.2.1 Synthesis of PS microspheres

(1) Assemble the reaction apparatus. Connect the flask with the condenser. Immerse the flask into the oil bath on the electromagnetic stirrer with a heating plate.

(2) Add 2 g PVP, 150 mL ethanol, and a stirring bar into the flask.

(3) Increase the reaction temperature to 70 °C under stirring.

(4) Add 150 mg AIBN into the flask after the solution temperature reaches 70 °C. Then slowly add 15 mL styrene monomer using the constant pressure funnel within 5 to 10 min.

(5) Keep stirring at 70 °C for 3 h. The temperature must be controlled carefully in a range of 68–72 °C.

(6) During stirring, occasionally take out several drops of the reaction products from 0.5 to 3.0 h at intervals of 0.5 h. Drop them on glass slides for the following observation.

(7) Cool the solution to room temperature.

3.2.2 Purification of PS microspheres

(1) Separate the product by centrifugation at 10000 r/min for 15 min. Dump the supernatant.

(2) Add 30 mL 50% aqueous solution of ethanol into the centrifuge tubes. Disperse the products ultrasonically. Then centrifuge the suspension again and dump the supernatant.

(3) Repeat step (2) three times.

(4) Take a small amount of the purified microspheres and place them on a glass slide for the following observation. Dry the other product at 55 °C for about 2 h.

3.2.3 Characterization of the as-synthesized PS microspheres

(1) Dilute the PS microspheres taken out at different reaction times and after

purification and drying with 95% ethanol on a glass slide and dry them with infrared light.

(2) Observe the morphology and measure the size of the PS microspheres with the optical microscope.

(3) Compare the sizes of the microspheres synthesized with different reaction times, and compare the sizes before and after purification and drying.

(4) Measure the FTIR spectra of PVP, AIBN, and the as-synthesized PS microspheres. The measurement procedure has been described in Experiment 39.

4 Requirements for experimental report

(1) Discuss the effects of reaction time, purification, and drying on the size of the PS microspheres.

(2) Analyze the purity of the as-synthesized PS microspheres from the FTIR spectra.

5 Questions and further thoughts

(1) What are the features of suspension polymerization, emulsion polymerization, and dispersion polymerization?

(2) How can we control the size of the PS microspheres?

(3) How can we further functionalize the PS microspheres? Please give several examples and their applications.

6 Cautions

(1) Styrene monomer is a type of flammable and harmful liquid. Improper operations will irritate eyes and skin. Therefore, the experiment should be conducted in a fume hood, and students should wear gloves and face masks during the experiment.

(2) All the samples should be dried thoroughly before the measurement of FTIR spectra.

Experiment 41   Synthesis of Organic Semiconductor and Determination of Frontier Orbital Energies

Type of the experiment: Comprehensive
Recommended credit hours: 24
Brief introduction: Organic semiconductors have attracted much attention due to their potential application in electronic devices. This experiment introduces the principle and synthesis of organic semiconductors. An imide-functionalized polymer is synthesized, and its frontier orbital energies are measured. The experiment is divided into two experiments that can be conducted, respectively. The first experiment introduces the synthesis procedure of the naphthalenedicarboximide-base organic semiconductor. General organic synthesis method and purified techniques, such as anhydrous and oxygen-free reaction, recrystallization, column chromatography (CC), and Soxhlet extraction, are used. The second experiment introduces

the principle of the measurement of frontier orbital energies by cyclic voltammetry. The lowest unoccupied molecular orbital (LUMO) and highest occupied molecular orbital (HOMO) energies are then determined.

Several reactions during the synthesis of the organic semiconductor take a long time. Therefore, the experiment procedure should be reasonably arranged so that the reactions do not occupy the class time.

Experiment 41.1 Synthesis of Organic Semiconductor

Type of the experiment: Comprehensive
Recommended credit hours: 16

1 Objectives

(1) To learn the synthesis method of organic semiconductors under air-free and oxygen-free conditions.

(2) To master the method of monitoring reaction progress using thin layer chromatography (TLC).

(3) To master the comprehensive application of chemical purification methods, including CC, recrystallization, and Soxhlet extraction.

2 Principles

2.1 Organic semiconductors

Organic semiconductors are made up of polymers or π-bonded molecules and can conduct when charge carriers are injected into them. In contrast to inorganic semiconductors, where the atoms are held together by strong covalent or ionic bonds, organic semiconductors consist of discrete (macro) molecules, which are weakly bound via supramolecular van der Waals, dipole-dipole, π-π, and hydrogen bond interactions. The energies of covalent or ionic bonds in inorganic materials are usually one to three orders of magnitude greater than those of supramolecular interactions in organic π-conjugated systems. Those relatively weak supramolecular interactions result in distinctive differences in materials processability and the corresponding optoelectronic device performance.

Organic molecular and polymeric semiconductors are attractive for the fabrication of cost-effective, large-area, and mechanically flexible electronic devices via solution-based high-throughput patterning techniques such as slot-die coating or printing. Furthermore, the low-temperature processing enables device integration with flexible substrates, such as fabrics or plastics, without compromising substrate functionalities. To date, the applications of organic semiconductors include, but are not limited to, organic light emitting diodes (OLEDs), organic thin film transistors (OTFTs), organic solar cells (OSCs), organic electrochromic devices (OECDs), and organic sensors.

2.2 Tune the bandgap of organic semiconductors

The performance of organic optoelectronic devices strongly depends on the bandgap.

As illustrated in Fig. 8.7, the bandgap of a conjugated polymer is determined by five contributors: the energy related to Peierls instability which depends on the degree of bond length alternation, $E^{\delta r}$, the energy related to the mean deviation from planarity of successive units along the polymer backbone, E^{θ}, the aromatic resonance energy of the aromatic cyclics, E^{res}, the mesomeric and inductive electronic effects of substituents, E^{sub}, and the intermolecular or interchain coupling in the solid state, E^{int}. Thus, the bandgap of conjugated polymers can be expressed by

$$E_g = E^{\delta r} + E^{\theta} + E^{res} + E^{sub} + E^{int}, \tag{8.1}$$

which provides a basis for tailoring the bandgap of conjugated polymers.

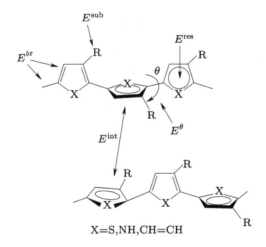

Fig. 8.7 Contributors to the bandgap of polymeric semiconductors.

The donor-acceptor (D-A) approach is one of the most promising approaches for tuning the bandgap of polymeric semiconductors, by which polymers are prepared with controlled sequences of electron-donating and electron-accepting units. Some D-A conjugated polymers have been reported with a bandgap lower than 1.0 eV. Moreover, the D-A strategy can tune not only the bandgap of the materials but also the energies of the frontier molecular orbitals (FMOs) or band edges relative to common electrode materials, which can significantly affect charge injection or extraction and also device stability.

Figure 8.8 illustrates molecular orbital formation in D-A polymers. The energy level of the resulting HOMO in a D-A polymer is a function of the energy levels of the HOMOs of both the donor and acceptor. When increasing the donating ability of donor monomers from donor "a" to donor "b" (higher E_{HOMO}), the E_{HOMO} and E_{LUMO} are both increased. Because E_{HOMO} is increased more than E_{LUMO}, the bandgap of the D-A polymer is decreased.

2.3 Naphthalenedicarboximide-base organic semiconductor

The great success of imide-functionalized small molecules serves as a testbed for using imide-functionalized polymers as semiconductors in organic electronics. Imide-

234　Chapter 8　Organic and Polymeric Materials

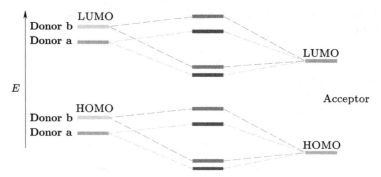

Fig. 8.8　Hybridization of the energy levels of donor and acceptor monomers to form molecular orbitals in a D-A polymer. (See color illustration at the back of the book)

containing polymers can be semiconductors or insulators, depending on the structural connectivity (Fig. 8.9). Polyimides have been widely used as electrical insulators due to their synthetic accessibility, mechanical property, thermal property, and supramolecular self-assembly. Thus some characteristics making polyimides superior insulators are also desired for organic semiconductors, such as the high electron deficiency and the capability to self-assemble into ordered structures.

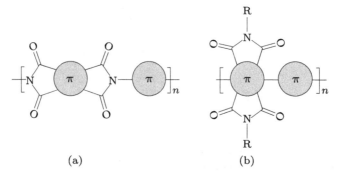

Fig. 8.9　General chemical structures of (a) polyimide insulators and (b) imide-functionalized π-conjugated polymer semiconductors.

　　The naphthalenedicarboximide-base is identified as the key polymer building block to ensure a strong electron-depleted electronic structure and, equally important, a highly p-conjugated polymeric backbone. In this experiment, the naphthalenedicarboximide-base polymer, [poly{[N,N'-bis(2-octyldodecyl)-1,4,5,8-naphthalenedicarboximide-2,6-diyl]-alt-5,5'-(2,2'-bithiophene)} :thiophene] (compound 9), will be synthesized from 1,4,5,8-naphthalenetetracarboxylic dianhydride (compound 1), as shown in Fig. 8.10 and Fig. 8.11.

　　For the synthesis of this polymeric semiconductor, the first step is monomer synthesis. The monomer, N,N'-bis(2-octyldodecyl)-2,6-dibromonaphthalene-1,4,5,8-bis(dicarboximide) (compound 3), can be synthesized by two-step reactions as shown in Fig. 8.10. First, compound 1reacts with dibromoisocyanuric in oleum at 85°C for 43 h to yield 2,6-dibromonaphthalene-1,4,5,8-tetracarboxydianhydride (compound 2). This is a selective substitution reaction, and no further purification is needed. Then it reacts with 2-octyldodecylamine, o-xylene, and propionic acid to form the monomer (compound 3). This is a nucleophilic substitution reaction. The monomer should have sufficient high purity to achieve high molecular weight for the resulting polymeric semiconductor. Therefore, the monomer will be subjected to flash column chromatography and recrystallization to improve the monomer purity (see Experiments 15 and 18).

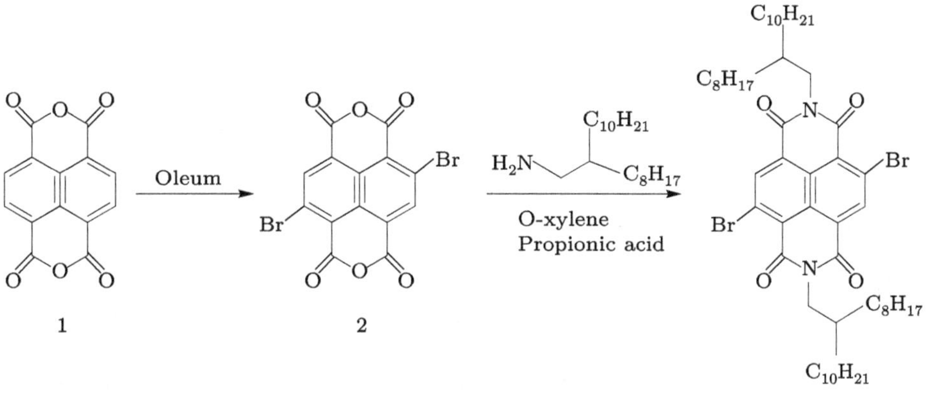

Fig. 8.10 Reactions for the synthesis of the monomer.

As shown in Fig. 8.11, the compound 3 reacts with 5,5'-bis(trimethylstannyl)-2,2'-bithiophene (compound 4) yield poly{[N,N'-bis(2-octyldodecyl)-1,4,5,8-naphthalene dicarboximide-2,6-diyl]-alt-5,5'-(2,2'-bithiophene)} (compound 5) via a Pd-catalyzed Stille polymerization. In the application of this polymer, it is reported that the polymer with uniform and controllable molecular has much better properties. The molecular weight of compound5 can be adjusted by the reaction time, and the molecular weight is increased with the reaction time. Trimethyl (thiophen-2-yl) silane (compound 6) and 2-bromothiophene (compound 8) are used as end-capping reagents. The as-prepared polymer will be purified by Soxhlet extraction and recrystallization (see Experiments 15 and 17).

3 Experimental

3.1 Preparation

(1) Chemicals: 1,4,5,8-naphthalenetetracarboxylic dianhydride (compound 1), oleum (20% SO_3), dibromoisocyanuric, 2-octyldodecylamine, o-xylene, propionic acid, chloromethane, hexane, silica gel (100–200 mesh), $Pd_2(dba)_3$, P(o-tolyl)$_3$, 5,5'-bis(trimethylstannyl)-2,2'-bithiophene (compound 4), trimethyl (thiophen-2-yl) silane (compound 6) and 2-bromothiophene (compound 8),hydrochloride acid, chloroform, methanol.

(2) Apparatuses: the reaction apparatus (the components have been listed in Experiment 29), vacuum filtration apparatus (the components have been listed in Experiment 15), TLC and CC apparatuses (the components have been listed in

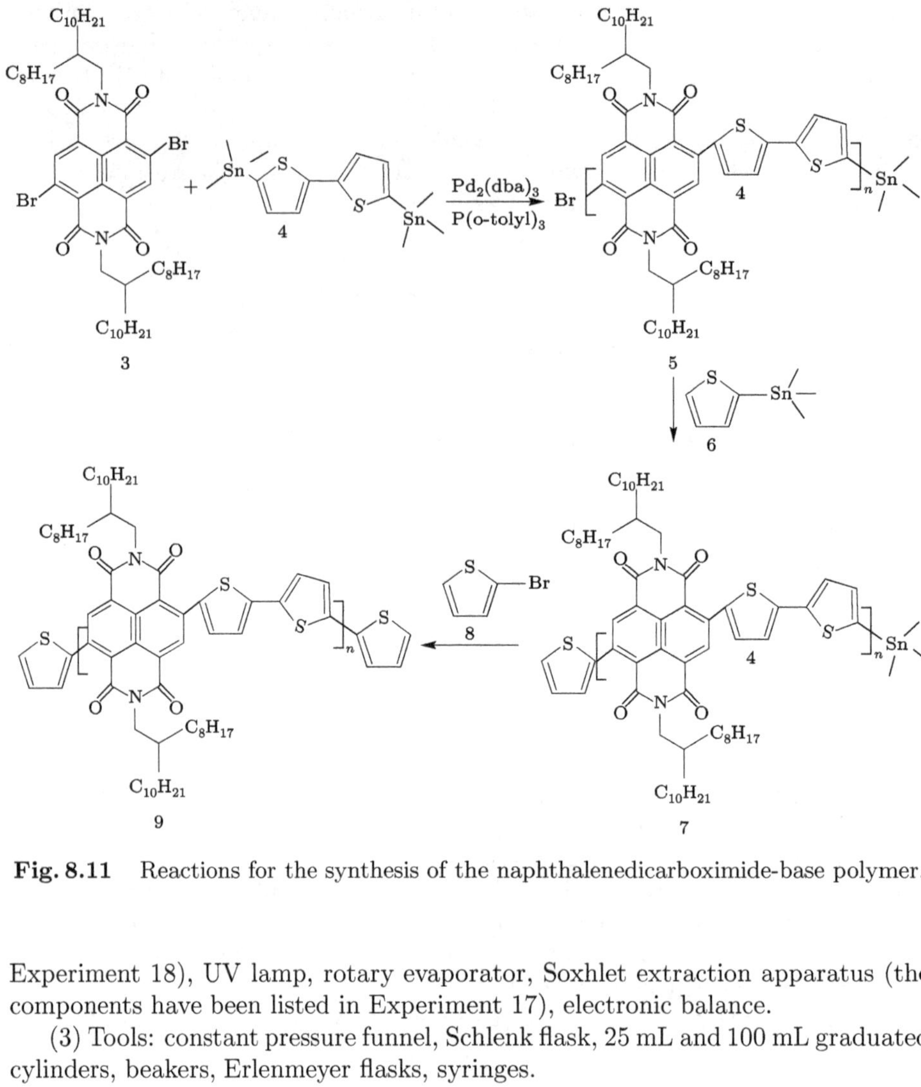

Fig. 8.11 Reactions for the synthesis of the naphthalenedicarboximide-base polymer.

Experiment 18), UV lamp, rotary evaporator, Soxhlet extraction apparatus (the components have been listed in Experiment 17), electronic balance.

(3) Tools: constant pressure funnel, Schlenk flask, 25 mL and 100 mL graduated cylinders, beakers, Erlenmeyer flasks, syringes.

3.2 Procedure

3.2.1 Synthesis of 2,6-dibromonaphthalene-1,4,5,8-tetracarboxydianhydride (compound 2)

(1) Assemble the reaction apparatus. Connect the flask with the condenser. Immerse the flask into the oil bath on the electromagnetic stirrer with a heating plate.

(2) Add 2.8 g (10.3 mmol) 1,4,5,8-naphthalenetetracarboxylic dianhydride and 100 mL oleum into the flask. Stir at 55 °C for 2 h.

(3) Prepare a solution of 3.0 g (10.5 mmol) dibromoisocyanuric in 50 mL oleum. Slowly add the solution into the flask using the constant pressure funnel in 40 min.

(4) Increase the temperature of the oil bath to 85 °C and stir for 40 h.

(5) Transfer the reaction mixture into 400 mL water in a 600 mL beaker. Stir at room temperature for 1 h.

(6) Separate the product by vacuum filtration (see Experiment 15).

(7) Dry and weigh the product.

3.2.2 Synthesis of N,N'-bis(2-octyldodecyl)-2,6-dibromonaphthalene-1,4,5,8-bis(dicarboximide) (compound 3)

(1) Assemble the reaction apparatus as that in Section 3.2.1.

(2) Add 2.34 g (5.49 mmol) compound 2, 4.10 g (13.78 mmol) 2-octyldodecylamine, 18 mL o-xylene, and 6 mL propionic acid into the flask.

(3) Stir at 140 °C for about 2 h.

(4) Monitor the reaction by TLC. The eluent is a mixture of chloromethane and hexane (1:1 in volume ratio). Visualize the spots with UV light. The procedure of TLC has been described in Experiment 18.

(5) When the synthetic reaction is completed, remove most of the solvents under reduced pressure using the rotary evaporator (see Experiment 16).

3.2.3 Purification of the monomer (compound 3) by CC and recrystallization

(1) Dissolve the as-synthesized compound 3 in a minimal amount of chloromethane. Separate the monomer by CC using the mixture of chloromethane and hexane as the eluent and silica gel as the adsorbent. The procedure of CC has been described in Experiment 18.

(2) Remove the solvent using the rotary evaporator. Collect the product in a three-necked glass flask.

(3) Add 3 mL hexane into the flask. Assemble the reaction apparatus and reflux the mixture at 85 °C for about 10 min. During the reflux, add more hexane (e.g., 0.5 mL) into the bottle if there is still some solid undissolved.

(4) Cool the solution to room temperature, and the desired compound will be crystallized.

(5) Separate the compound by vacuum filtration (see Experiment 15).

3.2.4 Synthesis of naphthalenedicarboximide-base polymer

(1) Add the monomer powders of 0.1 mmol compound 3 and 0.1 mmol compound 4 and the catalyst (1:8 molar ratio between $Pd_2(dba)_3$ and $P(o-tolyl)_3$) into an air-free Schlenk flask with a stirring bar.

(2) Pump and purge the Schlenk flask for 3 cycles with argon gas.

(3) Add 3 mL anhydrous degassed toluene by syringe under Ar protection.

(4) Stir the reaction mixture at 110 °C for 2 d under Ar protection.

(5) Add 0.02 mL compound 6 into the flask from the side pipe under Ar flow. Continue the reaction with stirring at 110 °C for 8 h.

(6) Add 0.04 mL compound 8 into the flask from the side pipe under Ar flow. Continue the reaction with stirring at 110 °C for 8 h.

(7) After cooling to room temperature, drip the deeply colored reaction mixture into a solution of 100 mL methanol and 5 mL 12 mol/L hydrochloride acid in an Erlenmeyer flask under vigorously stirring.

(8) Stir for 4 h, and then collect the precipitated solid by vacuum filtration.

3.2.5 Purification of the as-synthesized polymer

(1) Dissolve the solid polymer in chloroform and then add methanol to reprecipitate the polymer.

(2) Separate the solid polymer by vacuum filtration. Dry the polymer under reduced pressure.

(3) Extract the polymer with methanol, hexane, and chloroform orderly by Soxhlet extraction (see Experiment 17).

(4) Concentrate the polymer chloroform solution to approximately 20 mL using the rotary evaporator.

(5) Slowly dump the solution into 100 mL methanol under stirring.

(6) Separate the polymer by vacuum filtration. Dry the polymer under reduced pressure.

4 Requirements for experimental report

(1) Describe several purification methods used in this experiment in detail, including the principles, the operations, and the cautions in your own words.

(2) Describe and analyze the phenomenon of the experiment.

5 Questions and further thoughts

(1) How does it affect the separation of CC if a sample is dissolved in too much solvent?

(2) How can we judge the reaction is over by TLC? What is the substance corresponding to each spot?

(3) How can we decide the required volume of the solvent for recrystallization?

Experiment 41.2   Determination of Frontier Orbital Energies by Cyclic Voltammetry

Type of the experiment: Comprehensive
Recommended credit hours: 8

1 Objectives

(1) To understand the concepts of HOMO and LUMO of molecules.

(2) To understand the cyclic voltammetry method for determining the frontier orbital energies.

(3) To determine the HOMO and LUMO levels of the as-synthesized organic semiconductor by cyclic voltammetry.

2 Principles

2.1 HOMO and LUMO

In the molecular orbital theory, a molecular orbital gives the most probable location of an electron in an atom. Molecular orbitals are formed by the combination of atomic orbitals of separated atoms. Thus covalent bonds are formed by sharing electrons of neighboring atoms. These molecular orbitals split into FMOs, named HOMO and LUMO. The electrons in HOMO can donate to the LUMO type molecular orbitals because these molecular orbitals contain weakly attached electrons.

These molecular orbitals are the most available form for covalent chemical bonding. Because the energies of these orbitals are the closest of any orbitals of different energy levels, the HOMO-LUMO gap is where the most excitations can occur. Hence, it is the most important energy gap to be considered.

With the rapid development of organic electronics, designing high-performance devices requires that the frontier energies of organic semiconductor molecules can be accurately determined. Three methods are generally accepted to ascertain the frontier energies, which are ultraviolet photoelectron spectroscopy (UPS), inverse photoelectron spectroscopy (IPES), and cyclic voltammetry (CV). Unfortunately, UPS and IPES can be both time-consuming and complex. Moreover, different measurement conditions may result in variations of measured energies for the same material. Therefore, CV is the most common method to be used.

2.2 Determination of HOMO and LUMO levels by CV method

2.2.1 Three-electrode setup of CV

CV is an elegant and straightforward electrochemical technique for studying redox reactions at the electrode-solution interfaces. It is a simple, rapid, and powerful method for characterizing the electrochemical behaviors of materials that can be electrochemically oxidized or reduced.

As shown in Fig. 8.12, the CV method uses a three-electrode setup consisting of a reference electrode, a working electrode, and a counter electrode. The basic idea of CV is to apply a periodic potential that will alternately oxidize and reduce the material to be studied by extracting and injecting electrons, respectively. The potential of the working electrode is measured against a reference electrode that maintains a constant potential, and the resulting applied potential produces an

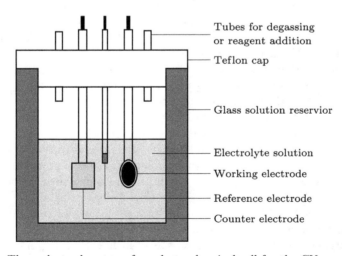

Fig. 8.12 Three-electrode setup of an electrochemical cell for the CV experiment.

excitation signal.

2.2.2 CV measurement of organic semiconductor

CV is recognized as an important technique for measuring the HOMO and LUMO levels and the bandgaps. Organic semiconductors are typically studied as thin films on the working electrode and using the non-aqueous solution as the electrolyte. Ideally, starting from any point, the sample is periodically oxidized and reduced when the periodic potential is applied. This gives rise to a pair of oxidation and reduction peaks for each electron transfer process. Figure 8.13 illustrates that the redox peaks are characterized by several potentials, namely the anodic and cathodic peak potentials, E_{peak}^{a} and E_{peak}^{c}, and the onset potentials, E_{onset}^{a} and E_{onset}^{c}, where the oxidation and reduction peaks start.

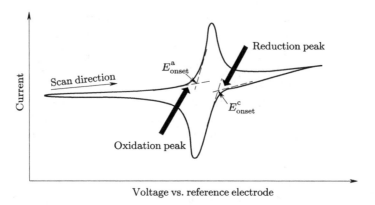

Fig. 8.13 Schematic current-voltage plot from the CV measurement.

HOMO represents the energy required to extract an electron from a molecule, which is an oxidation process, and LUMO is the energy necessary to inject an electron into a molecule, thus implying a reduction process. Therefore, the onset potentials of the corresponding oxidation and reduction peaks are taken to determine the HOMO and LUMO levels.

It is known that it is impossible to measure the absolute potential directly, and what can be determined is the potential difference. To obtain the HOMO and LUMO levels with respect to vacuum, one needs to know the potential of the reference electrode with respect to vacuum. The absolute potential of H^+/H_2 electrode is zero based on the standard hydrogen electrode (SHE) in aqueous media. However, it becomes unreliable when SHE is used in non-aqueous media for the complicated experimental system. Therefore, the ferrocenium/ferrocene (Fc/Fc$^+$) redox pair (4.8 eV versus vacuum) is recommended to be used as a reference system for its well-behaved electrochemically reversibility.

In practice, the Fc/Fc$^+$ redox couple of ferrocene in acetonitrile solution is often used as an internal or pseudo reference when acquiring the accurate potentials of inorganic/organometallic complexes. As shown in Fig. 8.14, Ag/Ag$^+$ (Ag/AgCl) is the reference electrode and Fc/Fc$^+$ is the external standard electrode. Then the energy levels can be calculated by

$$E_{\text{HOMO}} = -E_{\text{ferrocene}} + 4.8, \tag{8.2}$$

$$E_{\text{LUMO}} = -E_{\text{ferrocene}} + 4.8, \tag{8.3}$$

where $E_{\text{ferrocene}}$ is the energy level of Fc/Fc$^+$ with respect to the reference electrode. It is calculated using the average of the two peaks in the CV plot of the Fc/Fc$^+$ redox couple.

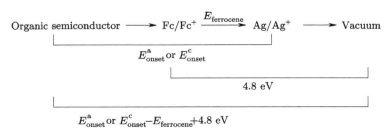

Fig. 8.14 The relationship of the energy levels among the organic semiconductor, external standard electrode (Fc/Fc$^+$), and reference electrode (Ag/Ag$^+$).

3 Experimental

3.1 Preparation

(1) Materials: naphthalenedicarboximide-base organic semiconductor synthesized in Experiment 41.1, Al$_2$O$_3$ powder.

(2) Chemicals: tetrabutylammonium hexafluorophosphate ($(n\text{-Bu})_4\text{N}^+\text{PF}_6$), acetonitrile (CH$_3$CN), chloroform (CHCl$_3$), ferrocene.

(3) Apparatus: electrochemical workstation (CH Instruments Ins., CHI604E).

(4) Tools: electrochemical glass cell with a PTFE lid, glass carbon electrode, Ag/AgCl electrode, Pt plate.

3.2 Procedure

(1) Prepare 30 mL electrolyte solution of $(n\text{-Bu})_4\text{N}^+\text{PF}_6$ in CH$_3$CN. The concentration is 0.1 mol/L.

(2) Prepare 10mL solution of $(n\text{-Bu})_4\text{N}^+\text{PF}_6$ in CHCl$_3$. The concentration is 1 mg/mL. Then dissolve the organic semiconductor sample in the solution.

(3) Prepare the working electrode. Polish the glass carbon electrode with Al$_2$O$_3$ powder and clean up. Then dip coat 25 μL of the sample solution onto the polished electrode twice.

(4) Put 15 mL electrolyte solution into the electrochemical cell. Assemble the three electrodes: Pt plate as the countering electrode, Ag/AgCl electrode as the reference electrode, and the glassy carbon electrode as the working electrode.

(5) Bubble the electrolyte with Ar for 10 min to ensure there is no oxygen on the working electrode.

(6) Perform the CV test at a scan rate of 0.1 V/s at room temperature under Ar. The scan scope is −1.5 to 1.0 V. Export the result as a ".txt" file.

(7) Replace the working electrode with an empty glass carbon electrode. Add 0.2–0.5 mg ferrocene to the electrolyte solution. Then perform the CV test and export the result.

4 Requirements for experimental report

(1) Plot the CV curves recorded in the experiment.
(2) Calculate the HOMO and LUMO energies and the HOMO-LUMO gap.

5 Questions and further thoughts

(1) How can we distinguish the reduction peak and the oxidation peak in the CV plot?
(2) Besides the present experiment, what are the typical applications of CV? Please provide one or two examples.

Extended readings

Elgrishi N, Rountree K J, McCarthy B D, et al. A practical beginner's guide to cyclic voltammetry. Journal of Chemical Education, 2018, 95: 197-206.

Guo X, Facchetti A, Marks T J. Imide- and amide-functionalized polymer semiconductors. Chemical Reviews, 2014, 114(18): 8943-9021.

Griffiths P R, de Haseth J A. Fourier Transform Infrared Spectrometry. 2nd Edition. Wiley-Interscience, 2007.

Leonat L, Sbârcea G, Brânzoi I V. Cyclic voltammetry for energy levels estimation of organic materials. UPB Scientific Bulletin, Series B, 2013, 75(3): 111-118.

Lutz J F, Hoth A. Preparation of ideal PEG analogues with a tunable thermosensitivity by controlled radical copolymerization of 2-(2-methoxyethoxy) ethyl methacrylate and oligo (ethylene glycol) methacrylate. Macromolecules, 2006, 39: 893-896.

Lynwood C. Polystyrene: Synthesis, Characteristics and Applications. Nova Science publishers Inc., 2014.

Roncali J. Molecular engineering of the band gap of π-conjugated systems: Facing technological applications. Macromolecular Rapid Communications, 2007, 28: 1761-1775.

Thalacker C, Röger C, Würthner F. Synthesis and optical and redox properties of core-substituted naphthalene diimide dyes. The Journal of Organic Chemistry, 2006, 71(21): 8098-8105.

Yan H, Chen Z, Zheng Y, et al. A high-mobility electron-transporting polymer for printed transistors. Nature, 2009, 457(7230): 679-686.

Chapter 9 Nanomaterials

Experiment 42 Synthesis of Metal Nanoparticles

Type of the experiment: Comprehensive and designing
Recommended credit hours: 16
Brief introduction: Metal-based nanocrystals with high-surface-energy facets are thermodynamically unstable in their growth, resulting in a big challenge in their synthesis. There is a growing interest in controlling the synthesis of metal nanocrystals and tailoring their properties toward various applications. This experiment introduces the synthesis techniques of Au and Ag particles, and the colloidal suspensions of the nanoparticles are synthesized. The Au@Ag core-shell nanocrystals are synthesis by a process designed by students. The experiment is divided into three experiments that can be conducted respectively. In the first experiment, the colloidal Au particles are synthesized by the classic citrate method. The effect of electrolytes, including Na_2SO_4, glucose, and sports drinks, on the agglomeration is examined by ultraviolet-visible (UV-Vis) spectroscopy. In the second experiment, the colloidal Ag particles are synthesized by the reduction method. The hinder effect of surfactant on the agglomeration is examined. In the third experiment, the students are required to design a synthesis process of Au@Ag core-shell composite nanoparticles by themselves based on a literature survey. Then the Au@Ag nanoparticles will be synthesized. Particularly, transmission electron microscope (TEM) will be introduced and used to observed the morphology of the as-synthesized particles.

Experiment 42.1 Citrate Synthesis of Gold Nanoparticles

Type of the experiment: Cognitive

Recommended credit hours: 4

1 Objectives

(1) To understand the principle of the citrate method.

(2) To synthesize the red colloidal suspension of gold nanoparticles by the citrate method.

(3) To evaluate the agglomeration of the nanoparticles by UV-Vis spectroscopy.

(4) To examine the effect of electrolyte addition on the stability of the colloidal nanoparticles.

2 Principles

2.1 Gold nanoparticles

Nanomaterials are defined according to their size of the structural unit, which is between 1 nm and 100 nm. They exhibit surface effects, small size effects, and macroscopic quantum tunneling effects because of their structure. Therefore, they are widely used in many fields, such as chemistry, physics, biomedicine, photology, and environics.

Metal nanoparticles are made of pure metals (e.g., gold, platinum, silver, titanium, zinc, cerium, iron, and thallium) or their compounds (e.g., oxides, hydroxides, sulfides, phosphates, fluorides, and chlorides). The research of metal nanoparticles has recently become the focus of intense work due to their unusual properties compared to bulk metals. Among the metal particles, gold nanoparticles feature a wide range of potential applications in medical imaging technology, tumor detection, optical probe, glucose sensor, and chemical metal coating because of their weak activity and insensitivity to air or light.

2.2 Synthesis of gold nanoparticles by the citrate method

Gold nanoparticles can be prepared via various synthesis routes, including chemical, electrochemical, sonochemical, and photochemical methods. The chemical method is the most common method for the synthesis of gold nanoparticles. The general procedure is the precipitation of gold nanoparticles in an aqueous solution from a dissolved gold precursor, such as $HAuCl_4 \cdot 3H_2O$, by a reducing agent, such as sodium citrate, ascorbic acid, sodium boron hydride, or block copolymers. Besides, in most cases, a further stabilizing agent, such as polyvinyl pyrrolidone and polyethylene glycol, is required to prevent agglomeration or further growth of gold particles. Some reducing agents, such as sodium citrate and sodium boron hydride, also act as stabilizers.

The classic citrate method is the reduction of a gold precursor with sodium citrate in the aqueous solution near the boiling point. It is also used as a model reaction to study the formation mechanism of gold nanoparticles because of its simple preparation procedure. In general, a detailed understanding of the mechanism and kinetics of precursor reduction and particle growth is necessary for controlling the size and shape of nanoparticles. However, a coherent explanation for the evolution of gold particles prepared via the citrate method has not been achieved yet. In a widely accepted multistep mechanism, the Au^{3+} ions are quickly reduced to Au^0 atoms that form clusters, followed by their assembly into larger polycrystalline particles and completed by further aggregation. Au^{3+} are reduced to neutral gold atoms, where citrate ions act as both the reducing agent and the capping agent. The reaction may produce nanoparticles with a diameter of about 13 nm. The layer of absorbed citrate anions on the surface of the nanoparticles produces an electrostatic repulsion that keeps the nanoparticles separated. The final product is in a state of colloidal suspension.

2.3 Assessment of stability of the nanoparticle suspension by UV-Vis spectroscopy

For many nanoparticle applications, it is necessary to form a stable colloidal nanoparticle suspension. However, the surface energy of nanoparticles is significantly higher than that of larger particles. Thus, nanoparticles tend to agglomerate in liquid suspensions. A stable nanoparticle suspension is often formed by adjusting the suspension ionic strength and pH value or by surface modification of the nanoparticles themselves. As mentioned above, the surface of the gold nanoparticles is decorated by a layer of citrate anions to separate the particles.

When a strong electrolyte is added to the suspension, the high concentration of ions will screen the repulsive electrostatic force between nanoparticles. Then the gold nanoparticles agglomerate due to the elimination of the repulsive force. In contrast, if a non- or weak electrolyte is added, the electrostatic repulsion among the nanoparticles is not disrupted.

The as-synthesized product of the colloidal suspension by the citrate method strongly absorbs green light with a wavelength of 520 nm, and the solution appears red. If the spacing between the nanoparticles decreases to less than their average diameter, the solution absorbs light at a longer wavelength (about 650 nm). Accordingly, the agglomeration state can be visibly detected by a change in the color of the suspension from red to blue or purple. If the nanoparticles aggregate into large precipitates, the suspension may even become clear.

Since the light absorption can be influenced by the agglomeration state, UV-Vis spectroscopy can provide a simple method to assess the stability of gold nanoparticles, which can be applied to monitor their quality over time and evaluate the integrity of the colloidal solution. The widening and shift of the absorption peaks in the spectra can be seen due to the agglomeration.

3 Experimental

3.1 Preparation

(1) Chemicals: gold chloride trihydrate ($HAuCl_4 \cdot 3H_2O$), trisodium citrate dihydrate ($Na_3C_6H_5O_7 \cdot 2H_2O$), sodium sulfate ($Na_2SO_4$), glucose, two types of sports drinks.

(2) Apparatuses: UV-Vis spectrometer (PerkinElmer, Lambda 25), electromagnetic stirrer with a heating plate, oil bath, electronic balance.

(3) Tools: 50 mL Erlenmeyer flask, graduated cylinders, droppers, test tubes, quartz cuvettes, laser pointer.

3.2 Procedures

(1) Add 20 mL of 1.0 mmol/L aqueous solution of $HAuCl_4 \cdot 3H_2O$ and 2 mL 10 mg/mL aqueous solution of $Na_3C_6H_5O_7 \cdot 2H_2O$ in an Erlenmeyer flask. Immerse the flask into the oil bath on the electromagnetic stirrer with a heating plate.

(2) Increase the reaction temperature to 85–90 °C under stirring. The colloidal nanoparticle suspension will gradually form as the citrate reduces the Au^{3+}. The reaction takes about 10 min. The suspension turns deep red.

(3) Use a laser pointer to examine the reflection of the laser by the suspension. The formation of the colloid can be confirmed by the Tyndall effect.

(4) Transfer 2 mL product into four test tubes, respectively.

(5) Add 0.5 mL 1.8 mol/L aqueous solution of Na_2SO_4 in one test tube. Observe the color change of the suspension.

(6) Repeat step (5) for 30wt% aqueous solution of glucose and two types of sports drinks.

(7) Measure the UV-Vis spectra of the samples in these four test tubes and the as-synthesized colloidal suspension. The measurement procedure has been described in Experiment 29.

4 Requirements for experimental report

(1) Describe the color change during the synthesis reaction.

(2) Describe the color change when the Na_2SO_4 solution, glucose solution, and sports drinks are added to the suspension.

(3) Plot the UV-Vis spectra of absorbance versus wavelength using the exported data.

(4) Discuss the effects of different additives on the color, the UV-Vis spectra, and the stability of the colloidal suspension.

5 Questions and further thoughts

(1) Why does the addition of the Na_2SO_4 solution and glucose solution produce different results of agglomeration?

(2) Describe the applications of gold nanoparticles with one or two examples.

Experiment 42.2 Synthesis of Silver Nanoparticles by Chemical Reduction Method

Type of the experiment: Cognitive
Recommended credit hours: 4

1 Objectives

(1) To understand the principle of the chemical reduction method.

(2) To synthesize the yellow colloidal suspension of silver nanoparticles.

(3) To evaluate the agglomeration of the silver nanoparticles by UV-Vis spectroscopy.

(4) To examine the hinder effect of additional surfactant on the agglomeration.

2 Principles

2.1 Synthesis of silver nanoparticles by chemical reduction method

Silver nanoparticles are particles of silver with a size ranging from 1 to 100 nm. They have been studied in many advanced technology applications, such as medical treatment, environment, conducting materials, wastewater treatment, and so on. Their unique properties depend on the size and morphology of the silver nanoparticles.

Many methods have been reported for the synthesis of the silver nanoparticles, such as chemical reduction, electron irradiation, laser ablation, and microwave processing. The chemical reduction is one of the most used methods, and it is accepted in the present experiment.

In general, the reactions of the reduction method are limited to those using silver nitrate ($AgNO_3$) as the precursor material, which is reduced by reducing agents. There are many choices of reducing agents, such as sodium borohydride ($NaBH_4$), sodium citrate, ascorbate, hydrogen, ascorbic acid, hydrazine, and ammonium formate. They are used for the reduction of the silver ions in the aqueous or nonaqueous solutions. In addition to the reducing agents, the surfactants, such as citrate, polyvinyl pyrrolidone (PVP), cetyltrimethylammonium bromide (CTAB), and polyvinyl alcohol (PVA), are required. They interact with particle surfaces to stabilize particle growth and protect particles from sedimentation and agglomeration. The selection of the surfactant varies with the choice of reducing agent, the concentrations of reagents, temperature, duration of reaction, and the diameters of the nanoparticles produced. Nearly all colloidal silver products by the reduction method are described as turbid and greenish-yellow or brown.

In this experiment, the yellow colloidal silver is produced by the reduction of silver nitrate with ice-cold sodium borohydride. The chemical reaction is

$$2AgNO_3 + 2NaBH_4 \rightarrow 2Ag + H_2 + B_2H_6 + 2NaNO_3. \quad (9.1)$$

The reaction may produce nanoparticles with a diameter of about 12 nm. The sodium borohydride acts as both the reducing agent and the surfactant. Therefore, a significant excess of sodium borohydride is required to reduce the ionic silver and stabilize the silver nanoparticles.

2.2 Optical property of silver nanoparticles

The distinctive color of colloidal silver is due to a phenomenon known as plasmon absorbance. Incident light creates oscillations in conduction electrons on the surface of the nanoparticles, and electromagnetic radiation is absorbed. In the UV-Vis spectrum, the wavelength of the plasmon absorption maximum is near 400 nm. The widening and shift of the absorption peaks in the spectra can be seen due to the increase in particle size and agglomeration.

As shown in Experiment 42.1, the addition of strong electrolyte results in agglomeration because the ions will eliminate the repulsive electrostatic forces between nanoparticles. The optical properties of colloidal metal nanoparticles will be influenced accordingly. Such an effect of the electrolyte can be avoided by adding a surfactant, which can screen the interaction between the ions of electrolyte and surface ion layer decorated on the nanoparticle surface.

3 Experimental

3.1 Preparation

(1) Chemicals: silver nitrate ($AgNO_3$), sodium borohydride ($NaBH_4$), sodium sulfate (Na_2SO_4), PVP.

(2) Apparatuses: UV-Vis spectrometer (PerkinElmer, Lambda 25), electromagnetic stirrer, electronic balance.

(3) Tools: magnetic stirring bar, culture dish, Erlenmeyer flask, graduated cylinders, droppers, test tubes, quartz cuvettes, laser pointer.

3.2 Procedure

3.2.1 Synthesis of colloidal silver

(1) Prepare 30 mL 2 mmol/L aqueous solution of $NaBH_4$ in an Erlenmeyer flask.

(2) Add a magnetic stirring bar and place the flask in the culture dish with ice on the electromagnetic stirrer. Cool the liquid for about 20 min.

(3) Drip 2 mL of 1mmol/L aqueous solution of $AgNO_3$ into the $NaBH_4$ solution at approximately one drop per second under stirring. Stop stirring when all the $AgNO_3$ solution is added.

(4) Examine the presence of a colloidal suspension by the reflection of a laser beam from the laser pointer. The Tyndall effect should be seen.

(5) Transfer 2 mL product into two test tubes, respectively.

(6) Prepare and add 0.5 mL 1.8 mol/L aqueous solution of Na_2SO_4 in one test tube. Observe the color change of the colloid.

(7) Add a drop of 0.3wt% aqueous solution of PVP into another test tube. Then add 0.5 mL 1.8 mol/L Na_2SO_4 solution. Observe the color change of the colloid.

(8) Measure the UV-Vis spectra of the samples in these two test tubes and the as-synthesized colloidal suspension. The measurement procedure has been described in Experiment 29.

4 Requirements for experimental report

(1) Describe the color change during the synthesis reaction.

(2) Describe the color change when the Na_2SO_4 solution or both PVP and Na_2SO_4 solutions are added into the colloidal suspension.

(3) Plot the UV-Vis spectra of absorbance versus wavelength using the exported data.

(4) Discuss the effects of the different additives on the color of the colloidal suspension and the UV-Vis spectra.

5 Questions and further thoughts

(1) Suggest one or two other synthesis methods of silver nanoparticles.

(2) Describe the applications of silver nanoparticles with one or two examples.

Experiment 42.3 Synthesis and Characterization of Au@Ag Core-Shell Composite Nanoparticles

Type of the experiment: Designing
Recommended credit hours: 8

1 Objectives

(1) To understand the common synthesis methods of core-shell metal composite nanoparticles.

(2) To synthesize the Au@Ag core-shell composite nanoparticles by designing the synthesis process.

(3) To understand the principle of TEM.

(4) To observe the morphology of the Au@Ag core-shell composite nanoparticles using TEM.

2 Principles

2.1 Metal composite nanoparticles

Composite nanoparticles are advanced materials that have recently gained increasing attention due to their scientific and technological importance. They have various applications such as catalysts with high activity and specificity, metal-semiconductor junctions, optical sensors, and modifiers of polymeric films for packaging.

Metal composite nanoparticles are based on metals, including two metals or more than two metals. They exhibit excellent mechanical properties, unique catalytic, electronic, and optical properties, leading to their extensive applications. The synthesis methods of the metal composite nanoparticles are categorized into two types: One is that the two metal ions are reduced simultaneously to synthesize a tightly bound metal-metal mixture; The other is that one metal deposits on a different metal nucleus surface to form a core-shell structure.

2.2 Au@Ag core-shell nanoparticles

Among metal composite nanoparticles, core-shell and alloy bimetallic nanoparticles are especially interesting because they provide opportunities to tune the optical and catalytic properties and are potentially useful as taggants for security applications.

Core-shell Au@Ag nanoparticles have been investigated most extensively because of their facile preparation and oxidation resistance. Besides, the surface plasmon resonance (SPR) band can absorb and scatter visible light because the SPR band is tunable between about 520 nm for Au and about 410 nm for Ag.

Several synthesis methods for Au@Ag core-shell nanoparticles have been developed, such as chemical reduction, seed-mediated growth method, photocatalytic reduction, and physical method. In this experiment, the students are required to select a synthesis method and design the detailed synthesis process based on a literature survey.

2.3 Principle of TEM

TEM is an instrument or technique in which a high-energy beam of electrons is transmitted through a sample to form an image. The sample is often an ultrathin section less than 100 nm thick or a suspension on a grid. A TEM utilizes energetic electrons to provide morphologic, compositional, and crystallographic information on samples, such as the crystal structure, dislocations, grain boundaries, and composition distribution.

The TEM operates on the same basic principle as the optical microscope but uses electrons instead of visible light. Because the wavelength of electrons is much smaller than that of light, the optimal resolution attainable for TEM images is many orders of magnitude better than that of an optical microscope. Thus, TEM can reveal the most delicate details of the internal structure at a maximum potential magnification of about 1 nm.

The electro-optical system of TEM is similar to that of SEM described in Experiment 5. As shown in Fig. 9.1, it consists of the following essential systems:

(1) The electron beam generation and convergence system produces the electron beam and focuses the beam. It consists of an electron gun and several condenser lenses.

(2) The image-producing system focuses the electron beam passing through the sample to form a real, highly magnified image. It consists of objective, intermediate, and projector lenses, and a movable sample stage.

(3) The image-recording system converts the electron image into some form perceptible to the human eyes. It consists of a fluorescent screen for viewing and

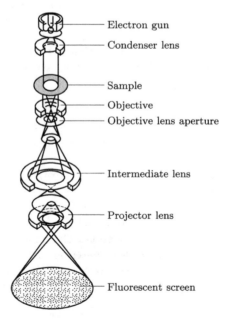

Fig. 9.1 Schematic illustration of the essential components of TEM.

focusing the image and a digital camera for permanent records.

(4) The vacuum system consists of pumps and their associated gauges and valves.

2.4 Preparation of TEM samples

Successful TEM observation strongly depends on the quality of the samples. The characterization of materials using the TEM requires that the samples used are thin enough with a thickness between 20 and 100 nm so that they are transparent to electrons. In addition, it is often necessary to make the samples small enough to fit in the electron microscope sample holder, which is usually 3 mm in diameter.

In most cases, the sample preparation can be executed in a few hours using straightforward steps. Electropolishing, chemical polishing, and replication methods have been used for decades as the principal preparation methods. Modern methods, including ion milling, tripod polishing, and focused ion beam (FIB), have also been widely used.

In the case of fine particle materials, the TEM sample preparation is relatively easy. The method consists of dispersing fine particles of a material. Generally, the material is suspended in various liquids. Ultrasonic can be used to improve the dispersion of the particles. Then the suspension is deposited onto the film of a support grid. This technique can be applied to any fine particles, whether they are micrometric or nanometric in size.

❸ Experimental

3.1 Preparation

(1) Chemicals: prepare according to the experiment designed by students.

(2) Apparatuses: UV-Vis spectroscopy (PerkinElmer Lambda 25), TEM (FEI Tecnai F30).

(3) Tools and others: prepare according to the experiment designed by students.

3.2 Procedure

3.2.1 General procedure

(1) Conduct a literature survey before the experiment.

(2) Design the preparation process of the colloidal suspension of the Au@Ag core-shell nanoparticles.

(3) Evaluate the agglomeration by UV-Vis spectroscopy.

(4) Observe the nanoparticles by TEM.

3.2.2 Example: Citrate synthesis of Au@Ag core-shell nanoparticles and characterization

(1) Synthesize Au nanoparticles. Dissolve 1.25 mL of 10 mmol/L $HAuCl_4 \cdot 3H_2O$ in 46.25 mL ultrapure water in a three-necked round-bottomed flask. Heat the solution to boil in the oil bath under stirring for 45 min. After that, dropwise add 2.5 mL 1 wt% $Na_3C_6H_5O_7 \cdot 2H_2O$ and hold the temperature for 10 min under stirring. The size of 10–12 nm Au particles can be obtained.

(2) Synthesize Au@Ag core-shell composite nanoparticles. Mix 0.65 mL of

0.1wt% PVP, 1.5 mL 0.05 mmol/L β-cyclodextrin, 1.5 mL 0.05 mmol/L cysteine, and 7 mL Au particles colloid. Stir the mixture for 10 min. The pH is adjusted to 10 by adding NaOH solution. Heat the solution to 85 °C in the oil bath. Dropwise add 3 mL 0.1 mmol/L $AgNO_3$ to the solution under stirring.

(3) Measure the UV-Vis spectrum of the product and compare it with those of the Au and Ag nanoparticles obtained in previous experiments.

(4) Ultrasonically disperse the product (about 1 mL) in ethanol (about 5 mL) for 10–20 min. Deposit the suspension onto the film of a support grid. Observe the morphology of the Au@Ag nanoparticles by TEM.

4 Requirements for experimental report

(1) Describe the synthesis procedure designed for Au@Ag core-shell nanoparticles in detail.

(2) Provide the characterization results of the UV-Vis spectroscopy and TEM.

5 Questions and further thoughts

(1) How can we confirm that the core-shell structure has been successfully prepared from the TEM image?

(2) Describe the applications of Au@Ag core-shell nanoparticles with one or two examples.

Experiment 43 Synthesis of Cadmium Sulfide Nanoparticles by Microemulsion Method

Type of the experiment: Cognitive
Recommended credit hours: 4
Brief introduction: In this experiment, the cadmium sulfide (CdS) nanoparticles are synthesized by the water-in-oil microemulsion method with cetyltrimethylammonium bromide (CTAB) as the surfactant. The method for determining the nanoparticle size based on the quantum-size effect is introduced. Then, the bandgap energy and the size of CdS nanoparticles are measured by ultraviolet-visible(UV-Vis) spectroscopy.

1 Objectives

(1) To synthesize CdS nanoparticles by the microemulsion method.

(2) To understand the principle of determination of nanoparticle size based on the quantum-size effect.

(3) To measure the bandgap energy and the size of the as-synthesized CdS nanoparticles by UV-Vis spectroscopy.

2 Principles

2.1 CdS semiconductor nanoparticles

A semiconductor is a material whose electrical conductivity us between a conductor and an insulator. Semiconductor nanomaterials exhibit extraordinary fluorescence properties and high quantum efficiency among the semiconductor materials because of the quantum-size effect. Some well-known semiconductor nanomaterials such as ZnS, CdS, Si, Si-Ge, GaAs, InP, GaN, SiC, and ZnSe have exhibited excellent applications in computers, palm pilots, laptops, cell phones, CD players, TV remotes, and solar cells.

CdS is a typical semiconductor nanomaterial. It possesses a room-temperature direct bandgap of 2.42 eV so that it is extremely photosensitive throughout the entire spectrum from infra-red down to ultraviolet. This makes it a potential and attractive semiconductor in the field of optoelectronics.

2.2 Synthesis of CdS nanoparticles by the microemulsion method

There are many methods to prepare the CdS nanoparticles, such as chemical precipitation, sol-gel process, microwave heating, microemulsion, and solid-state reaction. In this experiment, the CdS nanoparticles will be synthesized by the microemulsion method, which has been used to synthesize Fe-Au nanoparticles in Experiment 29.

The microemulsion is a thermodynamically stable dispersion of two immiscible liquids, such as a nonpolar liquid and a polar liquid stabilized by a surfactant. The dispersed domain diameter varies approximately from 1 to 100 nm, usually 10 to 50 nm. If the dispersion phase is oil and the dispersion medium is water, the system is called oil-in-water (O/W) microemulsion, and the reverse is called water-in-oil (W/O) microemulsion.

The O/W system is the normal micelle microemulsion, which is accepted in this experiment. CTAB dissolved in heptane will be used as the surfactant. CTAB has a long hydrophobic chain and a polar head group, as shown in Fig. 9.2. The molecule does not dissolve well in either aqueous or organic solvents. In an organic solvent containing a small amount of water, the CTAB traps the aqueous portion in a micelle sphere with the polar heads facing in and the non-polar tails facing out. In addition, pentanol will be added as the additional surfactant. The relative amount of pentanol cosurfactant controls the size of the micelle. The pentanol also acts as a capping agent to stabilize the CdS nanoparticles.

Mixing $CdCl_2$ and Na_2S with CTAB and pentanol micelles can produce CdS nanoparticles. The formation of CdS nanoparticles can be detected by UV-Vis spectroscopy since the quantum size effect make the visible absorption spectra different from that of bulk CdS.

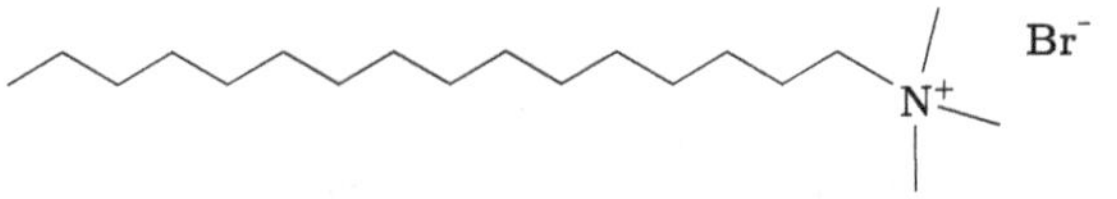

Fig. 9.2 The molecular structure of cetyltrimethylammonium bromide.

2.3 Determination of nanoparticle size based on the quantum-size effect

Nanometer-sized CdS particles have a wider bandgap than the bulk material due to the quantum confinement of the electron-hole pair that forms upon absorption of a sufficiently energetic photon. The larger energy difference causes a shift in the visible absorbance spectrum of CdS and a corresponding color change from orange to yellow. Therefore, the nanoparticle size can be calculated by the effective mass model that relates the change in bandgap energy to the radius of the particles. The bandgap energy of nanoparticles is expressed as

$$E_g^{nano} = E_g^{bulk} + \frac{h^2}{8m_0 r^2}\left(\frac{1}{m_e^*} + \frac{1}{m_h^*}\right) - \frac{1.8r^2}{4\pi\varepsilon\varepsilon_0 r}, \qquad (9.2)$$

where E_g^{nano} is the bandgap energy of the nanoparticle, E_g^{bulk} is the bandgap energy of bulk CdS at room temperature, h is Planck's constant, r is the particle radius, m_e^* is the effective mass of conduction-band electron in CdS, m_h^* is the effective mass of valence-band hole in CdS, e is the elementary charge, ε is the relative permittivity of CdS, and ε_0 is the permittivity of vacuum.

The E_g^{nano} and E_g^{bulk} can be obtained from the UV-Vis absorbance spectra as introduced in Experiment 30. Then, the particle radius r can be calculated by Equation (9.2).

3 Experimental

3.1 Preparation

(1) Chemicals: CTAB, heptane, pentanol, cadmium chloride ($CdCl_2$), sodium sulfide (Na_2S)

(2) Apparatuses: UV-Vis spectrometer (PerkinElmer, Lambda 25), electromagnetic stirrer, ultrasonic cleaner, electronic balance.

(3) Tools: 25 mL Erlenmeyer flasks, pipettes, quartz cuvettes.

3.2 Procedure

(1) Add 0.30 g CTAB, 1.5 mL pentanol, 6 mL heptane in two Erlenmeyer flasks, respectively. Stir the two mixtures to give even solutions.

(2) Prepare 0.6 mL 0.012 mol/L $CdCl_2$ solution in one flask, and 0.6 mL 0.012 mol/L Na_2S solution in the other flask. Stir the mixtures until the solutions are transparent.

(3) Transfer the two solutions into one Erlenmeyer flask. Ultrasonically disperse the solution for about 10 min until the light yellow suspension is formed.

(4) Add 15 drops of the suspension (about 1 mL) into a glass cuvette and dilute with 3–4 mL solution of heptane and pentanol (4:1). Then, measure the UV-Vis spectrum.

4 Requirements for experimental report

(1) Plot the UV-Vis spectrum of absorbance versus wavelength using the exported data.

(2) Calculate the bandgap energy of bulk CdS by $E = hc/\lambda$. The wavelength λ of the absorption peak is 512 nm.

(3) Calculate the bandgap energy of nanoparticles E_g^{nano} as introduced in Experiment 30.

(4) Calculate the particle radius r. The constants for the calculation are: $m_e^* = 0.19$, $m_h^* = 0.80$, $\varepsilon = 5.7$, $e = 1.602 \times 10^{-19}$ C, $\varepsilon_0 = 8.854 \times 10^{-12}$ F/m, $m_0 = $

9.110×10^{-31} kg, $h = 6.626 \times 10^{-34}$ J·s, and $c = 3.0 \times 10^8$ m/s.

5 Questions and further thoughts

(1) Is the bandgap energy of the CdS nanoparticles larger or smaller than that of bulk CdS? Why?

(2) Do you know other methods to determine the nanoparticle size? Please describe two or three methods.

Experiment 44 Preparation of Nickel Nanowires by Template Electrodeposition

Type of the experiment: Comprehensive
Recommended credit hours: 12
Brief introduction: Nickel nanowire is one of the typical metal nanowires having different properties from bulk metal. In this experiment, Ni nanowires are prepared by the template electrodeposition method in the anodic alumina membrane, which is easy to achieve and control the morphology of the products. Then, the growth morphology and the crystal structure are characterized by scanning electron microscope (SEM) and X-ray diffraction (XRD), respectively.

1 Objectives

(1) To understand the principle and process of the template electrodeposition method.

(2) To prepare Ni nanowires by the template electrodeposition method.

(3) To characterize the Ni nanowires using SEM and XRD.

2 Principles

2.1 Nickel nanowires

Nanowires are defined as solid materials in the form of wires with an average diameter of 10–100 nm and a length of up to 200 μm. The ratio of the length to the diameter is generally greater than 1000. At such a scale, quantum mechanical effects are important, which have attracted much investigation for potential applications in many fields, including optics, electronics, and genetics.

Nanowires can be classified by their constituent material as metal nanowires, semiconductor nanowires, oxide nanowires, multi-segment nanowires. Among these nanowires, Ninanowire is a prominent example for having different properties from the bulk metal. Its one-dimensional geometry and large shape anisotropy can enhance the coercivity and prevent it from becoming super paramagnetic. Based on the unique properties of Ni nanowires, numerous promising applications have been achieved, such as optical sensing, plasmonic resonance sensors, perpendicular magnetic storage, functional devices, negative permeability, biological separation, wire-grid type micro polarizer, trenches, and capacitors.

2.2 Template electrodeposition method

It is important to develop a facile method to synthesize high-purity Ni nanowires. Different techniques generating Ni nanowires include, but are not limited to, chemical vapor deposition (CVD), electrochemical deposition, electros pinning, bacterial system approach, microwave-assisted process, hydrothermal synthesis, and template synthesis method.

In this experiment, the Ni nanowires are prepared by the template electrodeposition method in the anodic alumina membrane. Template synthesis is a promising approach since it can control the shape and size of nanowires by altering pores in the template. It has been used to prepare nickel nanowires of regular and controllable rod sizes. In addition, the electrodeposition process is usually combined with the template method. The electrodeposition process involves electrolysis, in which there are a conductive cathode and an anode electrode in the solution. During electrolysis, the positively charged cations move towards the cathode, and the negatively charged anions move towards the anode. A reduction reaction and an oxidation reaction respectively take place at the cathode and anode, resulting in the formation of a metal layer deposited on the cathode.

3 Experimental

3.1 Preparation

(1) Materials: anodic alumina filter discs with 0.02 μm pores and a diameter of 2.5 cm (Whatman Co.), Ni sheet, Ag metal for evaporation coating.

(2) Chemicals: nitric acid, 6 mol/L aqueous solution of NaOH, nickel plating solution (300 g/L $NiSO_4 \cdot 6H_2O$, 45 g/L H_3BO_3, and 45 g/L $NiCl_2 \cdot 6H_2O$).

(3) Apparatuses: electrodeposition apparatus (consists of a Teflon supporting plate, a copper plate, a rubber O-ring, a Teflon tube, a Teflon cover, a DC power supply), evaporation coating machine (Shenyang Kejing Auto-Instrument Co., Ltd., GSL-1800X-ZF2), SEM (TECAN Vega3 LM), X-ray diffractometer (Rigaku, Miniflex 600).

(4) Tools: magnet, 50 mL beaker, alligator-clip wires, cotton swabs.

3.2 Procedures

3.2.1 Preparation of Ni nanowires

(1) Coat Ag conducting layer on one side of two alumina filter discs using evaporation coating machine. The thickness of the layer is about 100–200 nm.

(2) Assemble the electrodeposition apparatus as shown in Fig. 9.3. The order from bottom to top is the Teflon supporting plate, copper plate, alumina filter disc, rubber O-ring, Teflon tube, and Teflon cover. The Ag conducting layer on the filter disc faces the copper plate. Firmly clamp these components.

(3) Fill in the Teflon tube with the Ni plating solution. Insert the Ni sheet in the solution.

(4) Connect the positive and negative ends of the DC power supply to the Ni sheet and Cu plate, respectively.

(5) Conduct the electrolysis process at a constant voltage of 2 V for 15–20 min.

(6) Take out the alumina filter disc. Rinse it with water.

(7) Repeat steps (2) to (6) to prepare another sample.

3.2.2 Characterization

(1) Use cotton swabs dipped in nitric acid to remove the Ag coating layer on the two alumina filter discs.

(2) Take one disc. Acquire the XRD profile. The 2θ angle is from $20°$ to $100°$.

(3) Place another disc in a 50 mL beaker. Add 20 mL 6 mol/L NaOH solution to react with the alumina for at least 10 min so that Ni nanowires can be separated from the template.

(4) Place a magnet on the bottom of the beaker. The Ni nanowires will be attracted to the magnet. Dump the NaOH solution.

(5) Add water to rinse the nanowires twice. Use the magnet to attract the nanowires and dump the water.

(6) Collect and dry the nanowires. Observe the morphology of the dispersed nanowires as well as the other disc by SEM.

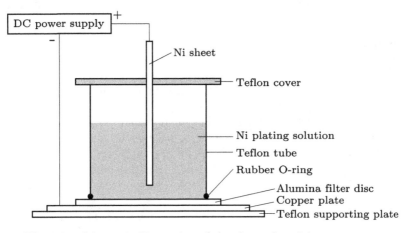

Fig. 9.3 Schematic illustration of the electrodeposition apparatus.

▌4 Requirements for experimental report

(1) Provide the SEM images of the dispersed Ni nanowires and the alumina filter disc. Estimate the average diameter and length of the nanowires.

(2) Plot the XRD profile and index the peaks. Determine the preferred orientation of the Ni nanowires.

▌5 Questions and further thoughts

(1) What can you see when the magnet is placed at the bottom of the beaker? Why?

(2) Please describe one or two other methods for preparing Ni nanowires.

Experiment 45 Synthesis and Characterization of Rod-like Silica Nanoparticles

Type of the experiment: Designing
Recommended credit hours: 8
Brief introduction: Rod-like silica nanoparticles will be synthesized in this experiment. The students will design the synthesis method and process by themselves. Then the morphology and structure of the as-synthesized nanoparticles will be characterized by scanning electron microscope (SEM), X-ray diffraction (XRD), and other apparatuses. If the reaction takes a long time, as the example introduced, the experiment can be arranged flexibly so that the reaction does not occupy the class time.

1 Objectives

(1) To understand the synthesis methods of rod-like silica nanoparticles.

(2) To synthesize the rod-like silica nanoparticles by designing the synthesis process.

2 Principles

2.1 Silica nanoparticles

Silica nanoparticles are commonly used as semiconductors in electronics because of their anisotropic structure, small size, high surface area, and quantum effect. There are many methods for the synthesis of silica nanoparticles, including both chemical and physical methods. Chemical methods mainly include hydrothermal synthesis, precipitation, template reaction technique, microemulsion method, spray pyrolysis technique, and electrochemistry method. Physical methods mainly include vacuum condensation, physical crushing, physical vapor deposition, ion sputtering, and freeze-drying.

Different morphologies of the silica nanoparticles can be produced by these synthesis techniques depending on the detailed synthesis route and condition. The shapes of the silica nanoparticles can appear as sphere, nanowire, rod-like, flower, and polyhedron.

2.2 Rod-like silica particles

Silica nanorods have attracted significant attention due to their small scale, high ratio of length to diameter, and readily functionalized surface. They are widely used in the field of biological detection, drug transportation, and disease treatment. Therefore, the synthesis and characterization of rod-like silica nanoparticles have been extensively investigated.

The rod-like silica particles can be synthesized by various methods, including template method, gas-liquid-solid growth method, hydrothermal synthesis, and sol-gel method. In this experiment, the students are required to synthesize rod-like silica nanoparticles. They need to select a synthesis method and design the detailed synthesis process based on a literature survey. Then, the morphology and crystal structure should be examined by SEM and XRD, respectively.

3 Experimental

3.1 Preparation

(1) Chemicals: prepare according to the experiment designed by students.

(2) Apparatuses: SEM (TECAN Vega3 LM), X-ray diffractometer (Rigaku, Miniflex 600).

(3) Tools and others: prepare according to the experiment designed by students.

3.2 Procedure

3.2.1 General procedure

(1) Conduct a literature survey before the experiment.
(2) Design the preparation process of the rod-like silica nanoparticles.
(3) Characterize the morphology and the crystal structure by SEM and XRD, respectively.

3.2.2 Example: Synthesis of rod-like silica nanoparticles by chemical reduction method

(1) Ultrasonically dissolve 30 g PVP in 300 mL 1-pentanol in a 500 mL flask for 2 h.

(2) Under electromagnetic stirring, add 30 mL absolute ethanol, 8.4 mL deionized water, and 2 mL 0.18 mol/L aqueous solution of sodium citrate dihydrate into the flask. Shake the flask by hand to mix the chemicals.

(3) Add 6.75 mL ammonia (25wt% aqueous solution) and 3 mL tetraethyl orthosilicate (TEOS) into the solution.

(4) After shaking the flask by hand, the flask is left to stand for about 12 h.

(5) The reaction mixture is centrifuged at 1500 g for 15 min. Dump the supernatant.

(6) Add 30 mL water or ethanal into the centrifuge tube. Disperse the powder ultrasonically. Centrifuge the suspension and dump the supernatant. Repeat this step 2 or 3 times.

(7) Add 30 mL ethanal into the centrifuge tube. Disperse the powder ultrasonically. Centrifuge the suspension at 700 g for 15 min and dump the supernatant. Repeat this step 2 or 3 times.

(8) Characterize the products by SEM and XRD, respectively.

4 Requirements for experimental report

(1) Describe the synthesis procedure designed for the rod-like silica particles in detail.

(2) Provide and analyze the characterization results of SEM and XRD.

(3) Describe the morphology and size of the as-synthesized particles.

5 Questions and further thoughts

(1) What are the advantages and disadvantages of your choice of synthesis

method?

(2) According to the characterization results, what improvements can you make in your experiment?

Experiment 46 Preparation of Surface Conductive Glass by Spray Pyrolysis Method

Type of the experiment: Comprehensive
Recommended credit hours: 8
Brief introduction: Surface conductive glass is optically transparent and electrically conductive in the thin coating layer on the glass. In this experiment, both the SnO_2 and Sb-doped SnO_2 conductive glass will be prepared using the spray pyrolysis method, which is a simple and commonly used method to form surface coatings. The sheet resistance and film thickness of the prepared coating layer will be measured using the four-point probe method and stylus profilometer, respectively. Then, the effects of the Sb dopant and film thickness on the conductivity will be examined.

■ Objectives

(1) To understand the principle and process of the spray pyrolysis technique.
(2) To prepare the SnO_2 and Sb-doped SnO_2 conductive glasses.
(3) To understand the effects of the Sb dopant and film thickness on the conductivity of surface conductive glass.

■ Principles

2.1 Surface conductive glass

It is well known that glass is an insulator. However, when the surface of a glass is coated with a conductive layer, it can have conductive properties. Hence, it is called surface conductive glass, which is optically transparent and electrically conductive in the thin layer on the glass. It is widely used in many applications, such as flat panel display, heterojunction solar cell, diathermic mirror, transparent electromagnetic shielding, and transparent conducting electrode because of its excellent transparent, conducting, reflective, absorptive, chemically stable, and anti-radiation properties.

The conductive layer can be made up of both inorganic and organic materials. Inorganic films typically are made up of transparent conducting oxides, such as In_2O_3, SnO_2, and ZnO. The films are doped, such as indium tin oxide (ITO), fluorine-doped tin oxide (FTO), and doped zinc oxide, to further improve the conductivity or stability. In addition, the film thickness, grain size, and crystallinity also affect the conductivity.

In this experiment, both the SnO_2 and Sb-doped SnO_2 conductive glass will be prepared. In most compounds, Sn can exist in either a divalent or tetravalent oxidation state. Thus the SnO_2 oxide layer is non-stoichiometric that consists of a mixture of SnO and SnO_2, and the structure can be considered as that for SnO_2

with a certain proportion of O^{2-} missing. The requirement of electro neutrality results in the formation of additional electrons, which contribute to the conduction of the non-stoichiometric oxide film. In the case of Sb-doped SnO_2, the dopant Sb has both trivalent and pentavalent oxidation states. The trivalent Sb ions, i.e., Sb^{3+}, also create oxygen deficiency in the same sense as partial reduction of SnO_2. Therefore, a small amount of Sb^{3+} increases the conduction.

2.2 Spray pyrolysis method

Various thin film deposition techniques can be used to apply a thin layer of semi-conducting materials on the glass surface. The spray pyrolysis technique is a common and simple process, which can be used to deposit Cu_2O, SnO_2, In_2O_3, CuO, CdS, and other materials on various substrates. This technique requires a simple instrument and can form a large size film. Besides, the reactants can control the structure and morphology of the film excellently.

In the spray pyrolysis process, the ingredients are first mixed with water, ethanol, or other solvents into solutions. Then, the reaction solution passes through the spraying device, making the solution evaporate quickly. The reactants undergo thermal decomposition or other simultaneous reactions and finally form nanoparticles on the substrate.

In the case of SnO_2 and Sb-doped SnO_2 as the conductive layer, the principal reaction involves the reaction of $SnCl_4$ with H_2O, i.e.

$$SnCl_4 + 2H_2O \xrightarrow{\triangle} SnO_2 + 4HCl \uparrow . \tag{9.3}$$

The non-stoichiometric oxide is formed by hydrolysis of a tin salt on the heated glass, while the organic constituent of the spray furnishes a reducing condition. If anhydrous tin salts are used, small amounts of water are added to enable the hydrolysis reaction. Thus, $SnCl_4 \cdot 5H_2O$ dissolved in methanol solution is commonly accepted.

2.3 Measurement of film thickness by the stylus profilometer

The stylus profilometer or step profiler is an instrument for measuring surface morphology. It delivers step height and surface roughness with high accuracy. It can contribute to optimizing production and machining processes. For example, it can be used to determine the type of machining used to manufacture a surface based on the detected roughness pattern. It can also be used tomeasure the thickness of a film or coating material.

As shown in Fig. 9.4, it can detect the surface morphology with a probe when there are tiny peaks and valleys on the surface.When the probe moves along the surface lightly, it moves up and down along the peaks and valleys. Consequently, the motion of the probe reflects the surface profile.

The stylus profilometer can acquire the high-precision morphology. However, this technique can be destructive to some surfaces because it involves physical movements in the X, Y, and Z axes while maintaining contact with the surface.

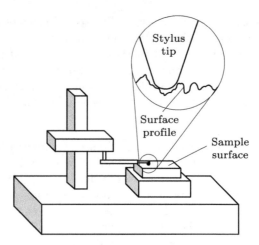

Fig. 9.4 Schematic illustration of the stylus profilometer.

3 Experimental

3.1 Preparation

(1) Chemicals: tin chloride pentahydrate ($SnCl_4 \cdot 5H_2O$), antimony oxide (Sb_2O_3), hydrochloric acid (HCl), methanol, ethanol.

(2) Substrates: glass substrates with a size of 20 mm × 20 mm.

(3) Apparatuses: spray gun and mini air compressor, muffle furnace, stylus profilometer (KLA-Tencor, Alpha-Step D-300), four-point probe system (Probes Tech, RTS-8), ultrasonic cleaner, electronic balance.

(4) Tools: graduated cylinder, beakers, ceramic tiles.

3.2 Procedure

(1) Dissolve 10 g $SnCl_4 \cdot 5H_2O$ in 10 mL methanol.

(2) Dissolve 0.20 g Sb_2O_3 in 2 mL concentrated HCl. Add 5 mL $SnCl_4 \cdot 5H_2O$ solution to obtain the Sb-doped $SnCl_4 \cdot 5H_2O$ solution.

(3) Ultrasonically clean 6 pieces of glass substrates in ethanol for 10 min, and then dry them.

(4) Heat the muffle furnace to 600 °C.

(5) Put the glass substrates on ceramic tiles and place them in the furnace for 10 min.

(6) Take out the glass substrates on the ceramic tiles from the furnace. Immediately spray the $SnCl_4 \cdot 5H_2O$ solutionor Sb-doped $SnCl_4 \cdot 5H_2O$ solution with the spray gun. Place the samples back to the furnace and reheat for 2 min.

(7) Repeat step (6) for 2, 4, and 6 times using the $SnCl_4 \cdot 5H_2O$ solutionor Sb-doped $SnCl_4 \cdot 5H_2O$ solution on different substrates. The SnO_2 and Sb-doped SnO_2 conductive layers with different thicknesses are then prepared.

(8) Measure the thicknesses of the conductive layers by the stylus profilometer.

(9) Measure the sheet resistances of the samples by the four-point probe system. The measurement procedure has been described in Experiment 27.

4 Requirements for experimental report

(1) Tabulate the layer thicknesses and the sheet resistances of SnO_2 and Sb-doped SnO_2 conductive layers.

(2) Discuss the effects of Sb dopant and film thickness on the conductivity.

5 Questions and further thoughts

(1) Please suggest several other techniques for preparing the surface conductive glass.

(2) The transparency of the conductive glass prepared in this experiment is generally not good. Why?

Experiment 47 Preparation and Characterization of Porous Thin Films

Type of the experiment: Designing
Recommended credit hours: 24
Brief introduction: Porous materials widely exist around us and play important roles in many aspects of our daily life. They can be used in energy conversion and storage, vibration suppression, photocatalysis, heat insulation, sound absorption, and fluid filtration. In this experiment, the students are required to select a porous thin film and design the preparation process based on a literature survey. The objective material can be an alloy or an alloy oxide and consists of at least two kinds of metal elements. The crystal structure and morphology of the as-prepared film will be confirmed by X-ray diffraction (XRD) and scanning electron microscope (SEM), respectively. The physical and chemical properties will be characterized according to the application of the film.

1 Objectives

(1) To understand the effects of porous structure, such as the size and shape of pores, on the properties of porous thin films.

(2) To design and prepare a kind of porous thin film based on a literature survey.

(3) To characterize the microstructure and related physical or chemical properties of the as-prepared film.

2 Principles

2.1 Porous thin films

Porous materials are materials containing pores. Due to the pores, which can be different in features and structures, they show peculiar characteristics which are much different from the conventional non-porous materials. The properties of porous materials depend largely on the pore morphology, size, and distribution. In addition, porosity is the essential parameter due to its significant influence on the mechanical, physical, and chemical properties of the materials.

Porous thin films are a kind of porous material whose thickness is much smaller than the lengths in the other two directions. They can be used for many applications, including protective coatings, optical coatings, membranes, gas sensors, electrode materials, and catalyst supports. These applications are also based on the pore structures, including total pore volume, porosity, mean pore size, pore shape, pore connectivity, and pore size distribution.

There are two reasons to make porous materials in the form of thin films. One is that thin films have a high ratio of surface area to volume, which results in an increased ability to interact with mediums. The other one is that thin films can be coated on existing devices, increasing added value.

2.2 Preparation and characterization of porous films

Various methods have been used to synthesize porous thin films, including the sol-gel method, electrochemical method, templated method, hydrothermal method, thermal evaporation, and the evaporation of source materials by the irradiation of energetic species or photons.

Since the properties of porous materials are significantly affected by their chemical and microstructure features, the film thickness, porosity, pore size, and pore shape are generally characterized after the preparation of porous films. Besides, the physical and chemical properties are also determined according to certain applications.

In this experiment, the students are required to prepare a kind of porous film based on a literature survey. They need to determine what kind of porous thin film will be synthesized, including the components, the structure features (e.g., pore size and porosity), and the potential applications. The detailed preparation method and the experimental scheme will be designed. Then, the microstructure and related properties will be characterized.

3 Experimental

3.1 Preparation

(1) Materials and chemicals: prepare according to the experiment designed by students.

(2) Apparatuses: SEM (TECAN Vega3 LM), specific surface area analyzer (Beishide 3H-2000PS1), X-ray diffractometer (Rigaku, Miniflex 600).

(3) Tools and others: prepare according to the experiment designed by students.

3.2 Procedure

3.2.1 General procedure

(1) Conduct a literature survey before the experiment.

(2) Determine the porous thin film to be synthesized. Design the preparation process. The method, which is simple, inexpensive, not time-consuming, and does not need expensive apparatuses, is recommended.

(3) Prepare the samples.

(4) Characterize the composition, structure, morphology, and related physical

and chemical properties.

3.2.2   Example: preparation of porous TiO_2/ZnO composite film using a simple hybrid sol-gel-powder method

(1) Dissolve 0.0045 mol zinc acetate dihydrate ($Zn(CH_3COO)_2 \cdot 2H_2O$, 99.5%) in 30 mL solution containing ethanol and monoethanolamine (MEA, C_2H_7NO, 99.0%) at room temperature. The molar ratio of MEA to zinc acetate is kept at 1.0.

(2) Stir the mixture at 60 °C for 3 h until it becomes a clear and homogeneous solution of ZnO sol.

(3) After aging for 1 d, ultrasonically disperse 0.04 mol TiO_2 powder into the ZnO sol to get the uniform TiO_2/ZnO composite sol.

(4) Mechanically polish 304 stainless steel plates to a mirror finish. Ultrasonically clean the plates in acetone, ethanol, and deionized water, respectively.

(5) Deposit the sol on the substrates by spin coating at 3000 r/min for 30 s.

(6) Dry the samples in the air at 80 °C for 30 min, and then calcine at 500 °C for 2 h.

(7) The structure and morphology of the samples are respectively characterized using XRD and SEM.

(8) The photocatalytic properties are examined by UV-Vis spectroscopy.

4 Requirements for experimental report

(1) Describe the reasons for selecting the porous thin film in the designed experiment.

(2) Describe the preparation process in detail.

(3) Analyze the experimental results and compare them with those reported in the literature.

5 Questions and further thoughts

(1) What are the advantages of porous materials comparing with general materials?

(2) Discuss the effects of porous structure on the properties of the prepared material.

6 Cautions

(1) The designed experiments should not involve harsh conditions, e.g., high temperature, high pressure.

(2) Avoid using dangerous chemicals, e.g., highly toxic, explosive, or flammable chemicals.

Extended readings

Bettotti P. Submicron Porous Materials. Springer, 2017.

Bradbury S. Transmission electron microscope. Available at: https://www.britannica.com/technology/transmission-electron-microscope. Accessed: 2 March 2021.

Cao G, Wang Y. Nanostructures and Nanomaterials: Synthesis, Properties, and Applications. 2nd Edition. World Scientific Publishing Company, 2010.

Haldar K K, Kundu S, Patra A. Core-size-dependent catalytic properties of bimetallic Au/Ag core-shell nanoparticles. ACS Applied Materials & Interfaces, 2014, 6: 21946-21953.

Kuijk A, Van Blaaderen A, Imhof A. Synthesis of monodisperse, rodlike silica colloids with tunable aspect ratio. Journal of the American Chemical Society, 2011, 133(8): 2346-2349.

Marcus L K, Lu M H, Zhang Y. Preparation of porous materials with ordered hole structure. Advances in Colloid and Interface Science, 2006, 121: 9-23.

McFarland A D, Haynes C L, Mirkin C A, et al. Color my nanoworld. Journal of Chemical Education, 2004, 4(81): 544A-544B.

Robbie K, Friedrich L J, Dew S K. Fabrication of thin films with highly porous microstructures. Journal of Vacuum & Technology A, 1995, 13(3): 1032-1035.

Solomon S D, Bahadory M, Jeyarajasingam A V, et al. Synthesis and study of silver nanoparticles. Journal of Chemical Education, 2007, 2(84): 322-325.

Tanaka J, Suib S L. Surface conductive glass. Journal of Chemical Education, 1984, 61(12): 1104-1106.

Winkelmann K, Noviello T, Brooks S. Preparation of CdS nanoparticles by first-year undergraduates. Journal of Chemical Education, 2007, 4(84): 709-710.

Xu H, Liu W, Cao L, et al. Preparation of porous TiO_2/ZnO composite film and its photocathodic protection properties for 304 stainless steel. Applied Surface Science, 2014, 301: 508-514.

Part V
Material Applications

Chapter 10 Energy Materials

Experiment 48 Fabrication and Characterization of Lithium-Ion Batteries

Type of the experiment: Comprehensive and designing
Recommended credit hours: 20
Brief introduction: Lithium-ion batteries are fabricated in this experiment. The operating principle, the characterization methods, and the influencing factors on the performance are introduced. The experiment is divided into two experiments that can be conducted respectively. The first experiment introduces the operating principle of Li-ion batteries and synthesis methods of cathode material $LiCoO_2$. Students are required to design the detailed synthesis process by themselves based on a literature survey. The phase structure of the synthesized cathode material should be confirmed using X-ray diffraction. The second experiment introduces the preparation method of electrodes and the assembly of coin-cell type Li-ion batteries. The fabricated batteries will be characterized by several widely used techniques. Besides, students are required to design an application using the batteries assembled in this experiment.

Several steps in the experiment, such as the drying process, the performance test, take a long time. Therefore, the experiment can be arranged flexibly so that these steps do not occupy the class time.

Experiment 48.1 Synthesis and Characterization of Cathode Material $LiCoO_2$

Type of the experiment: Designing
Recommended credit hours: 8

1 Objectives

(1) To understand the operating principle of Li-ion batteries.
(2) To know the common cathode materials for Li-ion batteries.
(3) To synthesizethe cathode material $LiCoO_2$ by both the solid-state and the sol-gel processes.

2 Principles

2.1 Development and application of lithium-ion batteries

It is well known that the advanced electrochemical storage technique is playinga more paramount role today.Compared with traditional batteries, such as Pb-acid batteries, Ni-MH, and Ni-Cd batteries, Li-ion batteries have attracted much more concern by academia and industry because of low weight, high potential, huge theoretical capacity, long life span, andsmall self-discharge.

One of the important applications of Li-ion batteries is used for hybrid and pure electric vehicles.Comparing with other traditional batteries, the introduction of Li-ion batteries reduces the total weight by approximately 40%–50% and the volume by about 20%–30% and meanwhile double the specific capacity. However, its development is entering the bottleneck. How to effectively prolong the cruising power, facilitate the rate of charging and strengthen safety are becoming hot issues.

2.2 Operating principle of lithium-ion batteries

Understanding the basic working principle of Li-ion batterieswill contribute to the selection of the electrode materials and the improvement of the electrochemical performance, such as energy density and power density.

Generally, Li-ion batteries mainly compose of cathode, anode, electrolyte, and separator. The working principle is illustrated in Fig. 10.1. During the charging

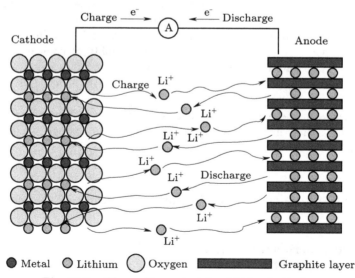

Fig. 10.1 Theworking principle of Li-ion batteries.

process, Li-ions are taken off the crystal lattice of the cathode, and correspondingly the cathode materials are oxidized. Li-ions are driven towards the anode through the separator in the electrolyte under the gradient of the electrical field. Then, they are inserted into the anode. The concentration difference of Li-ions between the anode and the cathode is formed in the opposite direction of the electrical field. The charge process will be terminated when the effects of concentration difference and the electrical field are counteracted. The discharge process is a contrary one. Li-ions are de intercalated from the anode and transferred through the separator to the cathode. They are intercalated into the cathode, and the cathode is reduced.

2.3 Electrode materials of lithium-ion batteries

Based on the working principle of Li-ion batteries, a promising electrode material should possess enough channels within its lattice structure to allow for the easy transportation of Li ions.

Graphite is the most popular anode material to be employed due to the attractive characteristics of low potential, cheap cost, excellent electrical conductivity, and high specific capacity (a theoretical capacity of about 372 mAh/g). The layered graphite exhibits a larger interlamellar spacing and a weaker binding force than the in-plane bonding force. Li-ions can be intercalated in the interlayer and formed LiC_n ($n \leqslant 6$) during the charging process under the potential window of 0.005 to 0.25 V (vs. Li^+/Li). Similarly, Li-ions can be de intercalated under the same voltage range.

Compared with the anode materials, the cathode materials with outstanding redox activity and electronic conductivity play a more important role in promoting the electrochemical performance of Li-ion batteries. Usually, the cathode materials are lithiated transition metal oxides with unique channels. Those with a layered structure, including $LiCoO_2$ (LCO), spinel Li_2MnO_4 (LMO), olivine $LiFePO_4$ (LFP), and ternary materials (NMC, NCA), have attracted more attention.

2.4 Synthesis of cathode material $LiCoO_2$

LCO is chosen as the cathode material for further preparation of Li-ion batteries in this experiment. The solid-state reaction route is the most widely used method to synthesize polycrystalline LCO powder from a mixture of solid raw materials. To successfully produce high-performance LCO material with desired properties, it is necessary to carefully control the reaction conditions, such as temperature, reaction duration, atmosphere, particle size, and mixing method of the raw materials.

Another widely accepted synthesis method is the sol-gel process, which is a promising method to prepare nano-structured chemical products, nano-sized membranes/films, and fibers. This process involves the formation of a colloidal solution (sol) and a gel-like diphasic system via hydrolysis or alcoholysis and polycondensation reactions. It is a cheap and low-temperature technique that allows for readily controlling the chemical composition, microstructure, particle size, and introducing dopants.

In this experiment, students will prepare LCO by both solid-state and sol-gel methods. They are required to design the detailed synthesis processes and conditions by themselves based on a literature survey. The structure of the as-synthesized products will be confirmed by X-ray diffraction (XRD). The products will be used to fabricate Li-ion batteries in the following experiment.

3 Experimental

3.1 Preparation

(1) Chemicals: lithium carbonate (Li_2CO_3), cobalt oxalate ($CoC_2O_4 \cdot 2H_2O$), lithium acetate ($CH_3COOLi \cdot 2H_2O$), cobalt acetate ($CoC_2H_6O_4Co \cdot 4H_2O$), citric acid ($C_6H_8O_7$), ethanol.

(2) Apparatuses: X-ray diffractometer (Rigaku, Miniflex 600), tube furnace, muffle furnace, differential scanning calorimeter (Nanjing Dazhan Institute of Electromechanical Technology, DSC-100), thermal gravimetric analyzer (Nanjing Dazhan Institute of Electromechanical Technology, TGA-105), drying oven, electronic balance, electromagnetic stirrer.

(3) Tools: beakers, crucibles, constant pressure dropping funnel, stirring bar, mortar and pestle.

(4) Others: prepare according to the experiment designed by students.

3.2 Procedure

3.2.1 General procedure

(1) Conduct a literature survey before the experiment.

(2) Design the solid-state and sol-gel synthesis processes of the cathode material LCO.

(3) Prepare the LCO powder by the two methods.

(4) Sift the product with a 300-mesh sieve.

(5) Confirm the structure of the as-synthesized powders by XRD.

3.2.2 Example 1: solid-state synthesis of LCO powder

(1) Weigh raw materials Li_2CO_3 and $CoC_2O_4 \cdot 2H_2O$ according to the stoichiometric ratio (0.55:1–0.5:1). The weight of the materials should make more than 2g of the final product.

(2) Thoroughly mix the raw materialswith a mortar and pestle. Ethanol is used as the dispersant.

(3) Dry the mixture in the drying oven.

(4) Determine the reaction temperature by TGA or DSC.

(5) Calcine the mixture in the furnace. Increase the furnace temperature to 300–600 °C and hold for 240–600 min. Then further increase the temperature to 600–950 °C and hold for 600–1200 min.

3.2.2 Example 2: sol-gel synthesis of LCO powder

(1) Weigh reagents $CH_3COOLi \cdot 2H_2O$ and $C_2H_6O_4Co \cdot 4H_2O$ according to the chemical stoichiometric ratio (1.05:1–1:1). The weight of the reagents should make more than 2g of the final product.

(2) Dissolve the reagents in ethanol or deionized water. The total reagent concentration is controlled to be about 1 mol/L. Drop citric acid into the solution

under stirring to form a sol.

(3) Let the sol stand for about 4 h to accomplish the gel process.

(4) Dry the gel in the drying oven.

(5) Determine the reaction temperature by TGA or DSC.

(6) Calcine the dried gel in the furnace. Increase the furnace temperature to 300–400 °C and hold for 60–120 min. Then further increase the temperature to 400-800 °C and hold for 240–480 min.

4 Requirements for experimental report

(1) Describe the solid-state and the sol-gel synthesis processes in detail.

(2) Calculate the yield of the LCO powder.

(3) Plot the XRD profiles of the synthesized powder and index the peaks. If there are extra peaks, analyze the possible reasons.

5 Questions and further thoughts

(1) What will happen if the raw materials are not mixed homogeneously in the solid-state reaction route?

(2) The amount of the ethanol dispersant has to be controlled as small as possible. Why?

(3) Generally, 10mol% excessive Li_2CO_3 is used in preparing the raw materials. Why?

(4) How does the calcination temperature affect the phase and composition of the product?

Experiment 48.2 Fabrication and Performance Test of Lithium-Ion Batteries

Type of the experiment: Comprehensive
Recommended credit hours: 12

1 Objectives

(1) To understand the basic structure of Li-ion batteries.

(2) To learn the assembly of coin-cell type Li-ion batteries.

(3) To understand the methods for the performance test of Li-ion batteries.

(4) To evaluate the performance of the as-prepared Li-ion batteries.

(5) To understand the influencing factors on the performance.

2 Principles

2.1 Structure of lithium-ion batteries

Generally, lithium-ion batteries consist of a cathode, an anode, an electrolyte, a separator, and a cell can. The coin-cell type batteries are assembled in this experiment, and the main components are illustrated in Fig. 10.2.

Normally, the cathode material is coated on aluminum film, which is used as the current collector for the cathode. The anode material is coated on copper film,

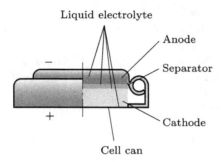

Fig. 10.2 The structure of a typical Li-ion coin-cell type battery. (See color illustration at the back of the book)

which is used as the current collector for the anode. The organic solvent solution of lithium salt is used as the electrolyte, which helps Li-ions transfer inside the battery. The PP microporous membrane is used as the separator, which can be passed through by Li-ions during the charging and discharging process. The stainless-steel cell can is used in coin-cell battery packaging.

2.2 Characterization of electrochemical properties

2.2.1 Cyclic voltammetry

Cyclic voltammetry (CV) is a type of potentiodynamic electrochemical measurement. In Experiment 41.2, it has been used to determine the frontier orbital energies. This technique can also be used to investigate the possible electrochemical reactions that occur in the electrode system, deciding the reversibility of the electrode process, determining the origins of the products, and investigating the adsorption/desorption process of the active electrochemical reactant and the electrochemical behavior of the electrode. In the case of Li-ionbatteries, the electrochemical behavior of the charge-discharge process can be investigated.

As shown in Fig. 10.3, the working electrode potential is scanned linearly versus timein the CV experiment.The potential is scanned to a set potential under a specific scan rate, and then the working potential is inverted at the same scan speed. At the same time, the response current is recorded. The current at the working electrode is plotted versus the applied voltage to give the CV curve.

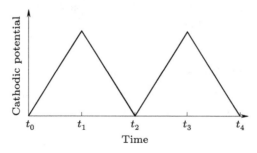

Fig. 10.3 CV potential waveform on the working electrode.

Important information about the kinetics of electrode reactions and transfer process could be provided by the CV curve. A typical CV curve of LiCoO$_2$(LCO) electrode is shown in Fig. 10.4. The redox couples are the potentials that the electrode reactions take place, and the peak currents present the intensity of the reaction. In this curve, there are three redox couples at 3.80V and 4.03V, 4.00V and 4.11V, 4.13V and 4.19V, respectively. The redox couple with the highest current peaks is related to the chemical valence change of Co^{3+}, which is the dominant reaction in batteries using LCOas the cathode material. The other two redox couples are related to the side reactions of the batteries, which may be the reactions from the electrolyte decomposition. The ratio of cathodic peak current i_{PC} to anodic peak current i_{PA} represents the reversibility of the electrode reaction, which is very important to the rechargeable batteries.

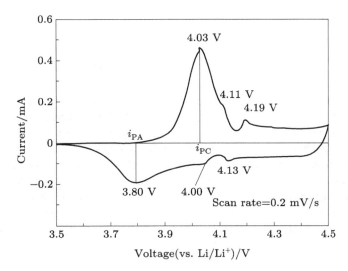

Fig. 10.4 TypicalCV curve of LCO. The counter electrode is Li metal.

2.2.2 Electrochemical impedance spectroscopy

Electrochemical impedance spectroscopy (EIS) is a technique to study the change of impedance versus the frequency under the specified current polarizing for the electrochemical system. It can be used to characterize and investigate the surface structure of solid, the metal corrosion system, metal electrodeposition, bio-system, and chemical power.

The Nyquist plot can be obtained from the EIS test, as shown in Fig. 10.5. Different parts of the plot show different kinds of the electrochemical impedance of the battery, which correspond to various electrochemical processes. The circular arcs at high and medium frequencies are mainly associated with the kinetics,reflecting the impedance of the charge transfer process. The straight sloping line at low frequencies is related to mass transfer,reflecting the impedance of the mass transfer process. Generally, the impedance can be fitted via an equivalentelectric circuit using software, such as Z-view. According to the various electrochemical system, the equivalentelectric circuit could contain resistance, capacitance, and other kinds of electronic components andprovide us a deep understanding of the electron transfer and mass transportation occurring in the battery system.

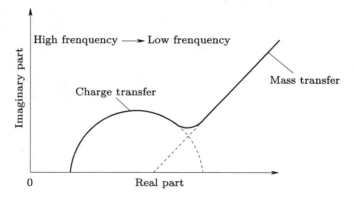

Fig. 10.5 Nyquistplot from the EIS test.

A typical simulated parallel circuit of Li-ion battery is shown in Fig. 10.6, anda set of parameters are determined. R_b represents the internal resistance of the cell, which reflects the overall ohmic resistance of the electrolyte and electrodes; R_{ct1} and C_{dl1} are the charge-transfer resistance and its related double-layer capacitance between the electrolyte and Li metal anode, which correspond to the semicircle at high frequencies; R_{ct2} and C_{dl2} are the charge-transfer resistance and its related double-layer capacitance between the electrolyte and cathode, which correspond to the semicircle at medium frequencies; W is the Warburg impedance related to the diffusion in the electrode, which is indicated by a straight slopping line at low frequencies.

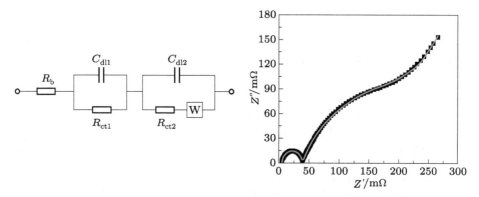

Fig. 10.6 An example of (a) simulated parallel circuit and (b) fitted Nyquist plot of a Li-ion battery. (See color illustration at the back of the book)

2.3 Performance of lithium-ion batteries

2.3.1 Theoretical capacity

The theoretical specific capacity of the electrode can be calculated by

$$C_0 = \frac{N_A e z}{t M_w}, \tag{10.1}$$

where N_A is the Avogadro constant (6.02×10^{23} mol^{-1}), e is the charge carried by electron unit (1.60×10^{-19} C), z is the electron number involved in the electrochemical reaction, M_w is the molar mass of electrode, t is time (3600 s for 1 h). Obviously, the result of $N_A e$ is the Faraday constant. For the LCO cathode, $z = 1$ and $M_w = 97.873$ g/mol. Since 1mAh $= 1 \times 10^{-3}$ A $\times$ 3600 s $= 3.6$ C, then $C_0 = 273$ mAh/g.

The max actual specific capacity is about half of the C_0. This is because while over 50% Li-ions de-intercalate for LCO, the structure of LCO may collapse. The collapse of the LCO structure will affect the reversibility of the electrode reaction leading to a decrease in the cycling stability.

2.3.2 Cycling stability

Cycling stability is an important parameter for rechargeable batteries, which stands for the service life of a rechargeable battery. Besides the collapse of the LCO structure, other factors, such as the electrolyte stability, charge and discharge rate, and temperatures, can also influence the cycling stability.

In the cycling stability test, a battery is charged and discharged under a specific current density, and the charging-discharging process is repeated dozens, hundreds, or even thousands of times. The capacity of the battery is recorded for each cycle. Figure 10.7 is a plot of discharge capacity versus cycle number showing the cycling stability.

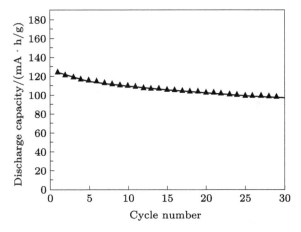

Fig. 10.7 Typical plot of discharge capacity versus cycle number for evaluating the cycling stability of a Li-ion battery.

2.3.3 Rate capability

The charging rate is also an important parameter for rechargeable batteries, which decides the charging time of a battery. It is known that the actual specific capacity of a battery under different charging current densities is different. Usually, when the charging current density rises, the actual specific capacity declines.

Figure 10.8 shows an example of discharge capacity influenced by the charging rate, i.e., by different rates of the current density C ($C = C_0(1h)$). Thus, this property is named rate capability. The decrease in the discharge capacity with the charging current density can be clearly seen in the figure.

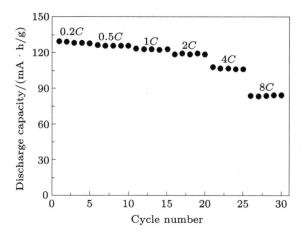

Fig. 10.8 Typical plot of discharge capacity versus cycle number for evaluating the rate capability of a Li-ion battery.

3 Experimental

3.1 Preparation

(1) Materials: Li metal, Al foil, microporous polypropylene membrane (Celgard), stainless steel plates, stainless-steel coin-cell can.

(2) Chemicals: LCO prepared in Experiment 48.1, carbon black, polyvinylidene fluoride (PVDF), n-methyl-2-pyrrolidone (NMP), $LiPF_6$ solution in ethylene carbonate (EC) and diethyl carbonate (DEC) (1 mol/L, EC:DEC=1:1).

(3) Apparatuses: electronic balance, vacuum drying oven, glove box, coating machine, sealing machine, electrochemical workstation (CH Instruments, CHI660E), battery testing system (Neware Technology, CT-3008-5V2mA-54).

(4) Tools: mortar and pestle, hole puncher.

3.2 Procedure

3.2.1 Preparation cathodelayer

(1) Weigh LCO powder prepared by solid-state or sol-gel methods, carbon black, and PVDF in a ratio of 8:1:1. The total weight is 0.5g. Mix them with a mortar and pestle.

(2) Add about 30 drops of NMP into the powder mixture to form a slurry with excellent mobility.

(3) Switchon the coating machine. Clean the working surface and scraper of the coating machine with ethanol.

(4) Place Al foil with a size of 10 cm × 15 cmon the working surface of the coating machine, and clean it with ethanol.

(5) Adjust the height of the scraper to 20 μm.

(6) Transfer the slurry on the Al foil. Settle the scraper, which sweeps over the Al foil to spread the slurry, resulting in the formation of the cathode layer on the Al foil.

(7) Dry the cathode in the vacuum drying oven at 120 °C for 12h.

(8) Punch the cathode into discs with a diameter of 16mm.

3.2.2 Assembly of coin-cell type Li-ion batteries

(1) Punch the separator (the microporous polypropylene membrane) with a diameter of 19mm.

(2) Transfer all the components, including the electrodes, electrolyte, separators, spacers, and cell cans,into the glove box.

(3) Assemble coin cells using the sealing machine in the glovebox. The assembly sequence of the components is shown in Fig. 10.9, and add 6–10drops of $LiPF_6$ solution between the separator and the electrodes.

(4) Rest the assembled coin cells for more than 12 h before the following tests.

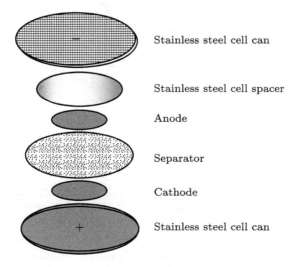

Fig. 10.9 Assembly sequence of coin-cell type Li-ion batteries.

3.2.3 CV and EIS test

(1) Switch on the electrochemical workstation and computer, and warmup for 10 min.

(2) Place acoin cell in the sample holder.

(3) Open the test software.
(4) Test the open-circuit voltage.
(5) Select the test module "Cyclic voltammetry" in the software and set the test parameters accordingly. The initial voltage is the open-circuit voltage;the voltage range is 2.5–4.5 V; the scanning rate is 0.1–2.0 mV/s; the segment is 6.
(6) Start the CV test. Export the test result as an Excel file.
(7) Select the test module "Electrochemical impedance spectroscopy" in the software and set the test parameters accordingly. The frequency range is 5 mHz–500 kHz, andthe amplitude is 5mV.
(8) Start the EIS test. Export the test result as an Excel file.

3.2.4 Performance test

(1) Switch on the battery testing system and computer. Open the testing software.
(2) Place the coin cell in the sample holder.
(3) Set the test parameters. The parameters for testing the specific capacity, rate capacity, and cycling stability are listed in Table 10.1.
(4) Start the test. Export the testing result as an Excel file.

Table 10.1 Parameters for the performance test

Specific capacity	Rate capacity	Cycling stability	Current density:
Current density: 54.6 mA/g (0.2C); Cut-off voltages: 2.5–4.2V; 10 cycles	Current density: 54.6 mA/g (0.2C), 273 mA/g (1C), 546 mA/g (2C); Cut-off voltages: 2.5V–4.2V; 10 cycles for each rate	273 mA/g (1C) Cut-off voltages: 2.5V–4.2V; 50 cycles	

3.2.5 Application design

Design an application using the as-prepared batteries according to the measured voltage and capacity.

4 Requirements for experimental report

(1) Draw the CV curve using the exported data. Identify the redox potentials of the CV curve, and evaluate the reversibility of the electrode reaction.
(2) Draw the Nyquist plot of the EIS test. Simulate the impedance into anelectric circuit using software Z-view, and determine the relative parameters.
(3) Draw the plots of charge capacity versus cycle number and discharge capacity versus cycle number for evaluating the rate capability and cycling stability. Analyze the specific capacity,rate performance, and cycling stability of the Li-ion batteries.
(4) Comparethe results of the LCO synthesized by solid-state and sol-gel methods.
(5) Analyze the factors that influence the performance of Li-ion batteries assembled in this experiment.

5 Questions and further thoughts

(1) Why is the specific capacity lower at a higher current rate? How can we improve the rate capability?

(2) Why does the specific capacity decrease with the cycle number?

(3) Can we use Cu foil as the current collector for the LCO cathode? Why or why not?

Experiment 49 Fabrication and Characterization of Polymer Solar Cell

Type of the experiment: Comprehensive
Recommended credit hours: 8
Brief introduction: Polymer solar cellsuse conjugated polymers as light absorbers, electron donors, acceptors, and hole-transporting materials. In this experiment, a polymer solar cell based on bulk heterojunction structure is fabricated using techniques of spin coating and thermal evaporating. The conjugated polymer is used as the light absorber and electron donor. The electro-optical properties of the as-prepared solar cell are determined by the solar simulator system.

1 Objectives

(1) To understand the structure and operating principle of solar cells.

(2) To understand the fabrication process of polymer solar cells.

(3) To understand the characterization method of the electro-optical properties of solar cells.

2 Principles

2.1 The principle of solar cells

Solar energy is used as renewable power, which is a sustainable solution to both energy crisis and environmental issues. Solar energy is abundant and free but intermittent and unstable. For most of the time, solar resource needs to be converted and stored for further application.

Photovoltaics is one of the important solar energy-related technologies. It is the direct conversion of light into electricity based on the photoelectric effect that has been introduced in Experiment 4. During the light absorption of solar materials, the electronic transition generally involvesan energy threshold below which no photons are absorbed. For a molecular system, this energy threshold corresponds to the energy difference between the highest occupied molecular orbital (HOMO) and the lowest unoccupied molecular orbital (LUMO), which is called the bandgap in a semiconductor.

Generally, a solar cell has a sandwiched structure constructed by an active layer, interfacial layers, and electrode layers. The active layer consists of n-type and p-type semiconductor layers as charge donor and electron acceptor. They form a p-n junction between them. The interfacial layers, i.e., the hole extraction layer

(HEL) or electron extraction layer (EEL), are located between the active layer and the anode or the cathode, respectively.

Under sunlight, electrons in the donor move from HOMO to LUMO, as shown in Fig. 10.10. The energy difference between the LUMO of the acceptor and the LUMO of the donor is much smaller than the difference between the LUMO and HOMO of the donor. Therefore, the electrons move from the LUMO of the donor to the LUMO of the acceptor easily, and they are transferred to the cathode by EEL rapidly. Meanwhile, the holes are transferred to the anode by HEL. The separation of electrons and holes is realized. Under the continuous illumination of sunlight, electricity is supplied by the solar cell.

As we know, the solar radiation that reaches the earth is a small but crucial part of the solar spectrum. Therefore, the materials in solar cells need a suitable energy threshold relative to the solar spectrum for high energy conversion efficiency. For maximizing the conversion efficiency, many solar materials with different energy thresholds, which are assembled in a solar cell, are used to absorb different parts of solar irradiation. However, a solar cell involving more than three or four different junctions with different optical absorption characteristics is relatively difficult to obtain.

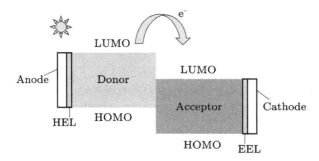

Fig. 10.10 The principle of electricity generation of solar cells.

2.2   Common types of solar cells

There are several common types of solar cells, including Si cells, thinfilm cells, multijunction cells, and organic cells. Most solar cells are constructed using crystalline silicon. The crystalline Si cells are single-junction devices, which are limited by thermodynamic considerations to a maximum theoretical power conversion efficiency of ~30% under direct sunlight with an intensity of 100 mW/cm^2. However, actual cells suffer from parasitic losses and exhibit lower efficiency. At present, the efficiency of Si cells ismoving close to the theoretical limit.

Organic photovoltaics (OPV) is a rapidly emerging technology with the potential for low-cost, non-vacuum processed devices. The heart of the cell is an active layer composed of an organic donor (such as small organic molecules and polymers) that absorbs most of the light and an electron acceptor (such as fullerene, inorganic quantum dot, and nanowire). The blend of the donor and acceptor at the molecular level contributes to the highefficiency of the devices.

In Experiment 4, an organic photodetector has been prepared base on the bulk heterojunction (BHJ) structure. A typical solar cell configuration with BHJ structure is shown in Fig. 10.11, which will be fabricated in the present experiment. Glass substrates coated with indiumtinoxide (ITO) are used as the cathode. The ZnO thin film is prepared by the spin-coating process and used asthe EEL. Molybdenum oxide (MoO₃) prepared by thermal evaporating technologyis used as the HEL. In this device, the Ag layer is also obtained by thermal evaporation, which serves as the anode.

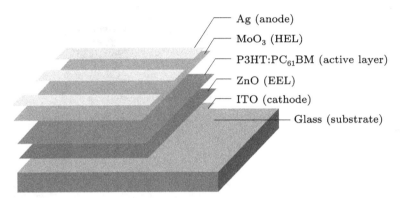

Fig. 10.11 The structure of a polymer solar cell. (See color illustration at the back of the book)

The active layer is composed of a poly (3-hexylthiophene) (P3HT) as electron donor and fullerene derivative([6,6]-phenyl-C_{61}-butyric acid methyl ester ($PC_{61}BM$)) as the electron acceptor.The P3HT and $PC_{61}BM$ are blended at the molecular scale. Thus, the P3HT:$PC_{61}BM$ bulk heterojunction contains plenty of interfaces, leading to more efficient separation of electrons and holes.

2.3 Performance of solar cells

The electro-optical properties of solar cells can be tested by the solar simulator system, which simulates the solar radiative spectrum of the sun and records the current-voltage characteristics, i.e., $I - V$ curve,of a solar cell. Usually, the $I - V$ curve of a solar cell is often tested by sweeping an external voltage or current source under certain light conditions.

The primarysolar cell performance is shown in Fig. 10.12. Quadrants I and III both consume power, corresponding to the operation as a rectifying diode and photodetector, respectively. Operation in quadrant IV generates negative power that can drive current in the external circuit. The curve in quadrant IV shown in Fig. 10.12 is a typical shape of a solar cell under solar light.

The following parameters determined from the $I - V$ curve shown in Fig. 10.13 are often used to describe the performance of a solar cell. Usually, current and power are measured as densities.

I_{sc} is short circuit current density at $V = 0$. It is the maximum value in the power quadrant, which occurs at the start of a forward-bias sweep. The maximum current value is the total current produced in the solar cell by photon excitation.V_{oc}

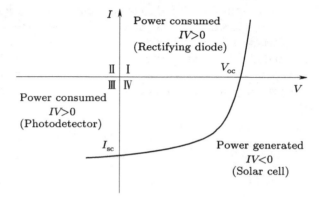

Fig. 10.12 $I - V$ curve of a solar cell.

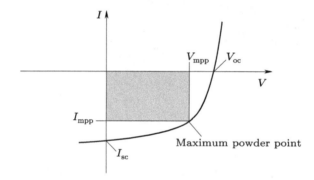

Fig. 10.13 The parameters of a solar cell determined from the $I - V$ curve.

is the open-circuit voltage at $I = 0$. It is the maximum voltage difference across the cell for a forward-bias sweep in the power quadrant. V_{mpp} and I_{mpp} are voltage density and current density at the maximum power point, respectively.

By using the above four parameters, the fill factor can be defined as

$$\text{FF} = \frac{V_{\text{mpp}} I_{\text{mpp}}}{V_{\text{oc}} I_{\text{sc}}}. \tag{10.2}$$

The fill factor is essentially a measure of the quality of solar cells. It is calculated by comparing the maximum power with the theoretical power that would be generated if the cell could simultaneously generate the open-circuit voltage and short-circuit current. Clearly, in terms of the efficiency of the solar-to-electricity conversion, the higher the fill factor, the better the solar cell. Graphically it will correspond to an $I - V$ sweep that is more rectangular and less curved. Typical fill factors range from 0.5 to 0.82. The fill factor is also often represented as a percentage.

The conversion efficiency of solar cell η is the ratio of the electrical power output P_{out} to the solar power input P_{in}. The P_{in} is the product of the irradiance of the incident light, measured in W/m^2. The maximum power output is

$$P_{\text{max}} = V_{\text{mpp}} I_{\text{mpp}}. \tag{10.3}$$

Then, the maximum conversion efficiency can be determined as

$$\eta_{\max} = \frac{P_{\max}}{P_{\text{in}}} = \frac{V_{\text{mpp}}I_{\text{mpp}}}{P_{\text{in}}} = \frac{V_{\text{oc}}I_{\text{sc}}\text{FF}}{P_{\text{in}}} \qquad (10.4)$$

The maximum measured conversion efficiency is the appropriate indicator of the performance of the device under test. It is usually reported as the device efficiency without noting that the measurement is made at the maximum-efficiency operating point.

3 Experimental

3.1 Preparation

(1) Chemicals: P3HT, $PC_{61}BM$, chlorobenzene, zinc acetate, ethanolamine, 2-methoxyethanol, sodium dodecyl benzene sulphonate (SDBS), acetone, ethanol.

(2) Substrate: ITO coated conductive glass with a size of 1.5 cm × 1.5 cm.

(3) Apparatuses: spin-coater (SETCAS Electronics Co., Ltd., KW-4A), evaporation coating apparatus (Shenyang Kejing Auto-instrument Co., Ltd., GSL-1800X-ZF2), solar simulator (Enli Technology, SS-F5-3A), source meter (Keithley 2400), drying oven, UV ozone cleaning system, electromagnetic stirrer, ultrasonic cleaner, electronic balance, heating plate.

(4) Tools: Petri dishes, beakers, pipettes, 2 mL screw-cap vial, washing holder, stirring bar.

3.2 Procedure

3.2.1 Cleaning of the ITO substrate

(1) Ultrasonically clean the ITO substrates in acetone, detergent SDBS, DI water (3 times), and ethanol for 20 min sequentially.

(2) Put the substrates into the drying oven at 80 °C for 2 h.

(3) Treat the samples with the UV ozone cleaning system for 20 min to remove the organic moieties.

3.2.2 Deposition of the ZnO thin film by the spin-coating method

(1) Prepare the ZnO precursor. Dissolve zinc acetate in 1 mL 2-methoxyethanol with a concentration of 0.5 mol/L, using ethanolamine as the stabilizer.

(2) Deposit about 40 nm ZnO thin film on the ITO substrate by spin coating. Spin at 800 t/min for 15 s followed by 3000 r/min for 10 s. Then anneal the sample in air at 200 °C for 10 min.

(3) Place the sample into a petri dish.

3.2.3 Deposition of the active layer by the spin-coating method

(1) Weigh 3 mg P3HT and 1.8 mg $PC_{61}BM$, and put them into a 2 mL screw-cap vial.

(2) Add 200 μL chlorobenzene into the vial. Stir the mixture for 2 h to get the solution of the active layer.

(3) Put the ZnO substrate on the spin-coater. Spread 40 μL solution on the substrate. Spin at 800 r/min for 60 s.

(4) Heat the sample on the heating plate at 90 °C for 10 min.

(5) Put the sample into a petri dish.

3.2.4 Deposition of the MoO_3 and Ag electrode by the thermal evaporating method

(1) Fix the substrate on the stageof the thermal evaporator and cover it with a shadow mask.

(2) Switchon the thermal evaporator. Vacuum the chamber.

(3) Set the film thickness and deposition rate, and start deposition.The thicknesses of the MoO_3 and Ag electrode layers are 10 nm and 100 nm, respectively, which are examined by a crystal oscillator placed in the chamber.

(4) While the evaporating is finished, vent the chamber and take out the sample.

3.2.5 Characterization of the electro-optical properties

(1) Switchon the solar simulator, source meter, and computer.Open the test software.

(2) Set the voltage from -1 to 1 V. The test area is 0.09 cm^2. The intensity of solar simulator radiation is 100mW/cm^2.

(3) Start the test and obtain the $I-V$ curve.Export the results as a ".txt" file.

4 Requirements for experimental report

(1) Plot the $I-V$ curve using the exported data.

(2) Determine the relative parameters of the solar cell from the $I-V$ curve.

(3) Compare the properties with those reported in the literature. Analyze the factors that influence the performance of the solar cell prepared in this experiment.

5 Questions and further thoughts

(1) Why does a solar cell work in Quadrant IV shown in Fig. 10.12?

(2) List several widely studied electron donor materials for polymer solar cells. Which monomers are better choices for efficient organic/polymer solar cells and why?

(3) List several common structures of polymer solar cells. Compare their advantages and disadvantages.

Extended readings

Ginley D S, Cahen D. Fundamentals of Materials for Energy and Environmental Sustainability. Cambridge University Press, 2011.

Nelson J. The Physics of Solar Cells. Imperial College Press, 2003.

Yoshio M, Brodd R J, Kozawa A. Lithium-Ion Batteries: Science and Technologies. Springer, 2010.

Chapter 11 Electronic Materials and Devices

Experiment 50 Magnetron Sputtering Deposition of ITO Thin Film and Hall Effect Test

Type of the experiment: Comprehensive
Recommended credit hours: 8
Brief introduction: In this experiment, the principles of the magnetron sputtering technique and the theory of the Hall effect are introduced. The indium tin oxide (ITO) thin film is prepared by radio-frequency (RF) magnetron sputtering on a silicon wafer. The film thickness will be determined by the stylus profilometer. Then the Hall effect test is conducted on the as-prepared ITO film, and the relevant parameters, such as the Hall coefficient, carrier concentration, and mobility, are determined.

■ Objectives

(1) To understand the principle of the magnetron sputtering technique.
(2) To understand the theory of the Hall effect.
(3) To prepare the ITO thin film by RF magnetron sputtering.
(4) To determine the relevant properties of the ITO film using the Hall effect.

■ Principles

2.1 ITO thin films

ITO thin films are wide-gap semiconductors with relatively low resistivity and transparent in the visible range of the spectrum. They are the most widely used and developed transparent conductive oxide material. Applications of ITO thin films include transparent electrodes for a range of display, photovoltaic and sensor applications, EMI shielding, low-E windows, transparent heaters, and transparent electronics.

Since ITO is essentially Sn-doped oxide, the dopant Sn plays a crucial role in ITO properties.There is an Sn content of about 20wt% at which the resistivity, visible transmittance, and electron mobility are optimized. At Sn levels below this value, electrical properties are determined primarily by doping concentration and scattering off oxygen vacancies for higher Sn levels.

2.2 Magnetron sputtering technique

Sputtering is a widely used technique for thin film fabrication. The DC sputtering technique has been introduced in Experiment 27. It is a plasma vapor deposition process in which a plasma is created, and positively charged ions from the plasma are accelerated by an electrical field superimposed on the negatively charged electrode or target. The positive ions collide with the negatively charged target material. Consequently, the atoms from the target surface are ejected or "sputtered" andthen deposit on a substrate.

In the case of magnetron sputtering, it uses a closed magnetic field to trap electrons. The magnetron uses strong magnetic fields, typically from permanent magnets, to keep secondary electrons spatially confined in the vicinity of the target surface. Thus, the residence time of the secondary electrons in the plasma is greatly lengthened, resulting in greater ionization of the sputter-gas atoms, a denser plasma, and higher plasma currents and deposition rates.

2.3 RF sputtering technique

When anargonion strikes the target, an electron is released from the surface and combines with the ion to neutralize it, returning it to the vacuum as an argon atom. If the target material is dielectric, this process rapidly causes a charge build-up at the surface until argon ions are no longer attracted. In response to address this issue, the RF sputtering technique is developed to sputter non-conducting materials. It applies alternating current (AC) or pulsed power to the target whereby the ion charge built up on the surface is expelled during the positive or neutral phase. Because the power supply is only negative half the time, the deposition rate is lower than that of direct current (DC) sputtering. Moreover, the power supply is more complex and therefore more expensive.

2.4 Hall effect

2.4.1 Theory of Hall effect

Hall effect develops a transverse electric field in a solid material when it carries an electric current and is placed in a magnetic field perpendicular to the current. A demonstration of the Hall effect is simple to set up, as illustrated in Fig. 11.1. It shows a thin plate of conductive material that is carrying a current I. A pair of probes connected to a voltmeter are positioned along the sides of this plate. In this case, the measured voltage is zero. When a magnetic field is applied to the plate so that it is at right angles to the current flow, as shown in Fig. 11.1(b), a small voltage appears across the plate. If the direction or the polarity of the magnetic field is reversed, the induced voltage will also reverse.

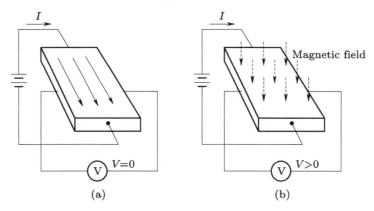

Fig. 11.1 The Hall effect in a conductive plate. (a) The voltage across the plate is zero; (b) a voltage appears when a perpendicular magnetic field is applied.

The Hall effect is based on the movement of charged carriers move in response to the electric and magnetic fields. The force exerted on a charged carrier by the electromagnetic field is described by the Lorentz force, which is calculated by

$$F_l = q_0 v B, \quad (11.1)$$

where v is the velocity of the charge along the plate length, B is the magnetic field along the plate normal, and q_0 is the charge of the charge carrier.

Although the magnetic field forces the charge carriers to one side of the plate, this process is self-limiting because the excess concentration of charges to one side and consequent depletion on the other side gives rise to an electric field across the plate. This field applied an electric force on the charge carriers, expressed as

$$F_e = q_0 E_H, \quad (11.2)$$

where E_H is the Hall electric field across the plate.

An equilibrium develops where the magnetic force pushing the charge carriers aside is balanced by the electric force trying to push them back toward the middle. Thus,

$$F_l = F_e \quad (11.3)$$

Solving for E_H yields

$$E_H = vB, \quad (11.4)$$

which means that the Hall field is solely a function of the velocity of the charge carriers and the strength of the magnetic field. For the plate with a given width w between sense electrodes, the Hall electric field can be integrated over w. Assuming it is uniform, the Hall voltage is

$$V_H = wvB. \quad (11.5)$$

The carrier velocity will be proportional to the current. Consider the case of a piece of conductive material with a given cross-sectional area of A. Assuming that the carrier density N is constant and the carriers behave like an incompressible fluid, the velocity will also be inversely proportional to the cross-section. A larger cross-section means lower carrier velocity. Then, the carrier drift velocity can be determined by

$$v = \frac{I}{q_0 N A}. \quad (11.6)$$

By combining Equations (11.5) and (11.6), the Hall voltage can be expressed as

$$V_H = \frac{IB}{q_0 N d} = R_H \frac{IB}{d}, \quad (11.7)$$

where d is the thickness of the conductor, and R_H is the Hall coefficient defined as

$$R_H = \frac{1}{q_0 N}. \quad (11.8)$$

2.4.2 The Hall effect in semiconductors

Since the charge carrier density N is relatively large in metal materials, the voltage resulting from the Hall effect is minimal in terms of Equation (11.8). For this reason, it is not usually practical to make Hall-effect transducers with most metals. Semiconductor materials have carrier concentrations that are orders of magnitude lower than those found in metals. Therefore, they exhibit the Hall effect more strongly than metals for a given current and thickness.

The conductivity of a semiconductor is a function of both carrier concentration and a property called carrier mobility. Carrier mobility measures how fast the charge carriers move in response to an electric field and varies with the type of semiconductor, the dopant concentration, the carrier type (n- or p-type), and temperature. The van der Pauw method, which combines the Hall effect with the four-point technique, allows one to measure both the carrier concentration and the mobility of carriers in a semiconductor.

Consider a square sample of a thin semiconductor having four small electrodes at the corners. As shown in Fig. 11.2, a current I_{MN} is applied via electrodes M and N using a voltage source. Then the voltage difference $(V_P - V_O)$ appearing across the electrodes O and P is measured. The resistance measured for this electrode arrangement is

$$R_{MNOP} = \frac{V_P - V_O}{I_{MN}}. \quad (11.9)$$

The resistivity of the semiconductor can be calculated by

$$\rho = \frac{\pi d}{\ln 2} \overline{R}, \quad (11.10)$$

where d is the thickness of the specimen, and $\overline{R}$ is the resistance averaged over the four possible measurement configurations.

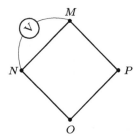

Fig. 11.2 Experimental setup for the van der Pauw four-point measurement.

The resistivity is related to the mobility by

$$\rho = \frac{1}{\mu N q_0} = \frac{R_H}{\mu}, \tag{11.11}$$

where μ is the carrier mobility. Thus, μ can be calculated from the equation if N or R_H is known.

The carrier density N or the Hall coefficient R_H can be determined using the Hall effect. The sample is placed in a uniform magnetic field B perpendicular to the semiconductor, and a current is applied to the sample via electrodes M and O. Then, the voltage across P and N is measured both before and after applying the magnetic field. The change in resistance ΔR_{MONP} before and after applying the magnetic field can be readily calculated. It has been derived that

$$R_H = \frac{d}{B} \Delta R_{MONP} \tag{11.12}$$

The average of the Hall coefficient can be calculated from the four possible measurement configurations. Then, the carrier density N and mobility μ can be determined using Equations (11.8) and (11.11), respectively.

3 Experimental

3.1 Preparation

(1) Materials: thermal oxidized p-type silicon wafer with a size of 30 mm × 30 mm (the thickness of the SiO_2 layer is about 200 nm), ITO target (SnO_2:In_2O_3 = 1:9).

(2) Chemicals: ethanol, acetone, silicon wafer detergent.

(3) Apparatuses: magnetron sputtering machine (Beijing Furuisi Ltd., TWS-350), stylus profilometer (KLA-Tencor, Alpha-Step D-300), Hall measurement system (Ecopia, HMS-3000), ultrasonic cleaner.

(4) Tools: washing holder, beakers.

3.3 Procedure

3.3.1 Cleaning of oxidized silicon wafer

(1) Place the oxidized silicon wafer into the PTFE washing holder and put it into the beaker with detergent solution. Clean the wafer ultrasonically for 10 min.

(2) Wash the silicon wafer twice in deionized water ultrasonically for 5 min each time.

(3) Mix ethanol and acetone solution in the volume ratio of 1:1. Wash the silicon wafer in the solution by ultrasonic cleaning for 10 min.

(4) Again, wash the silicon wafer twice in deionized water ultrasonically for 5 min each time.

(5) Dry the wafer with nitrogen.

3.3.2 Deposition of the ITO thin film

(1) Install the ITO target in the magnetron sputtering machine.

(2) Place the oxidized silicon wafer on the sample stage.

(3) Switchon the machine. Vacuum the sputtering chamber by the mechanical pump to a pressure below 10 Pa. Then switch on the molecular pump and vacuum the chamber to a pressure below 2×10^{-4} Pa.

(4) Pass argon into the chamber and control the pressure in the chamber at 1 Pa.

(5) Switch on the RF power supply working on a defined load impedance of 50 W. Pre-sputter for 10 min to remove the impurities on the target surface.

(6) Open the baffle. Deposit the ITO film for 30 min. The thickness is about 300 nm.

(7) Close the baffle. Switch off the power supply. Close the argon inlet valve and the valve between the chamber and the pumps. Switchoff the pumps and the machine.

(8) Vent the chamber and take out the sample.

3.3.3 Characterization of the ITO thin film

(1) Measure the thickness of the thin film by the stylus profilometer.

(2) Place the sample on the stage of the Hall measurement system. Locate the four probes at the border of the sample symmetrically.

(3) Switchon the Hall measurement system. Set the thickness value. Start the measurement and record the results.

4 Requirements for experimental report

(1) List the measurement results, including film thickness, resistivity, carrier concentration and mobility, Hall coefficient, and semiconductor type.

(2) Compare the results with those in the literature. Discuss the factors that influence the film quality.

5 Questions and further thoughts

(1) What is the difference between RF sputtering and DC sputtering?

(2) Does the geometric shape of the sample affect the Hall effect test?

(3) How can we identify the type ofsemiconductor by the Hall effect?

Experiment 51 Fabrication and Characterization of In-Ga-Zn-O Thin Film Transistor

Type of the experiment: Comprehensive
Recommended credit hours: 8
Brief introduction: Amorphous indium-gallium-zinc-oxide (IGZO) thin film transistor (TFT) has received broad attention due to their several advantages over other TFTs, such as high field-effect mobility, good short-range uniformity, and high electrical reliability. In this experiment, an IGZO TFT will be fabricated. Firstly, an IGZO layer and an indium tin oxide (ITO) layer are deposited on the SiO_2-Si wafer by magnetron sputtering successively. Note that these two layers will be patterned by manually aligning the shadow masks. Then, the electrical properties of the TFT will be tested.

■ Objectives

(1) To understand the working principle and fabrication methods of TFTs.
(2) To fabricate an IGZO TFT by sputtering with shadow masks.
(3) To understand the test method of the electrical properties of the transistors.
(4) To determine the carrier mobility, threshold voltage, subthreshold swing, and otheressentialparameters from the characteristic curves.

■ Principles

2.1 Structures of TFTs

A TFT is a type of field-effect transistor usually used in a liquid crystal display (LCD). This type of display features a TFT for each pixel. These TFTs act as individual switches that allow the pixels to change state rapidly, making them turn on and off much more quickly. Because these TFTs are arranged in a matrix, they are called active-matrix TFTs.

Figure 11.3 illustrates typical structures that are adopted in the fabrication of TFTs. The TFT structure can be specified by the stacking order of the gate, oxide semiconductor, and source/drain electrodes. They can be classified more precisely into combinations of top/bottom gate and top/bottom contact. The bottom-gate structure has the advantage that the components can be deposited on the highly doped silicon wafer with a thermal oxidation SiO_2 dielectric layer. The silicon substrate and SiO_2 layer act as the gate and the insulation layer, respectively. This structure can greatly simplify the preparation process. The top-gate structure has been employed to fabricate devices with epitaxial semiconductor layers, for which it is difficult to form bottom electrodes. This structure has another advantage that the upper gate insulator and electrodes may act as passivation, which protects the channel layer from external damage.

Since the amorphous IGZO TFT, in which the IGZO is the active layer, has received broad attention due to its advantages, such as high field-effect mobility, good short-range uniformity, and high electrical reliability, the IGZO TFT in the bottom gate and top contact structure is prepared in this experiment. Then, the electrical properties of the TFT are evaluated.

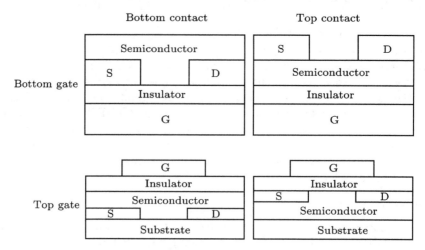

Fig. 11.3 Typical structures of TFTs. G, S, and D denote gate, source, and drain, respectively.

2.2 Working principle of TFTs

According to the different working mechanisms, TFTscan be divided into enhanced TFT and depleted TFT. The enhanced type means that when the gate voltage is zero, there is no conductive channel formed in the active layer, and only when the applied gate voltage is large enough, the TFT can be switched on. Therefore, the enhanced TFT is also called normally closed TFT. Depletion TFT means that the transistor can still have current even without an applied gate voltage, so it is also called normally open TFT.

In the TFT, an inversion layer at the interface between the semiconductor layer and the insulation layer acts as a conducting channel. For example, in an n-channel enhancement type TFT, and the substrate is p-type silicon. Since the insulation layer can be considered as the dielectric layer in a capacitor, the inversion charge consists of electrons that form a conducting channel between the source and the drain. Electrons flow from source to drain to make the channel conductive.

The inversion charge can be induced in the channel by applying a suitable gate-source voltage, V_{GS}, relative to other terminals. The onset of strong inversion is defined in terms of a threshold voltage V_T being applied to the gate electrode relative to the other terminals. When a drain-source bias V_{DS} is applied to an n-channel TFT in the threshold conducting state, electrons move in the channel inversion layer from source to drain. The relationship between drain-source current I_{DS} and drain-source voltage V_{DS} at constant gate-source voltage V_{GS} is known as output characteristics of TFT. Figure 11.4 shows the output characteristic curves of a typical n-channel enhanced TFT. It can be seen from the figure that the working region of the TFT can be divided into cutoff region, linear region (variable resistance region), saturation region (constant current region), and breakdown region.

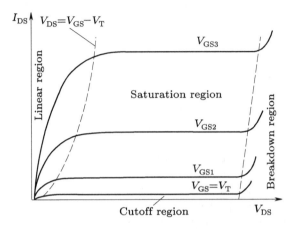

Fig. 11.4 Typical TFT output characteristic curves.

(1) Cutoff region ($V_{GS} < V_T$).

When the gate voltage V_{GS} is less than a threshold voltage V_T, the drain-source voltage V_{DS} cannot generate obvious drain-source current I_{DS}. It is because the surface of the active layer cannot form a conductive channel for small V_{GS}, leading to the high resistance state between the source and drain electrodes. Consequently, I_{DS} is close to zero, and the device is in the cutoff state.

(2) Linear region ($V_{GS} > V_T$ and $V_{DS} - V_{GS} - V_T$).

When $V_{GS} > V_T$, the electrons in the active layer move to one side of the insulating layer under the electric field force. Then the application of source-drain voltage V_{DS} gives a steady increase in the source-drain current I_{DS}. The current-voltage relationship of the gate, source, and drain is expressed as

$$I_{DS} = \frac{W}{L}\mu_n C_i \left[(V_{GS} - V_T)V_{DS} - \frac{1}{2}V_{DS}^2\right], \tag{11.13}$$

where W and L are respectively the width and length of the TFT channel, μ_n is the field-effect mobility of the device, and C_i is the capacitance of the gate insulator per unit area.

(3) Saturation region ($V_{GS} > V_T$ and $V_{DS} > V_{GS} - V_T$).

When gate-drain voltage V_{GD} is equal to V_T, i.e., $V_{GD} = V_T$, the inversion channel reaches a threshold at the drain, and the density of inversion charge vanishes. This is the pinch-off condition, which leads to a saturation of the drain current I_{DS}. Since $V_{GD} = V_{GS} - V_{DS}$, we may get $V_{DS} = V_{GS} - V_T$ under this condition, and the V_{DS} under this condition is called the saturation voltage. When $V_{DS} > V_{GS} - V_T$, the drain-source current I_{DS} in saturation remains approximately constant. In this case, the current-voltage relationship of the gate, source, and drain is expressed as

$$I_{DS} = \frac{W}{2L}\mu_n C_i (V_{GS} - V_T)^2. \tag{11.14}$$

(4) Breakdown region.

When the V_{DS} increases to a particular value, the voltage across the pinched-off region creates a strong electric field, which efficiently transports the electrons from the strongly inverted region to the drain. The I_{DS} increases sharply, and the TFT device is broken down.

2.3 Characterization of TFTs

The quality of a TFT device is generally evaluated by threshold voltage V_T, field-effect carrier mobility μ_n, current switching ratio I_{on}/I_{off}, and subthreshold swing S. These parameters can be calculated or fitted from the output characteristic curve and transfer characteristic curve. These curves can be obtained by the semiconductor characterization system, which usually consists of a semiconductor parameter tester and a prober. These parameters are explained in detail as follows.

(1) Field-effect carrier mobility.

In the absence of an applied electric field, the motion of the charge carriers in a semiconductor is an irregular thermal motion under an equilibrium state. When an electric field is applied, the equilibrium state is broken, and a part of the carriers moves directionally under the electric field. Carrier mobility μ_n refers to the mean value of the directional motion rates of these carriers under the electric field. The mobility is an important parameter to measure the performance of TFT. Together with the carrier concentration, it determines the resistivity of the semiconductor. When the carrier concentration is fixed, higher mobility will result in a better current carrying capacity of the semiconductor. Besides, the mobility affects the operating frequency of TFT. A higher mobility means a shorter time required for device switching and a faster switching speed.

The relationship between drain-source current I_{DS} and gate-source voltage V_{GS} at constant gate-source voltage V_{DS} is known as the transfer characteristic of TFT. The mobility can be determined from the transfer characteristic curve. When the TFT is working in the saturation region, from Equation (11.14), we get

$$\sqrt{D_{DS}} = \sqrt{K}(V_{GS} - V_T), \qquad (11.15)$$

where

$$K = \frac{W}{2L}\mu_n C_i. \qquad (11.16)$$

Thus, the mobility is

$$\mu_n = \frac{2KL}{C_i W}. \qquad (11.17)$$

When the transfer characteristic is expressed as a $\sqrt{I_{DS}} - V_{GS}$ curve, as shown in Fig. 11.5, the value of K can be obtained from the slope of the curve by linear fitting based on Equation (11.15).

(2) Threshold voltage.

The threshold voltage V_T is the minimum gate voltage required to turn on the TFT conductive channel. It can also be determined from the transfer characteristic curve in Fig. 11.5. It is the value at the intersection of the fitting line and the V_{GS} axis.

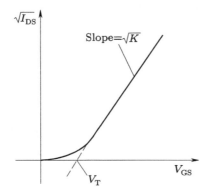

Fig. 11.5 Determination of the carrier mobility and threshold voltage from the $\sqrt{I_{\text{DS}}} - V_{\text{GS}}$ curve.

(3) Current switching ratio.

Under a specific source-drain voltage V_{DS}, we may define the current I_{DS} in the saturation region as on-state current I_{on} and the current when $V_{\text{GS}} = 0$ as off-state current I_{off}. The current switching ratio, $I_{\text{on}}/I_{\text{off}}$, is a parameter that reflects the TFT ability to control the current. For the LCD application, a higher $I_{\text{on}}/I_{\text{off}}$ means better performance since a higher I_{on} corresponds to stronger TFT driving ability and faster writing speed; lower I_{off} corresponds to stronger TFT turn off capability and smaller power consumption when the TFT is in the off state.

(4) Subthreshold swing.

The subthreshold swing S is defined as the increment of V_{GS} when I_{DS} rises by an order of magnitude at a constant V_{DS} in the subthreshold region, i.e., a region that $V_{\text{GS}} \leqslant V_{\text{T}}$. From the transfer characteristics expressed in semilog coordinates ($\log I_{\text{DS}} - V_{\text{GS}}$), S can be extracted using the equation

$$S = \frac{\mathrm{d} V_{\text{GS}}}{\mathrm{d} \log I_{\text{DS}}}. \tag{11.18}$$

That is the reciprocal of the slope of the linear segment in the subthreshold region.

The subthreshold swing S measures the changing speed of I_{DS} from the off-state to the on-state and hence reflects the controlling ability of the gate voltage on the device. Obviously, a smaller S means a faster switching from the off-state to the on-state of TFT.

3 Experimental

3.1 Preparation

(1) Materials: IGZO target (In:Ga:Zn=1:1:1), ITO target (SnO_2:In_2O_3 =1:9).

(2) Substrates: thermal oxidized p-type silicon wafers with a size of 30 mm × 30 mm (the thickness of the SiO_2 layer is about 200 nm).

(3) Chemicals: ethanol, acetone, silicon wafer detergent.

(4) Apparatuses: magnetron sputtering machine (Beijing Furuisi Ltd., TWS-350), semiconductor characterization system (Keithley, 4200SCS), prober(SemiProbe,

Lab Assistant), ultrasonic cleaner.

(5) Tools: washing holder, two shadow masks (customized) for deposition of the active and electrode layers.

3.2 Procedure

3.2.1 Fabrication of the IGZO TFT

(1) Ultrasonically clean the wafer by detergent solution, water, ethanol, and acetone sequentially. Then dry the wafer with nitrogen. The procedure has been described in Experiment 50.

(2) Install IGZO target in the magnetron sputtering machine.

(3) Align the shadow mask for the active layer on the oxidized silicon wafer. Place them on the sample stage.

(4) Deposit the IGZO active layer for 15 min. The thickness of the layer is about 30 nm. The procedure of the RF magnetron sputtering has been described in Experiment 50.

(5) Install ITO target in the magnetron sputtering machine.

(6) Align the shadow mask for the electrode layer on the wafer. Place them on the sample stage.

(7) Deposit the ITO electrode layer for 15 min. The thickness of the layer is about 200 nm.

3.2.2 Characterization of the electronic properties

(1) Place the TFT device on the prober. Contact the three probes with the gate, source, and drain of the TFT device. Conduct the test in a dark chamber.

(2) Switchon the semiconductor characterization system. Enter the corresponding test module.

(3) Test the output characteristics. Set the V_{DS} in the range of $0-30$ V. Set V_{GS} varies from 0 to 30 V at intervals of 5 V. Measure a series of $I_{DS}-V_{DS}$ curves with different V_{GS}. Export the data for further analysis.

(4) Test the transfer characteristics. Set the V_{GS} in the range of $-20-60$ V. Set a fixed V_{DS} to 10 V. Measure the $I_{DS}-V_{GS}$ curve. Export the data for further analysis.

4 Requirements for experimental report

(1) Plot the $I_{DS}-V_{DS}$ curves in one chart.

(2) Plot the $I_{DS}-V_{GS}$ curve.

(3) Calculate the field-effect carrier mobility μ_n, threshold voltage V_T, current switching ratio I_{on}/I_{off}, and subthreshold swing S for the fixed V_{DS} of 10 V. Describe the main calculation process and plot the $\sqrt{I_{DS}}-V_{GS}$ and $\log I_{DS}-V_{GS}$ curves.

5 Questions and further thoughts

(1) Why is the characterization of the electrical properties conducted in darkness?

(2) Compare the performance of the as-fabricated TFT device with those re-

ported in the literature. Suggest the possible factors that influence the performance.

Experiment 52   Micro-nano Processing by Lithography

Type of the experiment: Comprehensive
Recommended credit hours: 20
Brief introduction: Lithography is one of the most important micro-nano processing methods. This experiment introduces three widely used lithography techniques, including photolithography, soft lithography, and nanoimprint lithography. The experiment is divided into three experiments that can be conducted respectively. The first experiment introduces the principle and the general process of photolithography after describing the basic concepts of microfabrication and lithography. A logo pattern is prepared on a circuit board. The second experiment introduces the principle and techniques of soft lithography. The microstructures of a digital video disc (DVD) and a simple microfluidic device are fabricated by soft lithography using polydimethylsiloxane (PDMS). The third experiment introduces the principle and techniques of nanoimprint lithography. The structures of quartz grafting and DVD are transferred by thermal and ultraviolet (UV) nanoimprint techniques, respectively.

It is noted that lithography for micro-nano processing usually demands to be conducted in a cleanroom. However, structures with a larger scale instead of nanostructure are produced so that the experiment can be conducted without a cleanroom.

Experiment 52.1   Preparation of a Logo Pattern by Photolithography

Type of the experiment: Comprehensive
Recommended credit hours: 4

◼ Objectives

(1) To understand the principle of photolithography.

(2) To learn the general process of photolithography.

(3) To prepare a logo pattern on a circuit board using the photolithography method.

(4) To understand the application of photolithography.

◪ Principles

2.1   Microfabrication

Microfabrication is the process of fabricating miniature structures at micro and nano-scale. The earliest microfabrication processes were used for microelectronics, particularly forintegrated circuit fabrication. Applications in other areas are also rapidly emerging. These include microanalysis systems, micro-volume reactors, microelectromechanical systems (MEMSs), and optical components. In addition, the construction of the existing structures in a down-sized version and new types of microstructures also provides the opportunity to study fundamentalscientific phenomena that occur at small dimensions.

2.2 Lithography

Lithography is one of the most important microfabricationmethods. It is based on traditional ink-printing techniques and is used for patterning various layers, such as conductors, semiconductors, or dielectrics, on a surface.

The patterning is achieved by directly writing the pattern or transferring the pattern through a mask or stamp. The pattern is designed via computer-aided design (CAD) software, which contributes to the definition of the features on the substrate.The features are often formed with a resist, and they can be defined using light with a photoresist, an electron beam with an e-beam resist, or via physical stamping without resist. The features can be further transferred to another layer using techniques such as etching, electroplating, or lift-off.

2.3 Photolithography

Photolithography is the most commonly used microlithographic technique. In photolithography, a substrate, spin-coated with a thin layer of photoresist, is exposed to a UV light source through a photomask. The photoresist is a photo-sensitive polymer. The photomask is typically a quartz plate covered with a patterned microstructure of an opaque material (generally chromium) made by lithography. The photoresist exposed to UV light becomes either more or less soluble in a developing solution for positive or negative resist, respectively. In either case, the pattern on the photomask is transferred into the film of the photoresist, and the patterned photoresist can subsequently be used as the mask in doping or etching the substrate.

Photolithography is widely used to fabricate structures for microelectronic circuits, MEMS, microanalytical devices, and micro-optics. However, it has several disadvantages. It is a relatively high-cost technology, and the investment required to build and maintain photolithographic facilities makes this technique less accessible. It cannot be applied easily to curved surfaces, but the formation of micropatterns and microstructures on non-planar substrates is important in certain types of optical and MEMS devices. These limitations have motivated the development of alternative, low-cost microlithographic techniques for manufacturing microstructures as those introduced in the following experiments.

3 Experimental

3.1 Preparation

(1) Materials: transparent plastic sheet, UV pre-sensitized copper-clad circuit board.

(2) Chemicals: photoresist developer (DP-50), etching agent for PCB plate.

(3) Apparatuses: UV exposure machine (Boda, HS-2030S), optical microscope, printer.

3.2 Procedure

(1) Print a logo pattern, such as the logo of the university, on a transparent plastic sheet by the printer.

(2) Place the plastic sheet onto the circuit board. Fix the sheet using scotch tape.

(3) Expose the circuit board to UV light for 210 s.

(4) Prepare the developer solution of 18 g photoresist developer in 800 mL water. Heat the solution to 30-35 °C.

(5) Place the exposed board in the solution. The exposed side should be upward. Shake the board gently for several minutes until the logo pattern becomes clear.

(6) Wash the board with water, and then dry the board in the air.

(7) Prepare the etching solution of 100 g etching agent in 300 mL water.

(8) Put the board into the etching solution to etch off the copper of the exposed area. Then wash the board with water and dry it in the air.

(9) Observe the details of the pattern on the board by optical microscope and take micrographs.

4 Requirements for experimental report

(1) Give the micrographs of the details of the pattern.

(2) Discuss the factors that affect the precision of the transferred pattern.

5 Questions and further thoughts

(1) How is the photoresist layer prepared if the experiment starts from a blank circuit board?

(2) Does the photolithography need to be conducted in a cleanroom? Why?

(3) Is it possible to get a pattern that smaller than the original printed pattern? How?

Experiment 52.2 Fabrication of DVD Microstructure and a Simple Microfluidic Device by Soft Lithography

Type of the experiment: Comprehensive
Recommended credit hours: 8

1 Objectives

(1) To understand the principle and techniques of soft lithography.

(2) To learn the general process of soft lithography.

(3) To transfer pattern and fabricate a microfluidic device by replica molding method.

(4) To understand the application of soft lithography.

2 Principles

2.1 Soft lithography

Soft lithography includes a collection of methods that uses soft polymeric materials to fabricate stamps, channels, or membranes with micro-sized features. It isbased on several patterning methods, including imprinting, molding, and embossing with an elastomeric stamp.

Soft lithography extends the possibilities of conventional photolithography. It can process a wide range of elastomeric materials. It is well suited for polymers, gels, and organic monolayers. The term "soft" is used since these materials are mechanically soft. PDMS has been the most widely used material for the applications of soft lithography because of its valuableproperties, including low cost, biocompatibility, low toxicity, chemical inertness, versatile surface chemistry insulating, and mechanical flexibility and durability. PDMS can be easily manipulated, and the fabrication of PDMS devices requires only a few apparatuses and tools.

The soft lithography process includes fabrication of the elastomeric mold, usually in PDMS, and the use of that mold to create features with geometries defined by the relief structure on the mold. After fabrication, the master mold is filled with PDMS precursor and degassed in a vacuum. Then, the PDMS precursor solution is cured by baking. After cooling to room temperature, the PDMS mold can be peeled off from the substrate. It is further used for stamping or micromolding to generate structures or patterns by various techniques. Several widely used techniques shown in Fig. 11.6 are described as follows.

(1) Replica molding (Fig. 11.6(a)) is a simple and single-step replication technique used to crease micro- and nano-scale features. A patterned layer of PDMS is used as a soft mold where a polymer is poured. After curing, the polymer is separated from the PDMS mold. The replica has also been employed against rigid molds for the mass production of various materials, including DVDs, diffraction gratings, holograms, and micro-tools.

(2) Microcontact printing (Fig. 11.6(b)) is essentially a print process in which a PDMS layer is used as a stamp. The PDMS layer is first soaked in a chemical solution called ink and then brought into physical contact with a substrate to transfer the ink onto the substrate surface. Various inks, including small biomolecules, proteins, or suspension of cells, can be used.

(3) Micromolding (Fig. 11.6(c)) is another technique where a patterned PDMS is used as a mold. The pattern of the PDMS layer is first brought into contact with a substrate (e.g., a glass slide). A liquid polymer is then filled in the channelsformed between the PDMS mold and the substrate. Capillarity or suction can be used to fill the channels progressively. After curing, the PDMS can be gently removed, leaving a solid microstructure at the surface of the substrate.

(4) Microtransfer printing (Fig. 11.6(d)) is to fill the patterned surface of a PDMS layer with a liquid polymer. After removing the excess polymer, the PDMS layer is inverted and brought into contact with a substrate. After curing the polymer, the PDMS layer is cautiously peeled away, leaving a solid structure on the surface of the substrate.

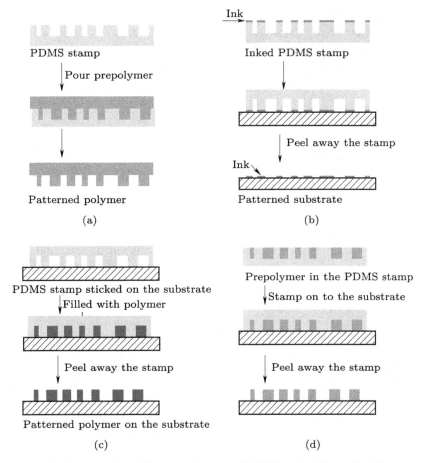

Fig. 11.6 Techniques of soft lithography using PDMS mold: (See color illustration at the back of the book)

2.2 PDMS soft lithography for microfluidics

Microfluidic devices consist of networks of channels at the microscale to perform specific functions. Miniaturization of the channels allows several functions to be integrated onto a single microfluidic device, allowing the device to perform very sophisticated tasks, such as sorting cells and worms, performing combinatorial chemistry, crystallizing proteins, detecting minute concentrations of DNA with ultra-sensitive PCR, as well as a host of other applications.

A microfluidic device can be readily obtained by the soft lithography technique. Typically, the microchannels imprinted in the PDMS layer are simply closed with a glass slide or PMMA plate. Alternatively, another piece of PDMS can be used. Permanent solid bonding between the layer of PDMS and the glass slide is required. This bonding may be facilitated by plasma bonding. When the microchannels are properly sealed, fluids can be pumped at pressures as high as about 350 kPa without failure.

In this experiment, a simple mimic microfluidic device dealing with precise control of fluids at the microscale is prepared. The fluidic channels are created by casting the PDMS on a Y-shaped master. The size of the master is about 6 cm, and the width and thickness of each branch are about 1 − 2 mm.

3 Experimental

3.1 Preparation

(1) Materials: DVDs

(2) Chemicals: PDMS base and curing agent (Dow Corning,Sylgard184 Silicone Elastomer Kit), isopropanol, red and green food color

(3) Apparatuses: drying oven, vacuum desiccator, optical microscope, electronic balance

(4) Tools: plastic Petri dishes, disposal cups, disposal plastic spoons, plastic droppers, stir sticks, PMMA plate, binder clips, Y-shaped master made by steel or plastic rods

3.2 Procedure

3.2.1 Replica of DVD structure

(1) Measure 25 g PDMS base using a disposable plastic spoon and 2.5 g curing agent using a disposable plastic dropper in a disposal cup.

(2) Mix them thoroughly using a stir stick for more than 5 min.

(3) Degas the mixture in the desiccator for 0.5 h to remove the bubbles.

(4) Cut a piece of DVD with a size of about 2.5 cm × 2.5 cm.

(5) Carefully peel apart the two plastic layers of DVD. Observe the surface of the part with a metal coating by the optical microscope. Take micrographs of the groove structure.

(6) Place the DVD specimen in a Petri dish with a diameter of 5 cm. The metal side with the groove structure should be upwards.

(7) Slowly pour about 5 g of the degassed mixture of PDMS and curing agent into the petri dish.

(8) Degas again for 0.5 h in the desiccator to remove the bubbles.

(9) Place the Petri dish into the drying oven at 80°C for 1 h.

(10) Cool down the Petri dish to room temperature and gently remove the DVD.

(11) Observe the surface of the curing PDMS sample by the optical microscope and take micrographs.

3.2.2 Fabrication of microfluidic device

(1) Put the Y-shaped master in the middle of a Petri dish with a diameter of 10 cm. Fix it on the bottom of the Petri dish.

(2) Slowly pour about 20 g of the mixture of PDMS and curing agent into the Petri dish.

(3) Evacuate for 0.5 h in the desiccator to remove the bubbles.

(4) Cure the PDMS in the drying oven at 80°C for 1 h.

(5) Cool down and remove the PDMS from the Petri dish.

(6) Cut the cured PDMS to show the end of the Y-shaped groove. Place it on the PMMA plate. Use bind clips to fix them.

(7) Mix isopropanol with red and green food color, respectively. Inject them into two channels. Observe the flow of the two liquids. Take a photo of the phenomenon.

4 Requirements for experimental report

(1) PDMS is the criticalmaterial in soft lithography. Describe its molecular formula, chemical properties, and physical properties.

(2) Give the micrographs of the DVD and the PDMS replica. Compare their details. Analyze the factors that influence the quality of the replica.

(3) Give the photo of liquid flow in the microfluidic device. Explain the experimental phenomenon.

5 Questions and further thoughts

(1) The track structure of the DVD is periodic. Can you design an optical experiment to judge how well the spacing of the copied tracks is?

(2) Please describe the applications of microfluidic devices with one or two examples.

Experiment 52.3   Pattern Transfer by Nanoimprint Lithography

Type of the experiment: Comprehensive
Recommended credit hours: 8

1 Objectives

(1) To understand the principle and techniques of nanoimprint lithography.

(2) To transfer the grating pattern onto PMMA by thermal nanoimprint technique.

(3) To transfer the groove structure of DVD by UV nanoimprint technique.

(4) To understand the applications of nanoimprint lithography.

2 Principles

2.1 Nanoimprint lithography

Nanoimprint lithography (NIL) is a method of fabricating nanoscale patterns with advantages of low cost, high throughput, and high resolution. It creates patterns by mechanical deformation of imprint resist and subsequent processes. The imprint resist is typically a monomer or polymer cured by heat or UV light during the imprinting. Adhesion between the resist and the template is controlled to allow proper release.

Compared with optical lithography, the difference in principles makes NIL capable of producing about 10 nm features over a large area with high throughput and low cost. Therefore, it can lead to a wide range of applications in electronics, photonics, data storage, and biotechnology.

2.2 Nanoimprint lithography techniques

There are many NIL techniques, but two of them are the most important and fundamental. They are thermoplastic nanoimprint lithography (T-NIL) and UV nanoimprint lithography (UV-NIL). All the other nanoimprint lithography techniques, such as roller nanoimprint lithography and jet and flash imprint lithography, are derivatives of these two techniques. Figure 11.7 shows the process of these two techniques.

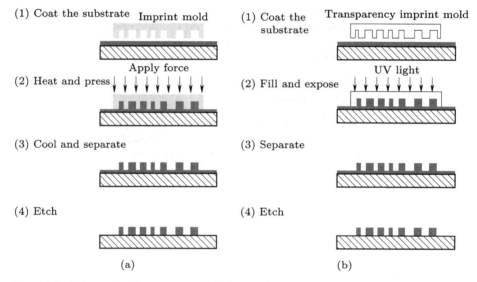

Fig. 11.7 Schematic illustrations of (a) thermoplastic nanoimprint lithography (T-NIL) and (b) UV nanoimprint lithography (UV-NIL).

In a standard T-NIL process, a thin layer of imprint resist, which is a thermoplastic polymer, is spin-coated onto the sample substrate. Then the mold, which has predefined topological patterns, is placed to contact the sample. The pattern on the mold is pressed into the softened polymer film at a temperature above the glass transition temperature of the polymer. After being cooled down, the mold is separated from the imprint. The acquired pattern in the imprint is the negative replica of the mold. The imprint can be further transferred to the substrate by other pattern transfer processes, such as photolithography.

In the UV-NIL process, a UV curable liquid resist is applied to the sample substrate, and the mold is typically made of transparent material like fused silica or PDMS. After the mold and the substrate are pressed together, the resist is cured in UV light and becomes solid. Then the mold is separated from the imprint. This step is also followed by a pattern transfer process to transfer the features from the imprint to the substrate.

Both these two techniques will be used in this experiment. The grafting pattern is transferred from a quartz glass grating to the thermoplastic polymer (PMMA) by T-NIL. Then the groove structure of the DVD is transferred by UV-NIL.

3 Experimental

3.1 Preparation

(1) Materials: quartz glass grating, thermal oxide silicon wafer with a size of 20 mm × 20 mm (the thickness of the SiO_2 layer is about 200 nm), glass slide, PDMS replica with DVD structure (prepared in Experiment 52.2).

(2) Chemicals: 1H,1H,2H,2H-perfluorodecyltrichlorosilicane (FDTS), poly(methyl methacrylate) (PMMA, MW=15000), toluene, heptane, UV resist (Norland optical adhesive, 6301), ethanol, acetone, isopropanol, silicon wafer detergent.

(3) Apparatuses: spin coater (SETCAS Electronics CO., Ltd, KW-4A), thermal press machine, ozone generator, ultraviolet curing box (Intelli-ray 400), optical microscope, stylus profilometer(KLA-Tencor, Alpha-Step D-300), electronic balance, heating plate, ultrasonic cleaner.

(4) Tools: washing holder, beakers, pipettes.

3.2 Procedure

3.2.1 Transfer grafting pattern by T-NIL

(1) Prepare a 7wt% PMMA toluene solution.

(2) Ultrasonically clean the silicon wafer. The procedure has been described in Experiment 50.

(3) Deposit a layer of the PMMA solution on a clean silicon wafer by spin coating. Spin at 500 r/min for 15 s and then at 3500 r/min for 30 s.

(4) Dry the coated wafer on the heating plate at 95 °C for 10 min to remove the solvent.

(5) Prepare a solution of 3 drops FDTS in 10 mL heptane.

(6) Clean a quartz glass grating ultrasonically with acetone, isopropanol, DI water, and ethanol for 10 min, sequentially.

(7) Soak the quartz glass grating in the FDTS solution for 10 min. An anti-stick layer of FDTS will form on the surface of the grafting.

(8) Place the quartz glass gratingonto the PMMA coated wafer carefully. Then, place them on the sample stage of the thermal press machine.

(9) Press under about 35 MPa at 175 °C for 10 min.

(10) After cooling, take out the sample. Remove the quartz glass grating.

(11) Observe the structure of the grafting and the transferred pattern on the PMMA using the optical microscope. Take micrographs.

(12) Measure the depth of the steps on the grafting mold and the transferred pattern by the stylus profilometer.

3.2.2 Transfer the groove structure of DVD by UV-NIL

(1) Clean a glass slide ultrasonically with acetone, isopropanol, DI water, and ethanol for 10 min, sequentially. Then treat it in an ozone generator for 20 min.

(2) Spincoat the UV resist layer on the glass slide at 500 r/min for 15 s and then at 3500 r/min for 30 s.

(3) Press the PDMS replica with DVD structure (prepared in Experiment 52.2) onto the resist layer.

(4) Cure the resist in the UV curing box at 100 W for 5 min.
(5) Carefully remove the mold.
(6) Observe the structure of the PDMS replica and the transferred pattern using the optical microscope. Take micrographs.

4 Requirements for experimental report

(1) Give the micrographs of the patterns observed under the optical microscope.
(2) Evaluate the quality of the transferred patterns by comparing the dimensions of the details in the transferred patterns with those in the grating and DVD. Discuss the factors that influence the quality.

5 Questions and further thoughts

(1) What are the advantages of NIL compared with photolithography?
(2) How can we choose the temperature and the pressure of the thermal press during T-NIL?
(3) Give an example to illustrate the application of the NIL technique.

Experiment 53   Fabrication of Enzyme-Based Glucose Biosensor

Type of the experiment: Comprehensive
Recommended credit hours: 16
Brief introduction: In this experiment, the working principle of enzyme base glucose biosensor is introduced. Nylon films coupled with glucose oxidase enzyme will be prepared as the biological component. YSI DO200 oxygen sensor is used as the transducer and detector component to detect the oxygen concentration. Then a glucose sensor is assembled by fixing the nylon film on the electrode of the oxygen sensor. It is used to detect the concentration of a solution with unknown glucose concentration based on the construction of the stand curve. Then the test conditions of pH value and temperature of the solution are optimized.

1 Objectives

(1) To understand the working principle and components of enzyme-based glucose biosensor.
(2) To master the assembly of the glucose sensor.
(3) To detect the glucose concentration and optimize the test conditions.

2 Principles

2.1 Biosensors

A biosensor is an analytical device used for the detection of an analyte. It combines a biological component with a physicochemical detector. In this device, the analyte under study is a biologically derived material or biomimetic component. It interacts with biologically sensitive elements that can be natural or made by biological engineering, such as tissue, microorganisms, organelles, cell receptors, enzymes, antibodies, and nucleic acids. The interaction of the analyte with the biological component will produce a signal, which will be further transformed by the transducer or the detector component into another signal that can be more easily measured and quantified.

2.2 Enzyme based glucose biosensor

The glucose oxidase enzyme (GOD) is also known as an oxidoreductase. It can catalyze the oxidation of glucose to hydrogen peroxide and D-glucono-δ-lactone. This enzyme is secreted by certain species of fungi and insects and shows antibacterial activity when oxygen and glucose are present.

Enzyme-based electrochemical biosensors utilize various oxidoreductases and are applied to many analytical instruments used in environmental, food, pharmaceutical, and clinical laboratories. The most successful application of electrochemical biosensors is glucose sensing, which helps millions of diabetic people monitor their blood glucose.

The first biosensor was described in 1962 by Clark and Lyons for quantifying glucose concentration in a sample directly. They immobilized GOD on an amperometric oxygen electrode surface covered with a semipermeable dialysis membrane. Diabetes mellitus is characteristic of hyperglycemia, a chronically raised concentration of blood glucose. As a result, it is critical for diabetic patients to monitor their glucose concentration in the blood frequently. In the presence of glucose, the oxidized form of GOD reacts with glucose and produces gluconic acid and reduced GOD, involving two electrons and two protons. This oxidation of glucose also consumes oxygen in solution since the dissolved oxygen reacts with the reduced GOD. Consequently, hydrogen peroxide and oxidized GOD, i.e., gluconic acid, are formed by the reaction

$$\text{Glucose} + O_2 \xrightarrow{\text{GOD}} \text{Gluconic acid} + H_2O_2. \qquad (11.19)$$

This reaction lowers the oxygen concentration. As a result, the electrode can sense the glucose by electrochemically sensing oxygen with the Clark oxygen electrode.

In this experiment, nylon films coupled with GOD are prepared as the biological component. Then the enzyme-based glucose sensor is assembled. YSI DO200 oxygen sensor will be used as the transducer and detector component to detect oxygen concentration.

3 Experimental

3.1 Preparation

(1) Materials: nylon mesh filters with a size of 30 mm × 30 mm.

(2) Chemicals: glucose oxidase, glucose, phosphate buffer (PB buffer), methanol, hydrochloric acid (HCl), boric acid (H_3BO_3), sodium tetraborate decahydrate ($Na_2B_4O_7 \cdot 10H_2O$), sodium chloride (NaCl), 50% glutaraldehyde ($C_5H_8O_2$), calcium dichloride ($CaCl_2$), deionized(DI) water.

(3) Apparatuses: dissolved oxygen meter (YSI DO200), oven, refrigerator, electromagnetic stirrer, electronic balance, pH meter, water bath.

(4) Tools: beakers, graduated cylinders, glass culture dishes, suction paper,

rubber strings.

3.2 Procedure

3.2.1 Preparation of solutions for the following usage

(1) $CaCl_2$ solution: dissolve 18.6 g $CaCl_2$ in 18.6 mL DI water in a 150 mL beaker; dilute it with methanol to 100 mL.

(2) 3.65 mol/L HCl solution: dilute 30.5 mL HCl with DI water to 100 mL in a 150 mL beaker.

(3) Boric acid salt solution (pH = 8.4): dissolve 0.858 g $Na_2B_4O_7 \cdot 10H_2O$ and 0.68 g H_3BO_3 with pure water in a 100 mL beaker; dilute it with DI water to 100 mL.

(4) 0.1 mol/L PB buffer (pH = 7.2): dissolve 5.12 g $NaH_2PO_4 \cdot 2H_2O$ and 4 g $Na_2HPO_4 \cdot 12H_2O$ with DI water in a 500 mL beaker; dilute it with DI water to 500 mL.

(5) 0.5 mol/L NaCl solution: dissolve 2.93 g NaCl with 0.1 mol/L PB buffer (pH = 7.2) in a 100 mL beaker; then dilute it with the PB buffer to 100 mL.

(6) 5% glutaraldehyde solution: dilute 10 mL 50% glutaraldehyde with H_3BO_3 solution (pH = 8.4) to 100 mL.

(7) 1 mg/mL glucosesolution: dissolve 100 mg glucose with 0.1 mol/L PB buffer in a 100 mL beaker; dilute it with the PB buffer to 100 mL.

3.2.2 The surface treatment of nylon and the grafting of glucose oxidase on nylon substrates

(1) Place 4 pieces of nylon mesh filters into a 150 mm glass culture dish. Wash the films with DI water 3 times, and 5 min for each time. Remove the residual water with suction paper.

(2) Fully immerse the nylon films with 50 mL $CaCl_2$ solution in the dish. Seal the dish and place it in the oven at 50 °C for about 30 min.

(3) Remove $CaCl_2$ solution and wash the films 3 times with DI water to remove the residual $CaCl_2$ solution. Remove the excess water with suction paper.

(4) Add 50 mL 3.65 mol/L HCl solution into the dish to hydrolyze the nylon films. Seal the dish and place it in the oven at 50 °C for about 45 min.

(5) Remove HCl solution and wash the films twice with DI water to remove the residual $CaCl_2$ solution. Wash the films with the 0.1 mol/L PB buffer once and then remove PB buffer.

(6) Couple the nylon films with 50 mL 5% glutaraldehyde solution for about 20 min. Then remove 5% glutaraldehyde solution and wash the films 3 times with the 0.1 mol/L PB buffer.

(7) Add 2 − 3 mL 1 mg/mL glucosesolution on the surface of nylon films. Then place the films for about 3.5 h at 4 °C in the refrigerator.

(8) Wash the films with 0.5 mol/L NaCl solution 3 times to remove the residual glucosesolution. Keep the films in the refrigerator at 4°C for the following experiment.

3.2.3 Assembly of glucose biosensor

(1) Cover the electrode of the oxygen sensor using a piece of nylon film.

(2) Fix the film using a rubber string. The film should paste on the electrode, and there must be no bubbles between the film and the electrode.

3.2.4 Plot the standard curve of glucose concentration against ΔO_2

(1) Prepare a series of glucose solutions in which the concentrations of glucose are 0.0625, 0.125, 0.25, 0.50, 1.00 mg/mL, respectively. The volume of each solution is 20 mL.

(2) Immerse the detector of the glucose biosensor into 20 mL PB buffer, and record the initial concentration of O_2 of PB buffer (in mg/L).

(3) Immerse the detector of glucose biosensor into glucose solution with different concentrations from 0.0625 to 1 mg/mL, respectively. The detector should be immersed for 5 min in each solution before reading the concentration. Rinse the detector with 0.1 mol/L PB buffer (pH = 7.2) after each measurement.

(4) Calculate ΔO_2 values, which is the difference between the concentration of O_2 in PB buffer and the concentration of O_2 in glucose concentration after immersed for 5 min.

(5) Tabulate the glucose concentration and ΔO_2. Plot the standard curve and the equation of curve fitting.

(6) Measure the glucose concentration of a solution with an unknown concentration. Detect the ΔO_2 value of the solution and calculation the concentration using the standard curve.

3.2.5 Optimize the pH value for the glucose sensor

(1) Prepare a series of 20 mL 1 mg/mL glucose solution with different pH values of 3, 5, 7, 9, and 11 by adding NaOH in the solution, respectively.

(2) The concentration of O_2 in 20 mL 0.1 mol/L PB buffer (pH = 7.2) is recorded as the initial concentration of O_2.

(3) Immerse the detector of glucose biosensor into the glucose solutions with different pH values, respectively. Record the O_2 concentration after immersed for 5 min. Rinse the detector with 0.1 mol/L PB buffer (pH = 7.2) after each measurement.

(4) Calculate the ΔO_2 values. Tabulate the pH and ΔO_2 values. The best pH value for the glucose biosensor is that with the largest ΔO_2.

3.2.6 Optimize the temperature for glucose biosensor

(1) Prepare a series of 20 mL 1 mg/mL glucose solution held at different temperatures of 0°C, 20°C, 35°C, 50 °C, and 70 °C in the water bath, respectively.

(2) The concentration of O_2 in 20 mL 0.1 mol/L PB buffer (pH = 7.2) is recorded as initial concentration of O_2.

(3) Immerse the detector of glucose biosensor into the glucose solutions at different temperatures, respectively. Record the O_2 concentration after immersed for 5 min. Rinse the detector with 0.1 mol/L PB buffer (pH = 7.2) after each measurement.

(4) Calculate the ΔO_2 values. Tabulate the temperature and ΔO_2 values. The best temperature for the glucose biosensor is that with the largest ΔO_2.

4 Requirement for experimental report

(1) Give the tables of the experiment results.
(2) Plot the standard curve and the equation of curve fitting.
(3) Analyze the results and suggest the optimum pH value and the temperature for the glucose sensor.

5 Questions and further thoughts

(1) Please explain the working principle of the Clark-type electrode of the oxygen sensor.
(2) What is the function of the $CaCl_2$ solution in this experiment?

Extended readings

Edward R. Hall-Effect Sensors. Theory and Application. 2nd edition. Newnes, 2006.

Guo L J. Nanoimprint lithography: Methods and material requirements. Advanced Materials, 2007, 19(4): 495-513.

Levinson H J. Principles of Lithography. 4tth Edition. SPIE, 2019.

Lundin D., Minea T., Gudmundsson J. T.,High Power Impulse Magnetron Sputtering: Fundamentals, Technologies, Challenges and Applications.Elsevier, 2020.

Luttge R. Nano- and Microfabrication for Industrial and Biomedical Applications. William Andrew, 2016.

Madou M J. Fundamentals of Microfabrication: The Science of Miniaturization. 2nd Edition. CRC Press, 2002.

Park J S, Maeng W J, Kim H S, et al. Review of recent developments in amorphous oxide semiconductor thin-film transistor devices. Thin Solid Films, 2012, 520: 1679-1693.

Swann S. Magnetron sputtering. Physics in Technology, 1988, 19: 67-75.

Taniguchi J, Ito H, Mizuno J, et al. Nanoimprint Technology: Nanotransfer for Thermoplastic and Photocurable Polymers.Wiley, 2013.

Wasa K, Hayakawa S. Handbook of Sputter Deposition Technology. Noyes Publications, 1992.

Wilbur J L, Kumar A, Kim E, et al. Microfabrication by microcontact printing of self-assembled monolayers. Advanced Materials, 2010, 6(7-8): 600-604.

Xia Y N, Whitesides G M. Soft lithography. AngewandtChemie International Edition, 1998, 37: 550-575.

Yoon J Y. Introduction to Biosensors. Springer, 2018.

Chapter 12 Biomaterials

Experiment 54 Preparation and Antibacterial Property of Silver-Loaded Activated Carbon Composites

Type of the experiment: Comprehensive
Recommended credit hours: 16
Brief introduction: This experiment introduces the preparation and antibacterial property of antibacterial silver-loaded activated carbon (Ag-AC) composites. The effect of Ag content on the property is analyzed. The experiment is divided into two experiments that can be conducted respectively. In the first experiment, Ag-AC composites with different Ag contents are prepared by soaking AC in $AgNO_3$ solution followed by calcination. Then, the morphology and crystal structure of the as-prepared composites are characterized by scanning electron microscope (SEM) and X-ray diffraction (XRD). In the second experiment, the *Escherichia coli* (*E. coli*) are cultured, and the antibacterial test is conducted for the as-prepared Ag-AC composites.

Experiment 54.1 Preparation and Structure Characterization of Antibacterial Silver-loaded Activated Carbon Composites

Type of the experiment: Comprehensive
Recommended credit hours: 8

■ Objectives

(1) To understand the basic features of activated carbon, silver nanoparticles, and silver-loaded activated carbon material.

(2) To prepare silver-loaded activated carbon composites with antibacterial property.
(3) To characterize the structure of the composite material by SEM and XRD.

2 Principles

2.1 Application of AC

AC is an allotrope of graphite and diamond. It is a porous material in the shape of powder, granules, or fibers, with a specific structure and a large surface area of (500–1000 m^2/g). Because of its massive surface area, AC can absorb gas, liquid, or colloidal substances onto its surface. This ability of absorption is partial since it prefers polar substances to non-polar substances. The amount of absorption depends on the pressure and temperature, i.e., the higherpressure and lower temperature, the higher absorption ability. AC has found widespread applications in removing pollutants from gas and liquid phases, and in wastewater treatment and water purification processes, due to their outstanding adsorption properties and high specific surface area.

The activated carbon is usually produced from popularly agricultural waste such as coconut husk and bamboo. Carbon materials are biocompatible with bacteria and suitablesupports for their development. An activated carbon material with antibacterial property can be obtained by being impregnated with silver nanoparticles.

2.2 Ag-AC composites

One way to foster new applications of AC involves the modification of its surface through chemical reactions or deposition of new materials with attractive properties, such as bactericidal agents. In this respect, silver is an excellent candidate that complies with the bactericidal requirements. Consequently, Ag-AC composites have been synthesized for this application.

Silver nanoparticles (AgNPs) are commonly utilized nanomaterial because of their antibacterial property and high electrical conductivity. Several antibacterial mechanisms of AgNPs have been proposed, and four widely accepted mechanisms are schematically illustrated in Fig. 12.1.

In the first mechanism, AgNPs can anchor onthe bacterial cell wall and subsequently penetrate it, thereby causing a structural change in the cell membrane. The permeability change of the cell membrane induces death of the cell. It is believed that the high affinity of AgNPs towards sulfur or phosphorous element is the criticalpoint of its antibacterial property. As sulfur and phosphorous elements are found in abundance throughout the cell membrane, AgNPs react with sulfur-containing proteins inside or outside the cell membrane, affecting cell viability.

The second mechanism is that free radicals induced by the AgNPscan damage the cell membrane and make it porous, ultimately leading to cell death.

In the third mechanism, the bacterial cells in contact with AgNPs take in silver ions, which inhibit several functions of the cell and damage the cells. Then, silver ions can produce reactive oxygen species (ROS) by inhibiting respiratory enzymes, which attack cells.

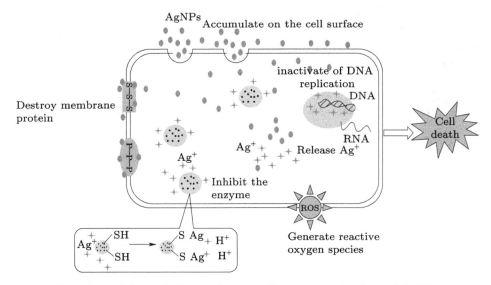

Fig. 12.1 Schematic illustration of antibacterial mechanisms of AgNPs.

The fourth mechanism proposed that the release of silver ions from the nanoparticles can interact with phosphorous moieties in DNA, which in turn inactivates DNA replication, or can react with sulfur-containing proteins to inhibit enzyme functions.

Whatever the mechanism, the antibacterial property of AgNPs is inevitable, which allows the incorporation of AgNPs into various matrixes such as activated carbon, polymeric networks, and textiles materials to prepare antibacterial composites.

3 Experimental

3.1 Preparation

(1) Materials: activated carbon washed twice with water and thoroughly dried in an oven before the experiment.

(2) Chemicals: silver nitrate ($AgNO_3$).

(3) Apparatuses: muffle furnace, oven, centrifuge, scanning electron microscope (TECAN Vega3 LM), X-ray diffractometer (Rigaku, Miniflex 600), vacuum filtration apparatus (the components have been listed in Experiment 15), electrical balance.

(4) Tools: graduated cylinder, beakers, glass Petri dishes, pipettes, mortar and pestle, tin foil.

3.2 Procedure

3.2.1 Preparation of silver-loaded activated carbon composite

(1) Prepare $AgNO_3$ solutions with concentrations of 0.025 mol/L, 0.050 mol/L, and 0.100 mol/L, respectively.

(2) Weigh 5.0 g activated carbon in four 100 mL beakers, respectively.

(3) Add 30 mL pure water and AgNO$_3$ solutions with different concentrations in each beaker. Then the activated carbon particles are impregnated in water or AgNO$_3$ solutions for 2 h. Cover these beakers with tin foil to avoid radiation of light.

(4) Collect the as-impregnated activated carbon by vacuum filtration and wash twice in the Büchner funnel with pure water. Collect the products in glass Petri dishes.

(5) Heat the samples in the Petri dishes in the muffle furnace at 250 °C for 1 h. The AgNO$_3$ is decomposed into Ag particles. The main reaction formula is

$$2AgNO_3 \rightarrow 2Ag + 2NO_2 \uparrow + O_2 \Uparrow \tag{12.1}$$

3.2.2 Structure characterization

(1) Grind the as-prepared Ag-AC composites and the AC with a mortar and pestle into fine powders, respectively. The crystal structures are then detected by XRD.

(2) The microstructures of the composites are observed by SEM. The contents of Ag are determined by X-ray energy dispersive spectroscopy (EDS).

4 Requirements for experimental report

(1) Plot the XRD profiles and index the peaks.

(2) Provide the SEM images and Ag contents of the Ag-AC composites. Describe the morphology of the composites.

5 Questions and further thoughts

(1) What are the states of Ag during the preparation of Ag-AC composites in the experiment?

(2) What is the type of bond betweenAg nanoparticles and activated carbon?

6 Cautions

Silver nitrate is a poisonous, corrosive, colorless crystalline compound. It becomes grayish black when exposed to light in the presence of organic matter. Once the skin is dipped in a silver nitrate solution, black spots appear due to the formation of black protein silver. Please wear gloves during the experiment.

Experiment 54.2 Antibacterial Test of Silver-Loaded Activated Carbon Composites

Type of the experiment: Comprehensive
Recommended credit hours: 8

1 Objectives

(1) To learn the culture method of the *Escherichia coli*.

(2) To understand the principle and method of antibacterial test.

(3) To analyze the effect of Ag content in Ag-AC composites on the antibacterial property.

Principles

2.1 Escherichia coli

Escherichia coli (*E. coli*) is a facultatively anaerobic, Gram-negative, coliform bacterium of genus *Escherichia*. It is commonly found in the environment, foods, and intestines of people and mammals. Most *E. coli* are harmless and actually are an essentialpart of a healthy human intestinal tract. However, some kinds are pathogenic, which means that they can make their hosts sick. Some can cause severefood poisoning, while others cause breathing problems, urinary tract infections, and pneumonia. *E. coli* can be transmitted through contaminated water or food, or it can be transmitted by humans or animals. In this respect, detection of the numbers of *E. coli* in drinking water is an important indicator of water quality testing.

The cells of *E. coli* are typically rod-shaped and are about 2.0 μm long and 0.25–1.0 μm in diameter. The structure of *E. coli* is illustrated in Fig. 12.2. It possesses adhesive fimbriae and a cell wall that consists of an outer membrane containing lipopolysaccharides, a periplasmic space with a peptidoglycan layer, and an inner, cytoplasmic membrane. Even though it has an extremely simple cell structure, with only one chromosomal DNA and a plasmid, it can perform complicated metabolism to maintain its cell growth and cell division.

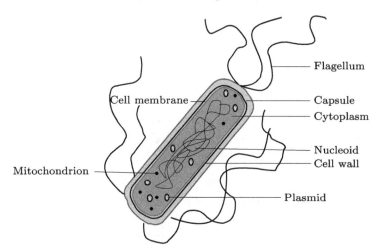

Fig. 12.2 The structure of *E. coli*.

2.2 The culture and isolation of E. coli

E. coli is a chemoheterotroph that medium must include a source of carbon and energy. It can grow in any medium that contains glucose, ammonium phosphate monobasic, sodium chloride, magnesium sulfate, potassium phosphate, dibasic, and water. *E. coli* can grow at 10–40 °C, but the optimum temperature for most of the strains is 37 °C. Besides, *E. coli* is a facultative anaerobe. It can grow best in the presence of oxygen or grow in a low oxygen environment.

Lysogeny Broth (LB) is the most common medium used for maintaining and cultivating *E. coli*. It usually contains three ingredients: tryptone that provides peptides and peptones, yeast extract that provides vitamins and certain trace elements, sodium chloride that provides sodium ions for transport and osmotic balance. From the physical state, LB can be divided into liquid and solid (contain 1%–2% agar) types used for culturing and isolating, respectively. Both the streak plate method and the coating plate method are used for isolating *E. coli*.

2.3 Plate colony counting method

In order to quantify *E. coli* in a culture, *E. coli* can be simply plated on a Petri dish with a growth medium. The plate count is linear for *E. coli* over the range of 30–300 CFU[①] on a standard-sized Petri dish. In microbiology, a CFU is a unit used to estimate the number of viable bacteria or fungal cells in a sample. Therefore, to ensure that a sample will yield colony numbers in this range, the solution will be diluted ten-fold series, and the dilution series is plated in replicates of 2 or 3 over the chosen range of dilutions. The colony numbers are read from a plate in the linear range, and then the bacteria concentration P (CFU/mL) of the original is calculated by

$$P = \frac{Z \times R}{V}, \tag{12.2}$$

where Z is the average value of the colony numbers in the plates, R is dilution rate, V is the volume plated.

3 Experimental

3.1 Preparation

(1) Materials: Ag-AC composites prepared in Experiment 54.1, bacterial glycerol stock.

(2) Chemicals: tryptone, yeast extract, sodium chloride (NaCl), agar, deionized water, sterile water, silver nitrate ($AgNO_3$).

(3) Apparatuses: super clean bench, incubator, autoclave, refrigerator, shaking incubator.

(4) Tools: 12 mL sterile tubes, 250 mL conical flasks, silicone stoppers, 9 cm culture dishes, 1000 mL beakers, alcohol burner, inoculating loop, liquid and solid waste containers.

3.2 Procedures

3.2.1 Prepare LB liquid medium

(1) Dissolve 1.0 g tryptone, 0.5 g yeast extract, and 1.0 g NaCl in deionized water in a 250 mL conical flask and dilute with deionized water to 100 mL.

(2) Plug the flask with a silicone stopper. This silicone stopper is specially designed with holes for vapor getting out readily.

---

① CFU is the colorvy-forming unit, the same belav.

(3) Put the LB liquid medium in an autoclave to sterilize at 121°C for 20 min.

3.2.2 Prepare LB solid medium (LB agar medium)

(1) Dissolve 9.0 g tryptone, 4.5 g yeast extract, and 9.0 g NaCl in deionized water in a 1000 mL beaker and dilute with deionized water to 600 mL.

(2) Transfer the medium into six conical flasks equally. Add 2.25 g agar to each conical flask, then dilute with deionized water to 150 mL.

(3) Plug the flask with a silicone stopper.

(4) Put the medium into an autoclave to sterilize at 121°C for 20 min.

3.2.2 Pour LB agar plates

(1) Sterilize the super clean bench by spraying with 75% ethanol and wiping with paper towels. Turn on the UV light to disinfect the super clean bench for 30 min.

(2) Take out the LB agar medium solution from the autoclave and place it near an alcohol burneronthe super clean bench to maintain sterility. Allow the solution to cool enough to touch by hand. Maintain sterility by working near the alcohol burner onthe super clean bench.

(3) Pour out LB agar (about 15 mL) solution into a 9 cm culture dish to form a thin layer (about 5 mm in thickness), as shown in Fig. 12.3. When the agar solution is poured into the culture dish, do not lift the cover off, and just open enough space for pouring the agar.

(4) Swirl the culture dish in a circular motion to distribute agar on the culture dish thoroughly. The resulting culture dishes are called the agar plates.

(5) Allow the agar plates to cool until solidification (about 20 min), and then invert the agar plates to avoid water condensation on the agar.

(6) Label these agar plates with the date. Put them in plastic bags and store them in a refrigerator.

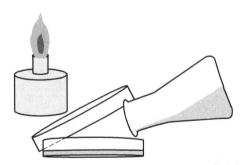

Fig. 12.3 Schematicillustration of pouring LB agar plate near an alcohol burner.

3.2.3 E.coli culture

3.2.3.1 Streaking and isolating bacteria on an agar plate

(1) Sterilize the lab bench by spraying with 75% ethanol and wiping with a paper towel. Maintain sterility by working near an alcohol burner.

(2) Take out an as-prepared agar plate from the refrigerator. Label the edge of the agar plate with the group number and the date.

(3) Take out the bacterial glycerol stock from the refrigerator.

(4) Sterilize an inoculating loop by placing it at an angle over a flame. The loop should turn red or orange before it is removed from the flame. Then cool it in the air.

(5) Gently dip the glycerol stock with the inoculating loop.

(6) Open the lid of the agar plate just enough to insert the loop.

(7) Streak the wire loop containing the bacteria at the top end of the agar plate moving in pattern #1 of three parallel lines, as shown in Fig. 12.4. The pattern covers about 1/4 area of the plate. When streaking the agar with the loop, be sure to move the loop horizontally to only streak the surface of the agar without destroying the agar layer.

(8) Sterilize the loop in the flame and cool it again.

(9) Rotate the agar plate at an angle and spread the bacteria from pattern #1 to a second area to obtain pattern #2.

(10) Repeat steps (8) and (9) to spread the bacteria to form patterns #3 and #4.

(11) Cover the lid and secure it with tape. Then invert the agar plates and incubate them overnight at 37°C in the incubator. Some white dots appear on the agar plates. They are colonies growing along the streaks and in isolated spaces. Each dot is composed of millions of genetically identical bacteria that arose from a single bacterium.

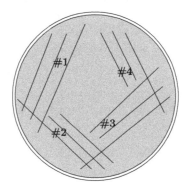

Fig. 12.4 Patterns of streaking bacteria on an agar plate.

3.2.3.2 E. coli inoculation

(1) Take out the as-prepared LB liquid medium and the *E. coli* agar plate that has been streaked from the refrigerator.

(2) Pour 3–5 mL of the LB medium into a sterile tube, label it. Using a sterile loop, transfer the colony to the liquid medium in the tube, as shown in Fig. 12.5. The tube is half-covered with the lid during the operation.

(3) Incubate the bacterial culture at 37 °C and 200 r/min for 12–18 h in a shaking incubator.

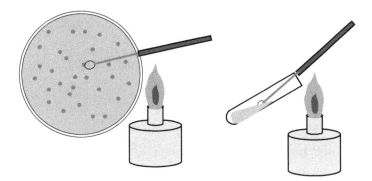

Fig. 12.5 Schematic illustration of *E. coli* inoculation.

3.2.4 Antibacterial test

(1) Sterilize all the glassware and materials in an autoclave at 121 °C for 20 min before the experiment.

(2) Prepare 5 mL distilled water containing 1.5×10^5 CFU/mL of *E. coli* in the sterile tube, and then shake it at 37 °C for about 1 h in the shaking incubator.

(3) Take 1.0 mL *E. coli* bacterial fluid and dilute with deionized water. The dilute rates are 10, 10^2, 10^3, 10^4, and 10^5. Figure 12.6 illustrates the procedure to prepare the diluted bacterial solution.

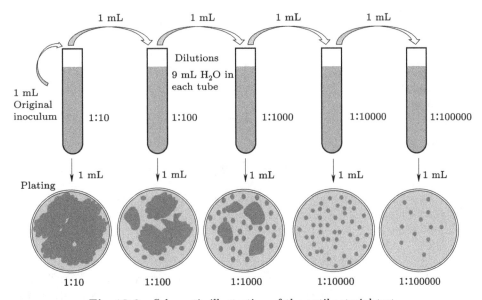

Fig. 12.6 Schematic illustration of the antibacterial test.

(4) Take 1.0 mL diluted solutions with different dilute rates to spread on three agar plates, respectively. These are the negative control group.

(5) Respectively add 0.1 g of Ag-AC composites with three different Ag contents

and 0.1 g of AC into 5.0 mL distilled water containing 1.5×10^5 CFU/mL of *E. coli*, and then shake them at 37 °C for about 1 h.

(6) Repeat steps (3) and (4).

(7) Place the agar plates in the incubator to cultivate at 37 °C for 24 h.

(8) Take out the plates and take photos.

(9) Count the numbers of bacterial colonies on the plates with different Ag contents, and different dilute rates. The valid count is in the range of 30–300 CFU. The count beyond this range is not recorded.

4 Requirements for experimental report

(1) Give the photos of the agar plates of the antibacterial test.

(2) Tabulate the number of bacterial colonies on the agar plates of the antibacterial test. Calculate the average value and standard deviation for each Ag content of Ag-AC composites and each dilute rate. Draw a column graph of the colony number against Ag content with error bars using Origin software.

(3) Calculate the antibacterial rate by

$$\text{Antibacterial rete} = \frac{(\text{Initial colony number} - \text{Colony number of termination})}{\text{Initial colony number}} \times 100\%, (12.3)$$

where the initial colony number is that of the negative control group, and the colony number of termination is that of the test group with Ag-AC composites. List the results in the table.

(4) Discuss the effect of Ag content of Ag-AC composite on the antibacterial behavior.

5 Questions and further thoughts

(1) *E. coli* are a group of Gram-negative bacteria. Do you know how to identify Gram-negative bacteria?

(2) Why do we usually shake the sterile tube when we culture *E. coli*?

6 Cautions

(1) Pouring LB agar plates and *E. coli* culture must be conducted onthe super clean bench.

(2) During growing bacterial colonies, millions of bacteria are treated. It is important to follow all lab safety rules, for example, wearing lab coats, gloves, and face masks to ensure that you do not inhale, ingest, or allow these germs to touch your skin.

(3) All experimental waste should be collected and processed properly. Waste bacterial plates should be disposed of properly by placing them in an autoclave to kill the bacteria before discarding them. Disinfectant is also used to destroy the bacterial colonies.

Experiment 55 Preparation and In Vitro Cytotoxicity of Pluronic F127-Encapsulated Curcumin Micelles

Type of the experiment: Comprehensive
Recommended credit hours: 20
Brief introduction: This experiment introduces the preparation and characterization of Pluronic F127-encapsulated curcumin micelles. The experiment is divided into two experiments that can be conducted respectively. In the first experiment, the Pluronic F127-encapsulated curcumin micelles are prepared by hydrating the polymer films of Pluronic F127 and curcumin in phosphate-buffered saline (PBS). The micelle powder is produced by the freeze-drying method. After that, the functional structure of the as-prepared powder of Pluronic F127-encapsulated curcumin micelles is studied by Fourier transform infrared (FTIR) spectroscopy. Several essential properties, including drug encapsulation efficiency, loading percentage, in vitro drug release, and storage stability, are characterized. In the second experiment, HeLa cells are cultured to investigate the toxicity of as-prepared powder of Pluronic F127-encapsulated curcumin micelles. The MTT (3-(4,5-dimethylthiazol-2-yl)-2,5-diphenyl tetrazolium bromide)assay is used for evaluating the cell toxicity of drug micelles.

Experiment 55.1 Preparation and Characterization of Pluronic F127-Encapsulated Curcumin Micelles

Type of the experiment: Comprehensive
Recommended credit hours: 12

■ Objectives

(1) To prepare Pluronic F127-encapsulated curcumin micelles.

(2) To understand the method of determining the solution concentration by ultraviolet-visible (UV-Vis) spectroscopy.

(3) To understand the methods of studying the in vitro release and drug stability.

(4) To determine the drug encapsulation efficiency and loading percentage.

■ Principles

2.1 Curcumin

Curcumin (1,7-bis (4-hydroxy-3-methoxyphenyl)-1,6-heptadiene-3,5 dione) is a polyphenol as shown in Fig. 12.7. It is a major constituent of turmeric, which is derived from the rhizomes of Curcuma longa. Curcumin has great potential in treatingvarious diseases, including cancer, cardiovascular diseases, arthritis, diabetes, psoriasis, and Alzheimer's, through modulation of numerous molecular targets.

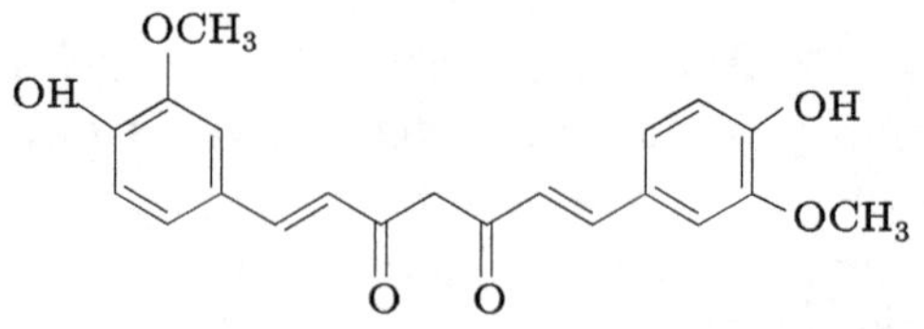

Fig. 12.7 The chemical structure of curcumin.

Due to the extremely low solubility of the compound in water, the clinical development of curcumin is limited, which restricts its use in intravenous administration and is associated with poor absorption in the intestine upon oral administration. Curcumin also undergoes rapid degradation by enzymes in the intestinal tract, resulting in low oral bioavailability. To overcome the problem of solubility, stability, and bioavailability of curcumin, the development of novel delivery systems has attracted more and more attention. One way to improve the aqueous solubility and stability of curcumin is to encapsulate the drug into polymeric micellar nanocarriers. In this experiment, we aimed to prepare Pluronic F127-encapsulated curcumin micelles for drug delivery application.

2.2 Pluronic F127-drug micelles

It has been proved that Pluronic block copolymer micelles are effective carriers for hydrophobic drugs with long-circulating characteristics. They are extensively used as emulsification, solubilization, dispersion, thickening, coating, and wetting agents. These block copolymers consist of hydrophilic ethylene oxide (EO), and hydrophobic propylene oxide (PO) blocks arranged in a triblock structure EO_x–PO_y–EO_x (abbreviated as PEO–PPO–PEO), where x and y chain lengths vary from 2–130 Units and 16–70 units, respectively.

One representative of these block copolymers is Pluronic F127, with a structure shown in Fig. 12.8. This arrangement results in amphiphilic copolymers, which self-assemble into micelles in the aqueous solution with core-shell structure above their critical micellar concentrations (CMC), as shown in Fig. 12.9. The core is formed by the hydrophobic interactions of the PPO blocks and remains segregated from the aqueous exterior by the hydrophilic corona formed by PEO blocks.

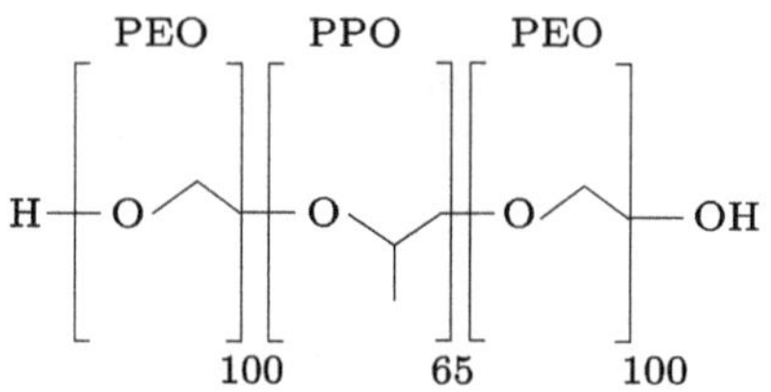

Fig. 12.8 The chemical structure of Pluronic F127.

The hydrophobic core of such micelles serves as cargo space to incorporate lipophilic therapeutic compounds, as shown in Fig. 12.10. At the same time, the hydrophilic corona contributes significantly to the pharmaceutical behavior of block copolymer formulations by maintaining the micelles in a dispersed state

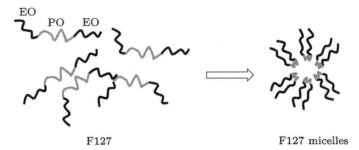

Fig. 12.9 Self-assemble of Pluronic F127 in aqueous solution. (See color illustration at the back of the book)

and decreasing undesirable drug interactions with cells and proteins through steric-stabilization effects. Therefore, the noncovalent incorporation of drugs into the hydrophobic PPO core of the Pluronic F127 micelles results in increased solubility, metabolic stability, bioavailability, and circulation time.

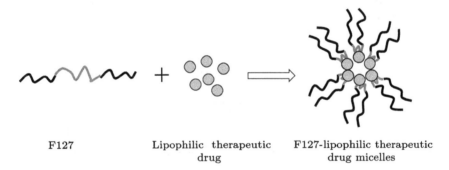

Fig. 12.10 Formation of Pluronic F127- lipophilic therapeutic drug micelles in aqueous solution. (See color illustration at the back of the book)

This biocompatible materialPluronic F127 has been used for the encapsulation of antifungal, neuroplastic, cycloplegic, and immune-suppressant. Moreover, Pluronic block copolymers have been proved to enhance the therapeutic activity of many anti-cancer agents such as carboplatin, camptothecin, and paclitaxel. The commercial availability, ease of preparation, well-studied physical properties, and strong safety profile makes Pluronic block copolymers appealing for drug delivery purposes.

Experimental

3.1 Preparation

(1) Chemicals: Pluronic F127 copolymer, curcumin, PBS(0.01 mol/L, pH = 7.4), dichloromethane (DCM, CH_2Cl_2), dimethyl sulfoxide (DMSO), KBr (IR grade)

(2) Apparatuses: rotary evaporator, freeze dryer, UV-Vis spectrometer

(PerkinElmer, Lambda 25), FTIR spectrometer (PerkinElmer Frontier), electromagnetic stirrer with a water bath, ultrasonic cleaner

(3) Tools: 100 mL round bottom flask, 1000 mL beaker, pipettes, dialysis tubing, dialysis clips, 20 mL syringe, 0.2 μm membrane filter, 10, 25, and 50 mL glass bottles, quartz cuvettes, tools for FTIR sample preparation listed in Experiment 39

3.2 Procedure

3.2.1 Prepare Pluronic F127-encapsulated curcumin micelles

(1) Prepare stock solutions of curcumin (1 mg/mL) and Pluronic F127 copolymer (10 mg/mL) in DCM, respectively.

(2) Transfer required amount of stock solutions as listed in Table 12.1 to separated round-bottom flasks to obtain the different drug to polymer ratios, 1:10, 1:25, and 1:50 by weight.

(3) The solvents are evaporated respectively by rotary evaporation at room temperature for about 10 min, and F127-curcumin films are obtained.

(4) Pre-warm the PBS solution at 37 °C in the water bath.

(5) Add 75 mL PBS solution in the flasks respectively to rehydrate the F127-curcumin films under ultrasonic vibration.

(6) The micelle suspensions are filtered by a 0.2 μm membrane filter. The filtered solutions are named stock micelle solutions.

Table 12.1 Mass ratios of curcumin to Pluronic F127 copolymer and the corresponding solutions

Curcumin-F127 weight ratio	Curcumin solution/mL	F127 solution/mL
1:10	5.0	5.0
1:25	5.0	12.5
1:50	5.0	25.0

3.2.2 Freeze-drying of Pluronic F127-encapsulated curcumin micelles

(1) Take 0.5 mL and 20 mL stock micelle solutions of each drug and polymer ratio. For convenience, these samples are denoted as samples A_{10}, A_{25}, and A_{50}, and B_{10}, B_{25}, and B_{50}, respectively (the subscripts correspond to the ratios).

(2) Freeze these solutions in the refrigerator.

(3) Place the frozen solutions in the freeze dryer. The chamber pressure is maintained at 0.035 mbar, and the temperature is -55 °C. For convenience, the lyophilized powders from 0.5 mL solutions are also denoted as A_{10}, A_{25}, and A_{50}, which will be used for determining the drug encapsulation efficiency, and those from 20 mL solutions are also denoted as B_{10}, B_{25}, and B_{50}, which will be used for FTIR analysis.

3.2.3 FTIR analysis of Pluronic F127-encapsulated curcumin powders

(1) Dry the F127-encapsulated curcumin powder samples B_{10}, B_{25}, and B_{50}, and Pluronic F127 and curcumin completely with the infrared heat lamp.

(2) Grind and mix samples (about 3 mg) and KBr powder (500 mg) with a mortar and pestle.

(3) Prepare pellets and acquire the FTIR spectra of these samples. The procedure has been described in Experiment 39.

3.2.4 Determination of drug encapsulation efficiency and loading percentage

(1) Plot the standard curve of absorbance at 430 nm (the wavelength at which absorbance of curcumin is the highest) against the curcumin concentration. The main operation steps are as follows:

(a) Prepare 20 μmol/L stock solution by dissolving 0.11 mg curcumin in 15 mL DMSO.

(b) Dilute 20 μmol/L stock solution to obtain series of drug solution with the concentrations of 1, 2, 5, 10, 20 μmol/L.

(c) Measure the UV-Vis absorption spectra of the solutions. Determine the absorbance at 430 nm.

(d) Plot a standard curve of the absorbance against the drug concentration. Fit the curve into a linear equation.

(2) Dissolve the samples A_{10}, A_{25}, and A_{50} with 20 mL DMSO, respectively, and then filter the solutions by a 0.2 μm membrane filter.

(3) Measure the absorbance at 430 nm of as-prepared solutions using the UV-Vis spectrometer. Then calculate the drug concentration in the micelles based on the standard curve.

(4) The drug encapsulation efficiency can be obtained by

$$\text{drug encapsulation efficiency} = \frac{\text{amount of encapsulated curcumin}}{\text{initial amount of curcumin}} \times 100\%. \tag{12.4}$$

The initial amount of curcumin is that listed in Table 12.1. The amount of encapsulated curcumin is calculated from the concentration determined by the absorbance.

(5) The loading percentage in micelle nanocarriers (i.e., Pluronic F127) can also be obtained by

$$\text{Drug loading percentage} = \frac{\text{Amount of encapsulated curcumin}}{\text{Initial amount of nanocarrier} + \text{Amount of encapsulated curcumin}} \times 100\%, \tag{12.5}$$

where the initial amount of nanocarrier is that of Pluronic F127 listed in Table 12.1.

3.2.5 In vitro release study of 1:50 Pluronic F127-encapsulated curcumin micelles

(1) Transfer 10 mL stock micelle solution with the weight ratio of 1:50 in a dialysis tubing. Place the dialysis tubing in a beaker with 1 L of the PBS solution at 37°C under gentle stirring.

(2) Samples of 400 μL are withdrawn from dialysis tubing periodically at the time points of 0 h, 1 h, 2 h, 3 h, 4 h, 8 h, 16 h, 24 h, 48 h for absorption measurement by UV-Vis spectrometer. 400 μL of PBS solution is replenished in the dialysis tubing after each withdrawal. The sample solutions are diluted with 4.6 mL DMSO for absorption measurement.

(3) After every 24 h, the PBS solution in the beaker should be changed with fresh PBS solution.

(4) Curcumin remaining in the micelles is quantified based on the standard curve. Then the percentage of curcumin released from the micelles at various time points is calculated by

$$\text{Drug release percentage} = \frac{\text{Amount of released curcumin}}{\text{Amount of encapsulated curcumin}} \qquad (12.6)$$

where the amount of encapsulated curcumin and the amount of released curcumin are calculated from the concentrations of the sample solutions at 0 h and other time points, respectively.

3.2.6 Storage stability of 1:50 Pluronic F127-encapsulated curcumin micelles

(1) Keep 10 mL stock micelle solution with the weight ratio of 1:50 at room temperature in the dark for the storage stability check.

(2) Every week, 400 µL solution is taken out and then diluted with 4.6 mL DMSO for absorption measurement.

(3) The absorbance at 430 nm will be recorded and plotted against time. This measurement needs to last for at least 10 weeks at room temperature, one data point per week.

4 Requirements for experimental report

(1) Plot the FTIR spectra of curcumin, F127, and Pluronic F127-encapsulated curcuminpowders in one coordinate system and separate them vertically using Origin software. Study the chemical structure of Pluronic F127-encapsulated curcumin by comparing with Pluronic F127 and curcumin to confirm curcumin encapsulation in Pluronic F127 micelles.

(2) Givethe standard curve of absorbance at 430 nm against the curcumin concentration and the equation of curve fitting.

(3) Tabulate the results of drug encapsulation efficiency and loading percentage. Give an example to illustrate the calculation process.

(4) Tabulate the drug release percentage. Plot drug release against time using the Origin software. Discuss the drug release of Pluronic F127-encapsulated curcumin micelles.

(5) Plot the curve of absorbance against time using the data from the storage stability test. Then analyze the stability of Pluronic F127-encapsulated curcumin micelles.

(6) Discuss how the drug-polymer ratio affects the drug encapsulation efficiency and the loading percentage of the Pluronic F127-encapsulated curcumin micelles.

5 Questions and further thoughts

(1) Why did we perform freeze-drying of Pluronic F127-encapsulated curcumin micelles for determining the drug concentration?

(2) How many times have we used PBS buffer in this experiment? What is the function of the PBS buffer each time?

(3) Why are the Pluronic F127-encapsulated curcumin micelles with the curcumin-F127 ratio of 1:50 selected for the studies of in vitro release, storage stability, and FTIR spectroscopy?

Experiment 55.2 In Vitro Cytotoxicity of Pluronic F127-Encapsulated Curcumin Micelles

Type of the experiment: Comprehensive
Recommended credit hours: 8

■ Objectives

(1) To understand the basic principle and method of cell culture.
(2) To understand MTT assay for evaluating the cell toxicity of drug micelles.
(3) To determine the cell viability of Pluronic F127-encapsulated curcumin micelles.

■ Principles

2.1 HeLa cell

HeLa cell is a cancerous cell belonging to a strain continuously cultured since it was isolated in 1951 from a patient suffering from cervical carcinoma. The designation HeLa is derived from the name of the patient, Henrietta Lacks. HeLa cells were the first human cell line to be established and have been used to test the effects of radiation, cosmetics, toxins, and other chemicals on human cells. They have been very important in gene mapping and studying human diseases, especially cancer. The most significant application of HeLa cells should be the development of the first polio vaccine. HeLa cells were used to supply hosts of poliovirus in human cells.

Now, there are many strains of HeLa, and all are derived from the same single cell. It is believed that the HeLa cells do not suffer programmed death. This is because they keep a special kind of enzyme telomerase that can prevent gradual shortening of the telomeres chromosomes since telomere shortening is a crucial-point of aging and death.

2.2 Cell culture

Cell culture is the technique by which cells are removed from an organism and placed in a fluid medium. Under proper conditions, the cells can live and even grow. The cells may be removed from the tissue directly and disaggregated by enzymatic or mechanical means before cultivation. They may alsobe derived from a cell line or cell strain that has already been established. In a broad sense, cell culture is the process by which prokaryotic, eukaryotic, insect, animal, or plant cells are grown under controlled conditions in vitro. In practice or the narrow sense, it refers to the culturing of cells derived from animal cells.

Culture conditions vary widely for each cell type. The artificial environment in which the cells are cultured consists of a suitable vessel containing a substrate or medium that supplies the essential nutrients (amino acids, carbohydrates, vitamins, minerals), growth factors, hormones, gases (O_2, CO_2), and regulates the physicochemical environment (pH, osmotic pressure, temperature). Most cells are anchorage-dependent and must be cultured when they are attached to a solid or semi-solid substrate (adherent or monolayer culture). In contrast, others can be grown floating in the culture medium (suspension culture).

Cells in culture can be divided into three basic categories based on their morphology. As shown in Fig. 12.11, they are fibroblast-like cells (Fig. 12.11(a)), epithelial-like cells (Fig. 12.11(b)), and lymphoblast-like cells (Fig.12.11(c)).

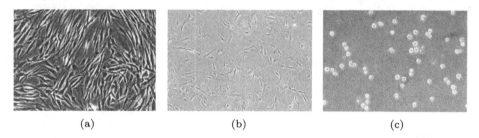

(a)   (b)   (c)

Fig. 12.11 Micrographs of (a) fibroblast-like cells, (b) epithelial-like cells, and (c) lymphoblast-like cells.

When the cells in adherent cultures occupy all the available substrate and have no room left for expansion, or when the cells in suspension cultures exceed the capacity of the medium to support further growth, they need to be subcultured. Subculturing, also referred to as passaging, is the removal of the medium and transfer of cells from a previous culture into a fresh growth medium. This procedure enables the further propagation of the cell line or cell strain.

Cell lines in continuous culture are prone to genetic drift, and finite cell lines are fated for senescence. Moreover, all cell cultures are susceptible to microbial contamination. Therefore, an established cell line is a valuable resource, and its replacement is expensive and time-consuming. It is vitally important that they are frozen down and preserved for long-term storage. The best method for cryopreserving cultured cells is storing them in liquid nitrogen with a complete medium contained a cryoprotective agent such as DMSO. Cryoprotective agents reduce the freezing point of the medium and also allow a slower cooling rate, significantly reducing the risk of ice crystal formation, which can damage cells and cause cell death. For some cells, DMSO can induce differentiation. In such cases, an alternative such as glycerol should be used.

Cryopreserved cells in the liquid N_2 can be thawed anytime for further applications. Thawing and harvesting cells are important for reviving cells after storage for a long time.The thawing procedure is stressful to frozen cells, and using good technique and quick operation ensures that a high proportion of the cells survive the procedure. Temperature and speed are essentialfactors of successful thawing. The temperature should not exceed 37 °C. Some cryoprotectants, such as DMSO, are toxic above 4 °C. Therefore, cryopreserved cells must be thawed quickly and diluted in the culture medium to minimize the toxic effects.

2.3 MTT assay

The MTT assay is a colorimetric assay for measuring cell metabolic activity. It is based on the ability of nicotinamide adenine dinucleotide phosphate (NADPH)-dependent cellular oxidoreductase enzymes to reduce the yellow MTT to its insoluble formazan, which has a purple color. The reaction of the MTT assay is shown in Fig. 12.12. A solubilization solution (DMSO or acidified ethanol solution, or a solution of the detergent, sodium dodecyl sulfate, in diluted hydrochloric acid) is added to dissolve the insoluble purple formazan product into a colored solution.

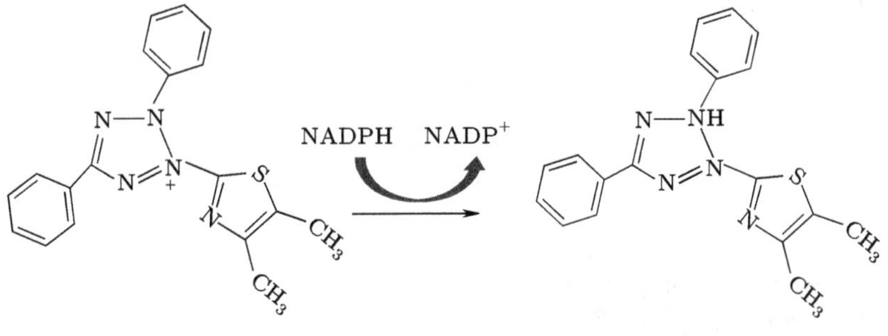

Fig. 12.12 Reaction of the MTT assay.

MTT assay is commonly used to analyze cell proliferation and viability. Viable cells with active metabolism convert MTT into a purple-colored formazan product with an absorbance maximum near 570 nm. When cells die, they lose the ability to convert MTT into formazan. Thus color formation serves as a useful and convenient marker of only the viable cells. It has been shown that the metabolism and exocytosis of MTT could dramatically damage cells. MTT could activate apoptosis-related factors, such as caspase-8, caspase-3 or accelerate the leakage of cell contents after the appearance of MTT formazan crystals. Therefore, the MTT method should be carefully chosen. Otherwise, the cell viability would be underestimated and incomparable. The main advantage of the MTT assay is the gold standard for cytotoxicity testing. The disadvantage is that the conversion to formazan crystals depends on metabolic rate and number of mitochondria resulting in many known interferences.

3 Experimental

3.1 Preparation

(1) Materials: HeLa cells, Pluronic F127-encapsulated curcumin micelles (drug and F127 ratio is 1:50) prepared in Experiment 55.1.

(2) Chemicals: MTT, PBS (0.01 mol/L, pH = 7.4), DMSO, Dulbecco's modified eagle medium (DMEM), fetal bovine serum (FBS), trypsin, penicillin and streptomycin (PS), 75% ethanol.

(3) Apparatuses: biosafety cabinet, CO_2 incubator, water bath, centrifuge, refrigerator, cell counter, autoclave, microplatereader, inverted phase-contrast mi-

croscope, liquid nitrogen container.

(4) Tools: sterileculture flask, pipettes, 15 mL and 50 mL sterile centrifuge tubes, liquid and solid waste containers, electric pipette, 3 mL sterile Pasteur pipet.

3.2 Procedure

3.2.1 Culture of HeLa cells

(1) All cell culture reagents, including PBS, DMEM, FBS, PS, and trypsin, are prewarmed in the 37 °C water bath for 30 min.

(2) Discard the spent cell culture media directly from the culture flask of HeLa cells to the liquid waste container because HeLa cells are adherent cells that stick firmly on the bottom of the culture flask.

(3) Add pre-warmed PBS solution tothe culture flask. Gently wash the cells by rocking the culture flask repeatedly to avoid disturbing the cell layer.

(4) Remove and discard the PBS wash solution from the culture flask.

(5) Add the pre-warmed trypsin dissociation reagent to the flask. Use enough reagents to cover the cell layer (approximately 0.5 mL per 10 cm^2). Gently rock the container to get complete coverage of the cell layer.

(6) Incubate the culture flask in the CO_2 incubator at 37 °C for approximately 2 min.

(7) Confirm the detachment of the cells from the flask bottom under the microscope due to the addition of the trypsin. If the detached cells are less than 90%, increase the incubation time for a few more minutes, and check the detachment every 30 s. You may also tap the flask lightly to expedite cell detachment.

(8) Add pre-warmed complete growth medium (DMEM-10%FBS medium containing 89% DMEM, 10% FBS, and 1% PS) with twice the volume of the previous trypsin dissociation reagent. Disperse the medium by electric pipetting over the undetached cell layer surface several times to detach the cells.

(9) Transfer the cell suspensionto a 15 mL conical tube and centrifuge at 1000 r/min for 3 min.

(10) Pour away the supernatant. Resuspend the cells in a minimal volume of the pre-warmed complete growth medium. Suck 10 μL cell suspension by a pipette. Determine the density of cells (in unit mL^{-1}) using the cell counter.

(11) Dilute cell suspension to the seeding density (1×10^5 mL^{-1}) by the complete growth medium. Seed the HeLa cells (1×10^4/well, i.e., 100 μL diluted cell suspension) in a 96-well cell culture plate and place the plate in the incubator.

3.2.2 Cell toxicity test of Pluronic F127- encapsulated curcumin micelles

(1) Prepare 20 mg/mL Pluronic F127-encapsulated curcumin micelles (curcumintoF127 ratio is 1:50) solution. Dissolve 120 mg of freeze-dried Pluronic F127-encapsulated curcumin into 6 mL DMEMas shown in Fig. 12.13. Sterilize the solution by passing through a 0.2 μm membrane filter. Take 3.6 mL of the filtrate and add 0.4 mL of FBS to obtain 20 mg/mL DMEM solution of Pluronic F127-encapsulated curcumin.

(2) Dilute 20 mg/mL DMEM solution of F127-encapsulated curcumin into 10,

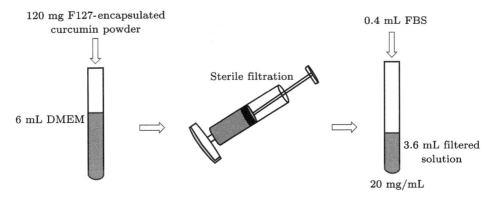

Fig. 12.13 Preparation of 20 mg/mL Pluronic F127-encapsulated curcumin solution.

5, 2.5 mg/mL respectively as illustrated in Fig. 12.14. Take 2 mL of the 20 mg/mL solution and mix it with 2 mL of the complete growth mediumto obtain 10 mg/mL F127-encapsulated curcumin solution. Similarly, 5 mg/mL and 2.5 mg/mL solution are prepared for subsequent MTT assay.

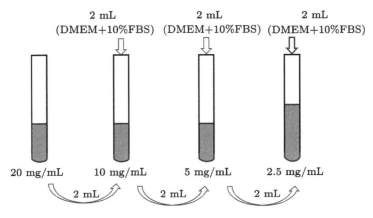

Fig. 12.14 Preparation of 10, 5, 2.5 mg/mL Pluronic F127-encapsulated curcumin solutions.

(3) The culture plate for the MTT assay is shown in Fig. 12.15. Add 100 μL Pluronic F127-encapsulated curcumin solutions with concentrations of 2.5, 5, 10, 20 mg/mL to the 96-well cell culture plate, in which HeLa Cells (1×10^4/well) have been seeded and cultured for 24 h. The control group is only the complete growth medium without the Pluronic F127-encapsulated curcumin. The background group is the complete growth medium without cells or drugs. PBS solution is added around the 96-well plate to avoid the evaporation of the medium. Each solution needs at least five replicates.

(4) After 24 h of treatment, remove the solutions by pipette. Add the PBS solution to wash the cells twice. Then, add 100 μL MTT solution of complete growth medium (0.5 mg/mL) to each well.

(5) Incubate the plate at 37 °C for 4 h. Remove the MTT solution. Add 100 µL DMSO to each well to solubilize the insoluble formazan crystals.

(6) The absorbance, which is proportional to cell viability, is subsequently measured at 570 nm in each well using the microplatereader.

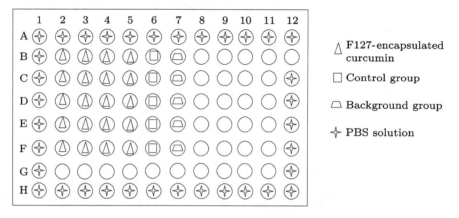

Fig. 12.15 The 96-well plate with HeLa cells for the MTT assay.

4 Requirements for experimental report

(1) Tabulate MTT data, including the absorbance values of the sample groups, the background group, and the control group, which are denoted as A_s, A_b, A_c, respectively.

(2) Calculate the cell viability by

$$\text{cell viability} = \frac{A_s - A_b}{A_c - A_b} \times 100\%. \tag{12.7}$$

Calculate the average value and standard deviation. List these results in the table. Draw a bar graph of cell viability against the concentration of Pluronic F127-encapsulated curcumin solution with error bars using Origin software.

(3) Discuss the effect of the concentration of Pluronic F127-encapsulated curcumin solution on cell toxicity.

5 Questions and further thoughts

(1) What is the principle of dissociating cells using trypsin?

(2) 5% CO_2 is commonly used to culture cells. What is the role of CO_2?

6 Cautions

(1) All solutions and apparatuses that come in contact with the cells must be sterile before cell culture. Always use proper sterile technique and work in a biosafety cabinet.

(2) Precautions should be taken to ensure that you do not inhale, ingest, or allow these cells to touch your skin.

(3) All experimental waste should be collected and processed properly.

Experiment 56 Preparation and Bioactivity of Polylactic Acid/Nano-Hydroxyapatite Composite

Type of the experiment: Comprehensive
Recommended credit hours: 24
Brief introduction: This experiment introduced the preparation and bioactivity of apolylactic acid/nano-hydroxyapatite (PLA/nHA) composite. The experiment is divided into two experiments that can be conducted respectively. In the first experiment, nHA particles are synthesized using the hydrothermal synthesis method and melt-compounded with PLA pellets followed by injection molding to round disks. Scanning electron microscope (SEM), X-ray diffraction (XRD), and Fourier transform infrared (FTIR) spectroscopy are employed to study the morphology and structure of as-synthesized nHA. In the second experiment, MC3T3-E1 cell lines are cultured to investigate the biocompatibility of as-prepared PLA/nHA composite. WST assay is used for evaluating the cell proliferation of the nanocomposite.

Experiment 56.1 Preparation and Characterization of Polylactic Acid/Nano-Hydroxyapatite Composite

Type of the experiment: Comprehensive
Recommended credit hours: 16

▮ Objectives

(1) To understand the principle of the hydrothermal synthesis method.
(2) To understand the principle of polymer melt compounding and injection molding techniques.
(3) To prepare the nHA and PLA/nHA composite using these techniques.

▮ Principles

2.1 Hydroxyapatite

Hydroxyapatite (HA, $Ca_{10}(PO_4)_6(OH)_2$) is acalcium phosphate compound. It is widely used for clinical applications due to its excellent biocompatibility, bioactivity, osteoconductivity, and osteoinductivity. HA is chemically similar to the mineral component of bones and hard tissues in mammals. It has been widely used in orthopedic and dental applications.

In general, synthetic nHA is the closest equivalent to bone apatite both in structure and chemical composition. Due to its large surface-to-volume ratio and synergistic effects, nHA generally possesses improved surface activity, exhibits active biological responses to the physiological environment, and promotes osteoblasts activities effectively. nHA can be synthesized by several techniques, including chemical co-precipitation, hydrothermalmethod, micelle-template, and hydrolysis. In this experiment, we use the hydrothermal method to synthesize nHA.

2.2 Hydrothermal method

Hydrothermal synthesis is a method to synthesize crystals that depend on the solubility of minerals in hot water under high pressure. The crystal growth is performed in an apparatus consisting of a steel pressure vessel called a hydrothermal reactor, in which a nutrient is supplied along with water. A temperature gradient is maintained between both ends of the growth chamber. At the hotter end, the nutrient solute dissolves in water, while at the cooler end, it grows into the desired crystal by deposited on a seed crystal. The hydrothermal process can be used to synthesize nanocrystalline materials from aqueous solutions at high temperatures, typically below 350 °C. It is effective and convenient in preparing nanocrystalline materials with controllable morphologies. The temperature can affect the morphology of crystal materials distinctly. The hydrothermal process can produce needle-shaped nHA with a size of 20–300 nm, a high degree of crystallinity, and a Ca/P ratio close to 1.67. The chemical reaction equation is

$$10CaCO_3 + 6(NH_4)_2HPO_4 + 2NH_3 \cdot H_2O \rightarrow$$
$$Ca_{10}(PO_4)_6(OH)_2 + 7(NH_4)_2CO_3 + 3H_2CO_3. \qquad (12.8)$$

However, the resulting nHAtends to agglomerate into clusters with a wide range of size distribution.

Possible advantages of the hydrothermal method over other types of crystal growth include the ability to form unstable crystalline phases at the melting point. Also, materials that require high-temperature dissolution can be grown by the hydrothermal method. The method is also particularly suitable for the growth of large, good-quality crystals while maintaining reasonablecontrol over their composition. Disadvantages of the method include the usage of expensive autoclaves and the impossibility of observing the crystal as it grows.

2.3 PLA/nHAcomposite

Biomaterials commonly used as bone implants include metal/alloys and polymer composites. They should be biocompatible and have sufficient mechanical strength to support human body weight. The mechanical properties of polymer composites can be tailored by the proper selection of the polymer matrix and fillers.nHA facilitates osteoblast attachment and proliferation effectively. It is an attractive filler for polymer composites designed for medical applications.

PLA is a biodegradable and bioactive thermoplastic derived from renewable biomass, typically from fermented plant starch such as corn, cassava, sugarcane, or sugar beet pulp. PLA is used as medical implants in the form of anchors, screws, plates, pins, rods, and as a mesh due to its advantage of being able to degrade into innocuous lactic acid. It breaks down inside the body within 6 months to 2 years, depending on the exact type used. This gradual degradation is desirable for a support structure because it gradually transfers the load to the body (e.g., the bone) as that area heals. The strength characteristics of PLA implants are well documented. This experiment aims to prepare a PLAnanocomposite filled with nHA and study the bioactivity of such composite.

3 Experimental

3.1 Preparation

(1) Chemicals: calcium carbonate ($CaCO_3$), diammonium hydrogen phosphate (($NH_4)_2HPO_4$), ammonia water($NH_3 \cdot H_2O$), hydroxyapatite, PLA, KBr (IR grade).

(2) Apparatuses: X-ray diffractometer (Rigaku, Miniflex 600), scanning electron microscope (TECAN, Vega3 LM), FTIR spectrometer (PerkinElmer, Frontier), electronic balance, electromagnetic stirrer, oven, torque rheometer with a mixer (Hapro, RM-200C), injection molding machine (Thermo Scientific, Minijet Pro), crusher, refrigerator, vacuum filtration apparatus (the components have been listed in Experiment 15).

(3) Tools: beakers, watch glass, glass rods, magnetic stirring bar, polytetrafluoroethylene (PTFE) autoclave, pH test paper, filter paper, tools for FTIR sample preparation listed in Experiment 39.

3.2 Procedure

3.2.1 Synthesis of nHA using hydrothermal method

(1) Weigh 1.0 g $CaCO_3$ powder and 0.792 g $(NH_4)_2HPO_4$ with a molar ratio of Ca/P 5:3.

(2) Dilute the $(NH_4)_2HPO_4$ in a 100 mL beaker with 60 mL deionized water under stirring.

(3) Add $CaCO_3$ into the above solution and stir the solution for 30 min.

(4) Adjust the pH value of the solution to 9–11 with ammonia water.

(5) Transfer the above solution to the PTFE autoclave.

(6) The solution is hydrothermally treated at 180 °C in an oven for 12 h and cooled in air.

(7) Separate the nHA product by vacuum filtration. Wash the product in the Büchner funnel with deionized water and absolute ethanol successively.

(8) Collect the filtrate in a watch glass and dry the powder in the oven at 80 °C for 3 h.

3.2.2 Characterization of as-prepared nHA

(1) Observe the morphology of the as-prepared nHA using SEM. The procedure has been described in Experiment 5.1.

(2) Determine the phase structure of as-prepared nHA and $CaCO_3$ using XRD. The procedure has been described in Experiment 28.2.

(3) Measure the FTIR spectrum of as-prepared nHA and $CaCO_3$ to confirm the chemical structure. The procedure has been described in Experiment 39.

3.2.3 The fabrication of PLA/nHA composite

PLA and nHA powders with the weight ratio of 7:3 are melt-compounded in the torque rheometer with blending temperature profiles of 170 °C, 190 °C, 190 °C, and 160 °C at a rotation speed of 30 r/min for 10 min.

(2) The compounded nanocomposite is broken into small pieces with a crusher.

(3) The as-prepared nanocomposite is injection molded into discs (more than

4 pieces) with a diameter of 17 mm and a thickness of 1 mm using the injection molding machine. The molding temperature and the injection temperature are maintained at 70 °C and 185 °C, respectively.The injection pressure is kept at 500 MPa. The discs are maintained at the pressure of 400 MPa for 10 s before being taken out. For comparison, pure PLA discs (more than 4 pieces) are also fabricated under the same processing conditions.

4 Requirements for experimental report

(1) Give the SEM photographs of the as-prepared nHA, and describe the morphology and particle size.

(2) Plot the XRD profiles of the as-prepared nHA and $CaCO_3$. Index the peaks by comparing them with the PDF card.

(3) Plot the FTIR spectrum of the as-prepared nHA and $CaCO_3$. Indicate the criticalfunctional groups shown in the FTIR spectra by comparing the peak positions with those reported in the literature.

5 Questions and further thoughts

(1) What is hydrothermal synthesis? What are the advantages and disadvantages of this method?

(2) Describe the chemical and crystallographic features of HA.

Experiment 56.2 Bioactivity Evaluation of Polylactic Acid/Nano-Hydroxyapatite Composite

Type of the experiment: Comprehensive
Recommended credit hours: 8

1 Objectives

(1) To learn the method of culturing cell line MC3T3-E1.
(2) To understand the WST-8 assay for evaluating the cell viability.
(3) To determine the cell viability of PLA/nHA composite.

2 Principles

2.1 MC3T3-E1 cell line

The osteoblastic cell line MC3T3-E1 has been established from a C57BL/6 mouse calvaria. MC3T3-E1 cells can differentiate into osteoblasts and osteocytes and have been demonstrated to form calcified bone tissue in vitro. Mineral deposits have been identified as hydroxyapatite. MC3T3-E1 cells secrete collagen and have been widely used as model systems in bone biology.

2.2 WST-8 assay

Some tetrazolium reagents can be reduced by viable cells to generate formazan products that are directly soluble in cell culture media. Tetrazolium compounds fitting this category include MTS (3-(4,5-dimethylthiazol-2-yl) -5-(3-carboxymethoxyphenyl)-2-(4-sulfophenyl) -2H tetrazolium), XTT (2,3-bis(2-methoxy-4-nitro-5-sulfophenyl)-5-[(phenylamino)carbonyl]-2H-tetrazolium hydroxide), and the WST-8 (2-(2-methoxy-4-nitrophenyl)-3-(4-nitrophenyl)-5-(2,4-disulfophenyl)-2H-tetrazolium monosodium salt). These improved tetrazolium reagents eliminate a liquid handling step during the assay procedure because the second addition of reagent to the assay plate is not needed to solubilize formazan precipitates, thus making the protocols more convenient. The negative charge of the formazan products, which contributes to solubility in the cell culture medium, is thought to limit the cell permeability of the tetrazolium. This set of tetrazolium reagents is used in combination with intermediate electron acceptor reagents, such as phenazine methyl sulfate (PMS) or phenazine ethyl sulfate (PES). These electron acceptor reagents can penetrate viable cells, become reduced in the cytoplasm or at the cell surface, and exit the cells where they can convert the tetrazolium to the soluble formazan product. In this experiment, we will use WST-8 reagent, and the reaction is shown in Fig. 12.16.

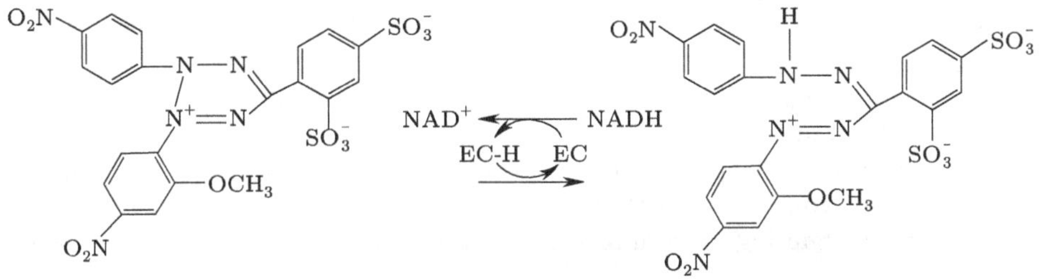

Fig. 12.16 Reaction of the WST-8 assay.

③ Experimental

3.1 Preparation

(1) Materials: MC3T3-E1 cells, PLA/nHA nanocomposite discs and PLA discs prepared in Experiment 56.1.

(2) Chemicals: phosphate-buffered saline (PBS, 0.01 mol/L, pH = 7.4), α-Minimum Essential Medium Eagle (α-MEM), fetal bovine serum (FBS), penicillin and streptomycin (PS), 75% ethanol, 95% ethanol, trypsin, WST-8.

(3) Apparatuses: biosafety cabinet, CO_2 incubator, water bath, centrifuge, refrigerator, cell counter, autoclave, microplatereader, inverted phase-contrast microscope, ultrasonic cleaner.

(4) Tools: stereculture flask, pipettes, 15 mL and 50 mL sterile centrifuge tubes, liquid and solid waste containers, electric pipette, 3 mL sterile Pasteur pipettes, beakers.

3.2 Procedure

3.2.1 Culture of MC3T3-E1 cells

(1) All cell culture reagents, including PBS, α-MEM, FBS, PS, and trypsin,

are preheated in the 37 °C water bath for 30 min.

(2) Discard the spent cell culture medium from the culture flask of MC3T3-E1 cells to the liquid waste container.

(3) Wash cells using pre-warmed PBS solution. Gently add the solution to avoid disturbing the cell layer, and rock the flask back and forth several times.

(4) Add the pre-warmed trypsin dissociation reagent to the flask. Use enough reagent to cover the cell layer (approximately 0.5 mL per 10 cm^2). Gently rock the container to get complete coverage of the cell layer.

(5) Incubate the culture flask in the CO_2 incubator at 37 °C for approximately 2 min.

(6) Confirm the detachment of the cells from the flask bottom under the microscope due to the addition of the trypsin. If detached cells are less than 90%, increase the incubation time for a few more minutes, and check the detachment every 30 s.

(7) Add pre-warmed complete growth medium (α-MEM-10%FBS medium) with twice the volume of the previous trypsin dissociation reagent. Disperse the medium by electric pipetting over the undetached cell layer surface several times to detach the cells.

(8) Transfer the cell suspensionto a 15 mL conical tube and centrifuge at 1000 r/min for 3 min.

(9) Pour away the supernatant. Resuspend the cells in a minimal volume of the pre-warmed complete growth medium and suck 10 μL of cell suspension by a pipette for cell counting. Determine the density of cells (in a unit of mL^{-1}) using the cell counter.

3.2.2 Determine the cell viability by WST-8 assay

(1) Ultrasonically wash the fourpure PLA discs and fourPLA/nHA composite discs twice with 95% ethanol for 5 min.

(2) Sterilizethe discs with 75% ethanol for 1 h. Then wash the discs with PBS solution to remove the residual ethanol. Expose both the front and back sides of the discs under the UV light for half an hour.

(3) Place the sample discs into the 12-well plate (plate 1) as shown in Fig. 12.17. FourPLA/nHA discs are placed in column A, and fourpure PLA discs are placed in column B.

(4) Dilute cell suspension to the seeding density 5×10^4 mL^{-1}. MC3T3-E1 cells are seeded in columns A and B of the 12-well cell culture plate 1 to 5×10^4/well. Add the same amount of cells with the culture medium in column C as the control group. Add the same amount of the culture medium in column C of plate 2 as the background group.

(5) Put the plates into the incubator at 37°C in a humidified atmosphere of 95% air and 5% CO_2 for 24 h.

(6) After incubating for 24 h, 100 μL WST-8 reagent is added into each well. Then, the plates are incubated at 37°C for another 2 h to form WST-8 formazan.

(7) After incubating, transfer 100 μL medium from each well of the 12-well plates to a 96-well plate. Then measure the absorbance at 450 nm using the microplate reader.

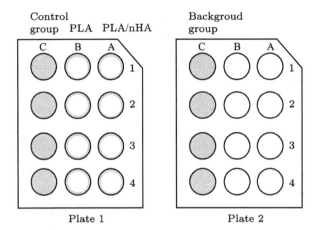

Fig. 12.17 The 12-well plates for the WST-8 assay. (See color illustration at the back of the book)

◪ Requirements for experimental report

(1) Tabulate the WST-8 assay data, including the absorbance values of the sample groups, the background group, and the control group.

(2) Calculate the average values and standard deviations of the cell viability. Draw a column graph of the cell viability of PLA/nHA, PLA, and the control group.

(3) Discuss the role of nHA on cell viability based on the results of the WST-8 assay.

◪ Questions and further thoughts

(1) What is the difference between the MTT assay and the WST-8 assay?
(2) Why are the MC3T3-E1 cells used in this experiment?

◪ Cautions

(1) All solutions, apparatuses, and tools that come in contact with the cells must be sterile before the experiment. Always use proper sterile technique and work in a biosafety cabinet.

(2) Precautions should be taken to ensure that you do not inhale, ingest, or allow these cells to touch your skin.

(3) All experimental waste should be collected and processed properly.

Extended readings

Alberts B, Hopkin K, Johnson A D, et al. Essential Cell Biology. 5th Edition. W. W. Norton & Company, 2019.

Danilczuk M, Lund A, Sadlo J, et al. Conduction electron spin resonance of small silver particles. Spectrochimica Acta Part A: Molecular and Biomolecular Spectroscopy, 2006, 63(1): 189-191.

Davis J M. Basic Cell Culture. Oxford University Press, 2002.

Feng Q L, Wu J, Chen G Q, et al. A mechanistic study of the antibacterial effect of silver ions on *Escherichia coli* and *Staphylococcus aureus*. Journal of Biomedical Materials Research, 2000, 52(4): 662-668.

Kim J S, Kuk E, Yu K N, et al. Antimicrobial effects of silver nanoparticles. Nanomedicine, 2007, 3(1): 95-101.

Matsumura Y, Yoshikata K, Kunisaki S, et la. Mode of bacterial action of silver zeolite and its comparison with that of silver nitrate. Applied and Environmental Microbiology, 2003, 69(7): 4278-4281.

Orange County Biotechnology Education Collaborative. Lab manual: Introduction to biotechnology. Available at: https://bio.libretexts.org/Bookshelves/Biotechnology/Lab_Manual%3A_Introduction_to_Biotechnology. Accessed: 2 March 2021.

Parker N, Schneegurt M, Tu A H T, et al. Microbiology. OpenStax, 2016.

Pollar T, Earnshaw W, Lippincott-Schwartz J, et al. Cell Biology. 3rd Edition. Elsevier, 2016.

Sahu A, Kasoju N, Goswami P, et al. Encapsulation of curcumin in Pluronic block copolymer micelles for drug delivery applications. Biomaterials Applications, 2011, 2(25): 619-639.

Sondi I, Salopek-Sondi B. Silver nanoparticles as antimicrobial agent: A case study on *E. coli* as a model for Gram-negative bacteria. Journal of Colloid and Interface Science, 2004, 275(1): 177-182.

Stoddart M J. Mammalian Cell Viability: Methods and Protocols. Humana Press, 2011.

Tortora G J, Funke B R. Case C L, et al. Microbiology: An Introduction. Pearson, 2018.

Wagner W R, Sakiyama-Elbert S E, Zhang G, et al. Biomaterials Science: An Introduction to Materials in Medicine. 4th Edition. Academic Press, 2020.

Xian W. A Laboratory Course in Biomaterials. CRC Press, 2010.

Chapter 13 Materials and Environment

Experiment 57 Passivation and Corrosion Behavior of Ferrous Alloys

Type of the experiment: Cognitive
Recommended credit hours: 4
Brief introduction: The corrosion and passivation behavior of ferrous alloys are evaluated by the electrochemical method in this experiment. Two alloys, Q235 mild carbon steel and 316 stainless steel, and two solutions, 0.5 mol/L H_2SO_4 solution and 0.5 mol/LH_2SO_4 with 0.01 mol/L NaCl solution, are selected to study the effects of alloying elements and Cl^- on the passivation behaviors of the alloys.

1 Objectives

(1) To understand the principles of corrosion and passivation.
(2) To learn the linear sweep voltammetry (LSV) technique on evaluating the passivation and corrosion behavior of alloys.
(3) To compare the corrosion behavior of Q235 mild carbon steel and 316 stainless steel in acid solution.
(4) To study the effect of Cl^- on the passivation and corrosion behavior of ferrous alloys.

2 Principles

2.1 Corrosion

Corrosion is the deterioration of a metal due to chemical reactions between it and the surrounding environment. Common corrosion types include uniform corrosion, pitting corrosion, crevice corrosion, galvanic corrosion, intergranular corrosion, and stress corrosion cracking.

In this experiment, the uniform corrosion and pitting corrosion that respectively takes place on Q235 mild carbon steel and 316 stainless steel will be examined. Uniform corrosion, also known as general corrosion, is the uniform loss of metal over an entire surface. It results in tons of waste, and, from this viewpoint, it is the most important form of corrosion. Pitting corrosion is a form of localized corrosion that leads to the creation of small holes or pits on the metal surface. Pitting corrosion is usually found on passive metals when the passive film is chemically or mechanically damaged and does not immediately re-passivate. Pitting corrosion is easy to occur in the presence of chloride because chloride is particularly damaging to the passive film.

There are other common corrosion types. Crevice corrosion is a form of localized corrosion occurring in confinedllocations to which the access of the working fluid from the environment is limited. Galvanic corrosion occurs when two different metals are located together in a corrosive electrolyte. A galvanic couple forms between the two metals, where one metal becomes the anode, and the other becomes the cathode. The anode or sacrificial metal corrodes and deteriorates faster than when used alone. Intergranular corrosion is a chemical or electrochemical attack on the grain boundaries of metals. These boundaries can be more vulnerable to corrosion than grain interiors. Stress corrosion cracking is a form of environmental cracking corrosion behavior. It occurs due to the external stress on the metals.

2.2 Investigation of corrosion behavior by electrochemical method

Electrochemical methodsare widely appliedfor investigating the corrosion process and determining the corrosion rate. The three-electrode setup of cyclic voltammetry described in Experiment 41.2 is generally used. As shown in Fig. 13.1, the

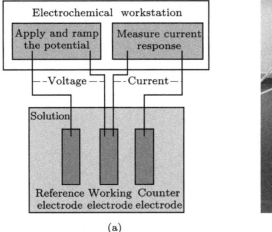
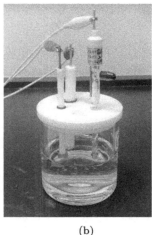

(a)   (b)

Fig. 13.1 (a) Schematic illustration of the experiment setup and (b) photograph of the electrochemical glass cell.

measurement can be processed in the electrochemical glass cell, in which the sample (working electrode, WE),Pt plate (counter electrode, CE), saturated calomel electrode (SCE, reference electrode, RE)are immersed in an electrolyte solution. In this experiment, 316 stainless steel and Q235 carbon steel are used as the working electrodes. The three electrodes are connected to anelectrochemical workstation, which is connected to a computer installed with software to run the test and record the results.

The corrosion behavior can be determined through a potential scan in a certain range, which is known as the potentiodynamic method or LSV. The potential applied on the electrode is changed at a constant rate,and the corresponding current is recorded. Figure 13.2 shows the anodic polarization curve, which is one of the basic tools for studying the electrochemical corrosion of metals. It is generally called the Tafel plot, which is plotted with logarithmic current density against potential.

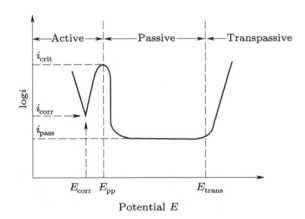

Fig. 13.2 Representative polarization curve of metals.

During potentiodynamic measurement, under applied potential, net oxidation and reduction reactions occur on the sample surface. Such a process breaks down the original equilibrium of the open circuit and causes a change in potential value to reach a constant intermediate value E_{corr}. It is the critical potential for the change from cathodic reaction to anodic reaction, i.e., from reduction to oxidation of the sample surface. The corresponding net current flowis corrosion current density i_{corr}. This stage is the active region in the polarization curve. The further increase in the potential leads to the oxidation or corrosion of the sample.

The linear part of the polarization curve, as shown in Fig. 13.3, is the Tafel region. β_a is the positive slope of anodic reaction, and β_c is the negative slope of cathodic reaction. The corrosion potential E_{corr} and the corrosion current density i_{corr} can be determined from the intersection of the extrapolation lines of the anodic and cathodic linear portions. At this point, the rate of metal dissolution is equal to the rate of hydrogen evolution. The polarization behavior of the straight lines follows by the Tafel equation

$$\eta = \pm \beta \log \frac{i}{i_0}. \tag{13.1}$$

In this equation, η is activation overpotential, which is the change in potential of an electrode from its equilibrium, i.e., $\eta = E - E_o$, E_o is the open circuit potential (OCP); β is Tafel constant, which is equal to β_a or β_c for the anodic or the cathodic reaction; i_0 is the exchange current that is the anodic or cathodic current under equilibrium conditions.

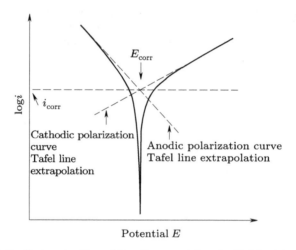

Fig. 13.3 Determination of the E_{corr} and i_{corr} by Tafel plot analysis.

After E_{pp} and the corresponding current i_{crit}, the sample starts to form passive film to protect the metal sample from being corroded. The i_{pass} is related to the passive stage, where the current density value is kept approximately stable. Passivation refers to the spontaneous formation of an ultrathin film of corrosion products, known as passive film, on the surface that acts as a barrier to further oxidation. The chemical composition and microstructure of a passive film are different from the underlying metal. The typical thickness of passive film on aluminum, stainless steels, and alloys is within 10 nm. The passive stage spans a certain potential range. The smaller values of i_{pass} and i_{crit} and the wider of the passive region imply better corrosion resistance.

The passive film will be destroyed with the continuous increase in the applied potential that is higher than E_{trans}. The current density increases correspondingly and thus leading to the increase in corrosion rate. This stage is the transpassive region in the polarization curve.

2.3 Determination of corrosion rate

Corrosion rate may be expressed as a current density, a rate of weight loss, or a rate of section loss for most engineering alloys. Since electron transfer occurs during the anodic and the cathodic reactions, the amount of metal uniformly corroded from an anode or electroplated on a cathode in a period can be determined by using Faraday's Law. Thus, the weight loss through the corrosion reaction is expressed as

$$W = \frac{ItM}{nF}, \tag{13.2}$$

where W is the weight of metal (in g) corroded or electroplated in time t (in s), I is the current flow (in A), M is the atomic mass of the sample (in g/mol), n is the number of electrons per atom produced or consumed in the process, and F is the Faraday's constant (96500 C/mol).

For convenience, the current flow is usually expressed in terms of a current density i. Then,

$$W = \frac{iAtM}{nF}, \tag{13.3}$$

where i is the current density (in A/cm^2), and A is the sample surface area (in cm^2).

Hence the corrosion rate of section loss at specified current density, which is the depth of corrosion per unit time and unit area, is defined as

$$\text{corrosion rate} = \frac{A}{A\rho t} = \frac{iM}{nF\rho} \tag{13.4}$$

where ρ is the density of the sample (in g/cm^3). Generally, the corrosion rate of section loss is expressed in a unit of mm/a, i.e., millimeters per year.

3 Experimental

3.1 Preparation

(1) Materials: 316 stainless steel(316SS) and Q235 carbon steel (235CS) rods with 5 mm in diameter mounted in PTFE and connected with the copper wire.

(2) Chemicals: 0.5 mol/L sulfuric acid solution, 0.5 mol/L sulfuric acid with 0.01 mol/L chloride solution.

(3) Apparatuses: electrochemical workstation (CH Instruments, CHI660E), grinding machine, optical microscope.

(4) Tools: electrochemical glass cell with a PTFE lid, SCE, Pt plate.

3.2 Procedure

(1) Switch on the electrochemical workstation and computer, and warmup for 10 min.

(2) Add 0.5 mol/LH$_2$SO$_4$ solution or 0.5M H$_2$SO$_4$ with 0.01mol/L NaCl solution into the electrochemical cell.

(3) Grind and polish the working electrode (316SS or 235CS) followed by rinsing with deionized water.

(4) Observe the surface morphology by optical microscope.

(5) Insert the working electrode (316SS or 235CS), the counter electrode (Pt plate), and the reference electrode (saturated calomel electrode) in the electrolyte cell, as shown in Fig. 13.1.

(6) Connect the electrodes to the electrochemical workstation. Green, white, and red clips connect with the working electrode, the reference electrode, and the counter electrode, respectively.

(6) Open the test software.

(7) Test the OCP E_o.

(8) Select the test module "Tafel plot" in the software and set the test parameters accordingly. The initial potential is $(E_o - 0.15)$ V; the final potential is 1.0 V or 1.6 V for 316SS or 235 CS, respectively; sweep segment is 1; the scan rate is 0.01 V/s^{-1}; the sensitivity is 1×10^{-2} A/V or 1×10^{-1} A/V for 316SS or 235 CS, respectively.

(10) Start test. A $\log i - E$ or $i - E$ curve is plotted by the computer. The plot can be changed by selecting the graphic options.

(11) Read the critical current density i_{crit}, the critical potential E_{pp}, the passivation current density i_{pass} directly from the curve.

(2) Determine the corrosion current density i_{corr} and the corrosion potential E_{corr}. Calculate the corrosion rate (in mm/a) at the current i_{corr}. The calculation can be conducted using the test software. The relative parameters for the calculation are listed in Table 13.1.

(13) Export the polarization curve as a ".txt" file.

(14) Observe the surface morphology after the test by optical microscope.

(15) Repeat steps (2) to (14) for the two working electrodes and the two solutions.

Table 13.1 Parameters of the working electrodes

Working electrode	Molecular weight/(g/mol)	Density /(g/cm^3)	Number of electrons	Area Φ5/ cm^2
316SS	56.84	8.00	2	0.1962
235CS	56.00	7.87	2	0.1962

4 Requirements for experimental report

(1) Plot four polarization curves ($\log i$ versus E) recorded in the experiment in one chart for comparison. The software Origin is recommended.

(2) Tabulate the results, including the critical current density i_{crit}, the critical potential E_{pp}, the passivation current density i_{pass}, the corrosion current density i_{corr}, and the corrosion potential E_{corr}, and corrosion rate at the current i_{corr}.

(3) Give the micrographs of the surface morphologies of the electrodes before and after the test.

(4) Compare the corrosion behaviors of 316SS and 235CS in acid and acid chloride solution. Analyze the effects of alloying elements and Cl^{-1} in the solutions.

5 Questions and further thoughts

(1) Why is the three-electrode system used in determining the polarization curve of metal? What are the functions of the counter electrode and the reference electrode?

(2) Why are the working electrodes required to be ground before each measurement?

Experiment 58 Salt Spray Test of Metals

Type of the experiment: Comprehensive
Recommended credit hour: 16
Brief introduction: The salt spray test for evaluating the corrosion behavior of metals is introduced in this experiment. Two pure metals, aluminum and copper, and two alloys, Q235 carbon steel and 304 stainless steel, are selected for the neutral salt spray test. The corrosion rates of the samples are measured.

During the experiment, the salt spray test takes about a week. Therefore, the experiment procedure should be reasonably arranged so that the salt spray process does not occupy the class time.

1 Objectives

(1) To understand the principle and technique of the salt spray test.
(2) To learn the operation of the salt spray corrosion test method.
(3) To determine the corrosion rate of common metals in salt spray test.
(4) To understand the corrosion mechanisms of metals in the salt spray atmosphere.

2 Principles

2.1 Salt spray test

The salt spray test is a corrosion test that uses the artificial simulated salt fog environment created by the salt spray equipment to evaluate the corrosion resistance of metal materials with or without corrosion protection.It is used as an accelerated means of testing the ability of materials to withstand atmospheric corrosion.It has long been astandardized corrosion test method because it is quick, repeatable, and relatively inexpensive.

The salt spray test can be classified into the neutral salt spray (NSS), acetic acid salt spray (AASS), and copper accelerated acetic acid salt spray (CASS) according to different salt spray conditions. The NSS is the base of the salt spray test. It is a neutral test which means the pH value of the test solution is in the range of 6.5 to 7.2. The AASS test uses NSS solution with the addition of glacial acetic acid in the pH value range of 3.1 to 3.3. The CASS uses AASS solution with the addition of copper chloride also in the pH value range of 3.1 to 3.3.

The selection of the test method depends on the material and coating of the samples. The NSS is generally used in metals, metal coating, conversion film, anode coating, and organic coating. The AASS and CASS methods are applicable for evaluating the corrosive performance of decorative copper/nickel/chromium or nickel/chromium coatings on steel, zinc alloys, aluminum alloys, and plastics designed for severe service. It is also applicable to the test of anodized aluminum. The selection of the method also depends on the specific application of the material. For instance, if the steel sample is tested in the salt spray with a fixed concentration of chloride ion, the test can simulate the corrosion rate of marine steel in the seawater environment.

In this experiment, the salt spray test is conducted using a salt spray chamber with a certain volume space, as shown in Fig.13.4. The corrosion rates of aluminum, copper, Q235 carbon steel and 304 stainless steel are tested by the NSS method.

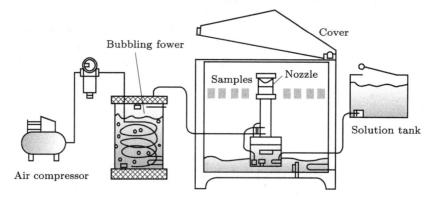

Fig. 13.4 Schematicillustrationof the salt spray test chamber.

2.2 Determination of corrosion rate

The corrosion rate of the salt spray test can be calculated by the weight gain or weight loss of asample under the corrosion of the salt spray atmosphere. If the corrosion product is easy to peel off, the weight loss method is recommended. Otherwise, the weight gain method is more suitable. The weight gain or weight loss per unit area can be used to express the corrosion rate as

$$v = \frac{|W_0 - W_1|}{At}, \tag{13.5}$$

where v is the corrosion rate (in g/(m^2· h)), W_0 is the mass of the sample before corrosion (in g), W_1 is the mass of the sample after corrosion (in g), A is the corroded area (in m^2), t is the corrosion time (in h).

Sometimes the experimental data show an apparentdiscrepancy from the value of the actual application because the experiment cannot completely simulate the actualenvironment. For example, the effect of ultraviolet light isignored in the general salt spray test, which may result in oxidation and crack of the coating layer.Therefore, it is necessary to consider the simplification of the test conditions whileanalyzing the experimental data.

3 Experimental

3.1 Preparation

(1) Materials: aluminum, copper, Q235 carbon steel, and 304 stainless steel plates with a size of 20 mm × 20 mm × 1 mm.

(2) Chemicals: hydrochloric acid (HCl), sodium hydroxide (NaOH), sodium chloride (NaCl), 95% ethanol solution, acetone.

(3) Apparatuses: salt spray chamber (Shanghai Precision Instrument Company, YWX/Q-150), grinding machine, electronic balance, optical microscope,

drying oven

(4) Tools: Vernier caliper, blow drier, pH test paper, transparent tape, polymer adhesive, 600, 800, and 1200 grit abrasive papers, brush.

3.2 Procedure

3.2.1 Preparation of salt spray solution

(1) Prepare 25 L salt spray solution of 5wt% NaCl in deionized water.

(2) Boil the solution to remove dissolved CO_2 before adjusting the pH value.

(3) Adjust the pH value of the solution to 6.5–7.2 using HCl or NaOH solution.

3.2.2 Sample preparation

(1) Grind one surface of the samples with 600, 800, and 1200 grit abrasive papers successively.

(2) Remove the surface oil by immersing them in the 95% ethanol solution, then dry them using the blow drier.

(3) Measure the dimensions of the samples with Vernier caliper. It is necessary to measure at least three times to obtain the average values.

(4) Cover one surface of the samples to be corroded with transparent tape, and then coat the other five surfaces with the polymer adhesive. Repeat several times to obtain a homogeneous coating.

(5) Peel off the tape, wipe the surface to be corroded with cotton dipped with acetone, and then dry the samples in the oven at 50 °C for half an hour.

(6) Weigh the samples with accuracy to 0.1 mg as the initial weight W_0.

(7) Take photos of the samples and micrographs of the surface morphology.

3.2.3 Salt spray test

(1) Switch on the power of the air compressor and the salt spray chamber.

(2) Adjust the air pressure to 0.7–1.8 kg/cm^3. Adjust the temperature of the salt spray chamber to 35 °C. Stabilize the temperature and the pressure for about 10 min.

(3) Place the samples in the chamber and ensure the test surface is 15–30° from the vertical direction.

(4) Select the intermittent mode (spray for 24 h and rest for 24 h). Record the start and the end time of the salt spray.

(5) Before closing the chamber, ensure an adequate amount of solution is in the trough. This also creates a seal when the chamber is closed and prevents the salt fog from escaping.

(6) Samples are often checked for rust development or coating degradation at regular intervals to monitor corrosion levels. Meanwhile, the spray conditions such as the temperature and humidity should be checked as well.

(7) After about a week, take out the samples and record the actual test time.

(8) Wash the samples with pure water, dry them in the oven, and weigh them as the final weight W_1.

(9) Remove the loose corrosion products with a brush.

(10) Take photos of the samples and micrographs of the surface morphology.

4 Requirements for experimental report

(1) Tabulate the experimental data, including the dimensions, the initial weights, and the final weights.
(2) Calculate the corrosion rates of the samples using Equation (13.5).
(3) Give the photos and micrographs of the surface morphology before and after the test.
(4) Compare and analyze the corrosion characteristics of the samples.

5 Questions and further thoughts

(1) Analyze the sources of error in this experiment. How can we improve the experiment to make the salt spray test closer to the actual situation?
(2) Please compare the advantages and disadvantages of the salt spray test withthe electrochemical corrosion test.

Experiment 59 Synthesis and Photocatalytic Activity of TiO_2 Powder

Type of the experiment: Comprehensive
Recommended credit hours: 16
Brief introduction: Titanium dioxide (TiO_2) is the most widely studied photocatalytic material. In this experiment, TiO_2 and Zn-doped TiO_2 photocatalysts are synthesized by the sol-gel method. Rutile and anatasephase structures are obtained by adjusting the calcination temperature. Then, the photocatalytic performance of the photocatalysts is evaluated by degradation measurement of methylene blue. The effects of the phase structure and Zn dopant on the photocatalytic performance will be examined.

1 Objectives

(1) To understand the principle of photocatalysis.
(2) To master the principle and procedure for the characterization of photocatalytic performance.
(3) To evaluate the photocatalytic performance of the as-synthesized powders.
(4) To understand the effects of phase structure and metal ion dopant on photocatalytic performance.

2 Principles

2.1 Photocatalysis of titanium dioxide

Photocatalysis is a kind of electrochemical reaction facilitated by photo-induced electrons and holes. Typically, a photocatalytic reaction involves the generation of an electron-hole pair. For instance, titanium dioxide (TiO_2), the most widely studied photocatalytic material, its electron-hole pair is generated through

$$TiO_2 + h\nu \rightarrow e^- + h^+, \qquad (13.6)$$

where h is the Planck constant and ν is the frequency of radiation, and hence $h\nu$ is the energy of a light quantum or a photon. The photons have to exhibit sufficient energy to excite an electron from the valence band to the conduction band. For TiO_2, the bandgap is about 3 eV. Consequently, the photons in the ultraviolet (UV) regime can contribute to charge carrier generation.

The lifetime of generated electrons and holes is of tremendous importance for high photocatalytic performance. The electron-hole pairs generated within the bulk have to diffuse through the bulk to the surface to initiate a photocatalytic reaction. All electrons and holes that recombine within the bulk or at the surface are no longer available for further reactions. Thus, the recombination of electrons and holes in the photocatalyst has to be suppressed. In addition, photocatalytic performance is also influenced by the surface area since it is directly related to the amount of absorbed organic molecules and available trapping centers.

It has been found that holes are the main oxidizing species in TiO_2 and get trapped at the surface within picoseconds after generation. Degradation of organic molecules can be achieved by a direct oxidation reaction with holes. Taken carboxylic compound as an example, the reaction is

$$RCOO^- + h^+ \rightarrow R + CO_2. \tag{13.7}$$

In addition, the $OH^{\cdot}$ radicals on the TiO_2 surface are formed by the reaction between the holes and water molecules. Instead of direct reaction with holes, successive attacks by $OH^{\cdot}$ radicals yield a degradation of organic compounds. The reaction is

$$R + OH^{\cdot} \rightarrow R' + H_2O. \tag{13.8}$$

2.2 Ion doping of TiO_2

The performance of TiO_2 is influenced by its crystallite size, surface area, phase structure, and dopant type and concentration. Among numerous approaches to improve the photocatalytic efficiency of TiO_2, ion doping has become the most efficient approach. Doping ions can narrow the bandgap of TiO_2 via an electronic coupling effect between doping ions and TiO_2. At the same time, they can extend the visible absorption by introducing impurity energy levels just above the valence band or below the conduction band of TiO_2. Besides, doping ions can trap the photogenerated electrons to suppress the hole-electron recombination effectively.

Various kinds of dopants, such as metals (Au, Ag, Pt, Fe, Co, Cu, Zn), nonmetals (C, N, F, P, S), and rare earth (La, Eu, Y, Er), have been investigated as dopants in TiO_2. Among various metals, zinc has been considered as one of the appropriate candidates for practical application due to the similarity in the ionic radii of Zn^{2+} (0.74 A) and Ti^{4+} (0.75 A). In this experiment, the effect of the Zn dopant on the photocatalytic performance will be examined.

2.3 Degradation measurement of methylene blue

Methylene blue (MB) is widely used as a chemical indicator and is generally used to investigate photocatalytic performance. Its constitutional formula is depicted in Fig. 13.5. By photocatalysis, methylene blue is completely degraded down to the final products SO_4^{2-}, CO_2, NH_4^+, and NO_3^-. Initially, a solution of MB shows a blue color. Discoloring occurs after degradation. Thus, the change in light absorbance related to the MB concentration is a direct measure for the degradation of MB molecules and consequentially relates to the photocatalytic performance of the investigated catalyst.

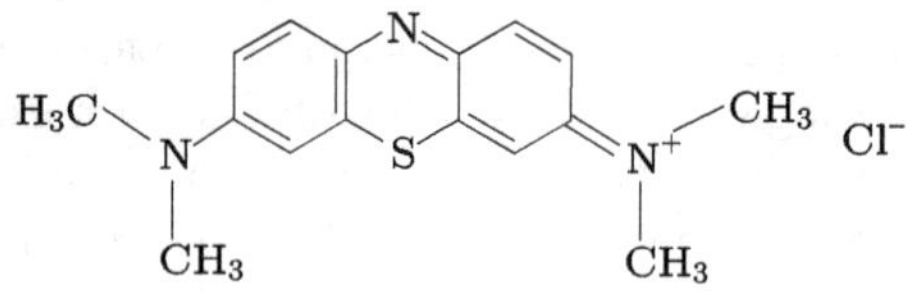

Fig. 13.5 The chemical structure of MB.

An idealized measurement curve of A/A_0 or C/C_0 for the UV-irradiated sample is shown in Fig. 13.6, where A_0 and A or C_0 and C are the absorbances or MB concentrations of the solution before and after UV irradiation. The curve may be divided into three different regimes: conditioning regime, linear degradation regime, and exponential degradation regime. Within the initial conditioning regime, the surface of the photocatalyst is not irradiated by UVlight and adsorbs MB molecules. As soon as the irradiation starts at t_0, MB degradation occurs. Within the linear degradation regime from t_0 to t_1, MB degradation is limited by charge carrier generation. At t_1, the measurement enters the exponential degradation regime. The degradation reaction rate does not depend on the generation of electrons and holes but on the availability of MB molecules at the catalystsurface.

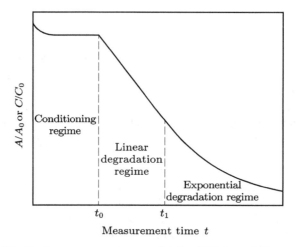

Fig. 13.6 Idealized measurement curve for the MB degradation measurement.

2.4 Determining solution concentration by UV-Vis spectroscopy

It is essential to determine the MB solution concentration to evaluate the photocatalytic performance. In this experiment, this is achieved by measuring the absorbance of the solutions using ultraviolet-visible (UV-Vis) spectroscopy.

The wavelength at which the absorbance of solute is highest should be found first, which is defined as λ_{max}. Whatever the concentration, the solutions of the same substance have the same value of λ_{max}. Then, after measuring the absorbance at λ_{max} of several solutions with different known concentrations, a standard concentration curve of absorbance versus concentration can be constructed. Consequently, the unknown concentration can be determined from the standard curve.

3 Experimental

3.1 Preparation

(1) Chemicals: tetrabutyl titanate ($C_{16}H_{36}O_4Ti$), acetic acid, nitric zinc, methylene blue, absolute ethanol.

(2) Apparatus: UV-Vis spectrometer (PerkinElmer, Lambda 25), UV lamp, X-ray diffractometer (Rigaku, Miniflex 600), centrifugalmachine (TG16-WS), electromagnetic stirrer, muffle furnace, oven, electronic balance.

(3) Tools: 150 mL beakers, 25 mL volumetric flask, constant pressure drop funnel, pipettes, mortar and pestle, 300-mesh sieve, quartz cuvettes, stirring bar

3.2 Procedure

3.2.1 Synthesis of TiO_2 powder by sol-gel method

(1) Mix 0.5 mL concentrated nitric acid and 2.6 mL deionized (DI) water in three 150 mL beakers, respectively. These solutions are named A, B, and C, respectively. Add 0.1 g and 0.7 g zinc nitrate in solutions B and C, respectively.

(2) Prepare three copies of solution D. Each contains 20 mL absolute ethanol, 2.0 mL acetic acid, and 10 mL tetrabutyl titanate.

(3) Transfer the solution D into the constant pressure dropfunnel. Dropwise add the solution D into the solutions A, B, and C under stirring.

(4) Keep stirring at room temperature for another 2 h to form a transparent gel.

(5) Stop stirring and let the gels stand for 4 h.

(6) Dry the gels at 100 °C in the oven for 24 h.

(7) Mill the dry gels into fine powders.

(8) Divide the three powders into two parts, calcine them at 400 °C for 2 h and 800 °C for 2 h, respectively.

(9) Mill the powders and sift them by the 300-mesh sieve. These powders are denoted as TiO_2-400, TiO_2-800, Zn1-TiO_2-400, Zn1-TiO_2-800, Zn7-TiO_2-400, Zn7-TiO_2-800, respectively.

(10) Determine the phase structures of the six powders by XRD.

3.2.2 Determine the λ_{max} of MB and plot the standard curve

(1) Prepare 1 L 5.5 mg/L aqueous solution of MB.

(2) Clean and dry two quartz cuvettes. They are filled 3/4 full with the MB solution (about 3 mL) and deionized water, respectively.

(3) Collect the UV-Vis absorption spectrum at the wavelength range from 350 to 850 nm.

(4) Determine the λ_{max} of MB.

(5) Quantitatively pipet 5, 9, 13, 17, 21 mL of the 5.5 mg/L MB solution into a 25 mL volumetric flask, respectively, and dilute them to 25 mL. Calculate the concentrations (in mol/L) of these diluted solutions.

(6) Repeat steps (2) and (3) for the diluted solutions.

(7) Determine the absorbance at λ_{max} of the solutions.

(8) Plot the standard curve of the absorbance at λ_{max} versus the MB concentration. Fit the curve into a linear equation.

3.2.3 Degradation measurement of methylene blue

(1) Pipette 80 mL 5.5 mg/L MB solution into six 150 mL beakers, respectively.

(2) Add 100 mg of each powder into the beakers, respectively.

(3) Stir the mixtures in the dark for 30 min.

(4) Take out 5 mL of each mixture, and centrifuge them at 10000 r/min for 10 min.

(5) Collect the UV-Vis absorption spectra of the supernatants. Determine the absorbance at λ_{max} and the concentration of MB of each supernatant using the standard curve.

(6) Expose the beakers to the UV light under stirring. Repeat steps (4) and (5) 6–8 times every 10 min.

4 Requirements for experimental report

(1) Plot the XRD profiles of the synthesized powders and index the peaks. If there are extra peaks, discuss the possible reasons.

(2) Give the plot of the standard curve and the equation of curve fitting. Tabulate the corresponding results of the MB concentrations and the absorbance at λ_{max}.

(3) Tabulate the results of degradation measurement of MB using photocatalysts TiO_2, Zn1-TiO_2, and Zn7-TiO_2, including the MB concentrations and measurement time.

(4) Plot the degradation measurement curves of C/C_0 versus measurement time for the six photocatalysts in one chart.

(5) Analyze the influence of phase structure and Zn dopant on the photocatalytic performance.

(6) Is the photocatalytic performance enhanced by Zn dopant in your experiment? Why?

5 Questions and further thoughts

(1) Please suggest several other techniques for preparing the TiO_2 powder.

(2) What are the main factors determining the photocatalytic performance of TiO_2? Suggest several strategies to improve the photocatalytic performance.

Extended readings

ASTM Standard practice for calculation of corrosion rates and related information from electrochemical measurements: G102-89(2015)e1. West Conshohocken: ASTM International, 2015.

Linsebigler A L, Lu G, Yates J T. Photocatalysis on TiO_2 surfaces-principles, mechanisms, and selected results. Chemical Reviews, 1995, 95(3): 735-758.

McCafferty E. Introduction to Corrosion Science. Springer, 2010.

Schneider J, Matsuoka M, Takeuchi M, et al. Understanding TiO_2 photocatalysis: Mechanisms and materials. Chemical Reviews, 2014, 114(19): 9919-9986.

Wang Q, Yun G, An N, et al. The enhanced photocatalytic activity of Zn^{2+} doped TiO_2 for hydrogen generation under artificial sunlight irradiation prepared by sol-gel method. Journal Sol-Gel Science Technology, 2015, 73: 341-349.

Yamashita H, Li H. Nanostructured Photocatalysts: Advanced Functional Materials. Springer, 2016.

Zhang H, Liang Y, Wu X, et al. Enhanced photocatalytic activity of (Zn, N)-codoped TiO_2 nanoparticles. Materials Research Bulletin, 2012, 47(9): 2188-2192.

Appendices

A1 Experimental Safety

Materials science and engineering is an interdisciplinary field that combines materials science, engineering, physics, chemistry, and biology. Work in the laboratory of materials science and engineering involves a number of inherent risks due to handling chemical reagents, operating mechanical equipment, and experiments related to high or low temperature and high pressure. There is always the possibility that an accident of varying severity can occur. Therefore, students are expected to develop safe working habits from the very beginning.

This appendix describes the most relevant aspects of safety in the process of experimental teaching. More detailed safety instruction dealing with all aspects of safety in the laboratory can be found in specialized monographs and regulations of universities.

A1.1 Safe behaviors and habits

Students should learn the best practices to safely conduct experiments that can be dangerous in case of carelessness or negligence. They should take appropriate precautions at all times and ask the instructor any questions regarding how to proceed with the experiment. The following general behaviors and habits can minimize the risk of accidents:
- Before using the laboratory, it is necessary to participate in safety, environment, protection, and storage education.
- To admittance into the laboratory, a laboratory coat should be worn. Gloves, masks, and goggles are also required according to specific requirements for the experiment. Slippers, shorts, short skirts, and high-heeled shoes are not allowed.
- It is strictly forbidden for miscellaneous personnel to enter the laboratory, bring the irrelevant articles into the laboratory, eat or drink in the laboratory, smoke or spit, store personal living articles, and take the laboratory articles

out of the laboratory.
- Before an experiment, study the techniques and principles. Know in detail all basic operations and manipulations that take place. Know the characteristics of drugs, reagents, and consumables.
- During an experiment, operate in strict accordance with the experimental operation procedure. Follow all safety instructions specific to each practice or experiment to be conducted. Without permission, it is not allowed to use or move apparatuses and reagents in the laboratory.
- Any newly prepared solution should be stored in a clean container and properly labeled or tagged.
- Keep flammable solvents away from heating elements such as heating plates, stoves, radiators, and furnaces.
- The electrical load of experimental apparatuses must match the circuit of the laboratory. When the high-power electrical apparatuses are working, students shall not leave without permission.
- Keep the laboratory quiet. Do not make any noise and gather together to chat. Keep the laboratory clean and sanitary.
- After an experiment, switch off apparatuses. Put apparatuses, articles, and reagents back to their original place. Clean up the experimental table. Carry out a safety inspection on water and electricity. Leave the laboratory with the permission of the instructor.

A1.2 In case of an accident

In order to minimize the consequences of an accident, some recommendations are given as follows:
- In case of an accident, immediately notify the instructor or head of the laboratory. In case of a severeaccident, call the emergency phone number. Warn anyone nearby about the nature of the emergency. Do not move any injured persons, except in case of ?re or chemical exposure.
- In case of burn or cut, wash well with water and cover with appropriate dressing. If the burn or cut is severe, go to a hospital immediately.
- In case of fire, evacuate following instructions. Efforts can be made to put it out with a fire blanket or fire extinguisher if the fire is small and localized. Never use water to extinguish a fire caused by organic solvents.
- In the case of chemical spills, act quickly for neutralization, absorption, and elimination. If chemical splashes on the skin, wash immediately with plenty of water.
- In case of inhalation of chemicals, evacuate the affected area and immediately go to a place with fresh air.

A2 Experimental Errors

The material analysis and research are directly affected by the accuracy of the measurement data, whereas various data obtained from experiments are subject to errors. Thus, it is necessary to study the experimental errors.

There are plenty of monographs about the experimental errors. Moreover, these generally are introduced in experimental courses of physics and chemistry. Hence, this appendix does not intend to give a detailed introduction to this topic. Instead, the knowledge that is closely related to the experiments in this book is summarized.

A2.1 Types of experimental errors

An essential thing to keep in mind when learning how to design experiments and collect experimental data is that our ability to observe the real world is not perfect. The observations we make are never precisely representative of the process we think we are observing. Mathematically, this is conceptualized as

$$\text{Measured value} = \text{Truevalue} \pm \text{Error}. \tag{A2.1}$$

The error is a combined measure of the inherent variation in the phenomenon we observe and the numerous factors that interfere with the measurement. It is impossible to eliminate all measurement errors.

Errors are generally classified as systematic and random errors, which are introduced in detail as follows.

A2.1.1 Systematic errors

Systematic errors, also known as bias errors, arise from the design and execution of experiments. The magnitude and signs of systematic errors are close every time the experiment is executed.

Systematic errors have many sources. The most often mentioned is a calibration error in a measuring instrument. Calibration errors may be a zero-o?set error that causes a constant absolute error in all readings or a scale error in the slope of output versus input that causes a constant percentage error in all readings. Some instruments have a systematic error associated with hysteresis leading to a sensitivity error, that is, the output di?ers depending on whether the input is increasing or decreasing. Also, there can be a systematic error due to local non-linearity in the instrument response. The environmental conditions, such as temperatures, humidity, and atmospheric pressure, may also affect the readout of some instruments.

In addition, the limitation of the experimental method and the invalidity of the theory or model applied to the system under study are intrinsic sources of systematic errors. Personal habits, such as being slow in recording time signals or insensitive to specific observations, can cause errors.

A2.1.2 Random errors

Random errors, also called precision errors, are that the magnitude and sign of random errors are different each time an experiment is conducted. Random errors arise from many undetermined ?uctuating experimental conditions, such as unpredictable mechanical and electrical fluctuations affecting the performance of instruments or unpredictable limitations of the experimenters. They are often associated with the "least count" of the scale on an instrument. For example, if a ruler has graduations at intervals of 1 mm, we can expect precision errors of about 1 mm in readings of the ruler.

A2.2 Reduction of errors

Systematic errors can be ascertained and eliminated through the improvement of experimental methods, change of instrument or experimental conditions, proper instrument calibration, and more careful experimental practices. Random errors cause scattering in experimental data. Whenever feasible, random errors should be estimated from repeated tests or observed scatter in graphed results. They are amenable to statistical analysis, and the theory of statistics has many powerful and valuableresults that facilitate a satisfactory treatment of random errors.

If the measurement of a property is repeated for n times under the same conditions, the sample average is calculated as

$$\bar{x} = \frac{1}{n} \sum_{i=1}^{n} x_i, \tag{A2.2}$$

where x_i represents the ith individual observation. The sample average is a statistic that estimates the mean or central tendency of the underlying random variable. The deviation of x_i from the mean value $\bar{x}$ represents the precision of each measurement.

The measured values scatter around the sample average. The commonly used measure of the scatter is the standard deviation calculated as

$$S_x = \left[\frac{1}{n-1} \sum_{i=1}^{n} (x_i - \bar{x})^2 \right]^{1/2}. \tag{A2.3}$$

The quantity $n-1$ is the number of degrees of freedom associated with the sample standard deviation.

It is often the case that we are more interested in the average value. If we take a number of samples, the resulting average values are also normally distributed, and it will be shown that the standard deviation of the average can be estimated as

$$S_{\bar{x}} = \frac{S_x}{\sqrt{n}}. \tag{A2.4}$$

Clearly, when the number of observations, n, is large, the uncertainty in the estimate of the mean is small. This relationship demonstrates that there is more uncertainty in an individual observation than in the estimated mean. Even if the underlying phenomenon is quite variable and there are significant measurement errors, it is still possible to reduce uncertainty in the estimate of the average value by making many measurements.

Often a comparison is made between theoretical and experimental results. When there are discrepancies between these results, abeginning student tends to believe that the experimental results are somehow less reliable. The student refers to errors in the experimental results when preparing a report. Of course, theoretical results do not have signi?cant precision errors. However, most often, the largest discrepancies are because the physical system does not exactly match the theoretical model, or conversely, the theoretical model does not exactly match the physical system. The student should always be careful to ascertain whether discrepancies between theory and experiment should be considered as errors in the theory, errors in the experiment, or perhaps in both.

A3 Data Presentation

The experiment results are often expressed as groups of numerical data. The data are required to be expressed scientifically, both clearly and concisely, to give a clearer indication of the experimental results. Then, reasonable analysis can be carried on. Most commonly, the data are presented in the form of tables, graphs, or equations.

A3.1 Present data in tables

The simplest way to present a list of data is to organize them into a table. It can show many numerical values in a small space, compare values or characteristics of related items, and show the presence or absence of specific characteristics. The guidelines for constructing a good table are as follows:
- The title of the table should be informative and tell something specific about the contents of the table. Generally, the title is placed above the table.
- A table consists of a matrix of rows and columns of data. The columns and rows should be labeled, and the correct units must be indicated.
- The primitive data should be written in the leftmost column, and the numerical data calculated from the primitive data in the columns to the right since the flow of information in a table is usually from left to right and from top to bottom.
- If the numbers are very large or very small, the data in a column or a row may be multiplied by an appropriate power often indicated in the column or row heading.
- The data should be reported with the correct number of significant figures. The decimal points of the numbers in a column should be aligned.

A3.2 Present data in graphs

Graphs directly illustrate the relationship between an experiment parameter (the independent variable) and the measured or calculated quantity (the dependent variable). When the trend is more important than the precise data values, they can show trends and relationships across and between data sets, visually explain a sequence of events, phenomena, characteristics.

There are many types of graphs, such as line graphs, bar graphs. For almost every numerical data set, there is a graph type that is appropriate for representing it. The general guidelines for properly constructing a graph are as follows:
- Like tables, a graph must also have a descriptive title. The word "Figure" is usually used in numbering the graph. The title is placed below the graph.
- The independent variable must be plotted along the abscissa and the dependent variable the ordinate.
- When choosing the scales of the variables, all the significant figures can be read out from the graph so that the precision of the variables in the graph is the same as that of the measurement.
- If more than one set of data is plotted on the same graph, different marks or line styles should be used for each set of data.
- After putting down all the data points on the graph, a straight line or a curve can be drawn using the points. If the line or curve is obtained by fitting the data, it does not need to pass through all the data points.

A3.3 Present data in equations

The large numbers of data obtained in a series of measurements often have to be manipulated or transformed into an equation to give the results sufficient meaning. Suppose the functional relation of the variables is already known as theorem expressed by an equation or empirical equation. In that case, the task is to determine the optimum values of the constants in the relation. In contrast, if the functional relation of the variables is unclear, the first thing is to make a graph of the data. The shape of the curve on the graph will certainly give you some hint for the relation. An equation, which is often an empirical equation, may be obtained by fitting the data.

Fitting the data into a linear equation is undoubtedly easier than fitting the data into a non-linear equation. Hence, when the independent variable is not related to the dependent variable in a simple linear manner, it is best to find a transformed equation that is linear. The least squares method can then be used to find the coefficients (slope and intercept) of the linear equation. This method is probably the most popular statistical method for determining the best line representing a linear relationship.

The least squares method provides the overall rationale for the placement of the line of best fit among the data points being studied. The most common application of this method aims to create a straight line that minimizes the sum of the squares of the errors that are generated by the results of the associated equations, such as the squared residuals resulting from differences between the observed value and the value anticipated, based on the model.

Given a set of data $\{(x_1, y_1), \cdots, (x_N, y_N)\}$, we may define the error of the ith data point associated to a linear equation, $y = mx + b$, as

$$r_i = y_i - (mx_i + b). \tag{A3.1}$$

The sum of the squares of r_i is expressed as

$$\sum r_i^2 = \sum (y_i - mx_i - b)^2. \tag{A3.2}$$

The goal is to find values of m and b that minimize the error. This requires us to find the values such that

$$\frac{\partial (\sum r_i^2)}{\partial m} = 0, \quad \frac{\partial (\sum r_i^2)}{\partial b} \tag{A3.3}$$

Then, the two derivatives yield

$$\sum 2(y_i - mx_i - b)(-x_i) = 0, \tag{A3.4}$$

$$\sum 2(y_i - mx_i - b)(-1) = 0. \tag{A3.5}$$

The solutions of these two equations are

$$m = \frac{N \sum x_i y_i - \sum x_i \sum y_i}{N \sum x_i^2 - (\sum x_i)^2}, \tag{A3.6}$$

$$b = \frac{\sum y_i \sum x_i^2 - \sum x_i \sum x_i y_i}{N \sum x_i^2 - (\sum x_i)^2}. \tag{A3.7}$$

Unfortunately, not all relations can be transformed into linear forms. In this case, we can compare the curve to a standard graph to determine which curve relationship is suitable for fitting. Of course, curve fitting is much more complicated in this case. The least squares method can also be used, but the solution of the curve relationship is generally solved by iteration.

A4 Experimental Report

In the process of scientific research and engineering practice, it is often necessary to write experimental reports or test reports. Therefore, the experimental report is an integral part of experimental teaching. Through a series of experimental courses, students not only need to master experimental knowledge and skills but also the ability to write reports. In addition, students can get training for writing research papers through writing experimental reports.

The experimental report needs to be detailed, clear, and concise. The formal experimental report has a structure similar to the experimental manual. It includes the following aspects: title, objectives, principles, preparation, procedure and results, analysis and discussion, conclusions, questions and further thoughts, references.

In order to ensure the experimental learning effect, students are usually required to preview the experiment and write a preview report. It also includes title, objectives, principles, preparation, and procedure. The request for the preview report is the same as those for the formal report, while it can be much simplified.

The specific writing points of each part of the experiment report are introduced as follows:

(1) Title.

The title must indicate what experiment is conducted. Generally, the title is the same as that given in the manual. Besides, it is recommended to give the title on a cover page and the name, student number, and date for the experiment.

(2) Objectives.

The content and objectives of the experiment have been described in the experimental manual. However, some contents have not been carried out in the specific experiment process, or the experiment contents have been changed. Accordingly, students cannot copy the manual completely but should write according to the actual requirements.

(3) Principles.

This part should clearly describe the principles, which are the basis of the experiment. The principles must be described in concise language. Usually, the principles in the relevant textbook and experimental manual are too much in detail. Therefore, you should summarize the principles in your own language. Sometimes students are unfamiliar with some principles that are simply described in the manual. In this case, they need to consult the literature and supplement the report.

Schematic diagrams of experimental apparatuses necessary to explain the principles can be given. If there is more than one figure, they should be numbered in turn and inserted after the corresponding text.

(4) Preparation.

In the experiment, the materials, chemicals, apparatuses, and tools are configured according to the requirements of the experiment. The names, models, specifications, and quantities need to be recorded.

(5) Procedure and results.

Write down the main steps of the experiment, especially the key steps and precautions. Record the phenomena during the experiment and original data truthfully. Record the experiment or measurement conditions if necessary.

Select appropriate methods to present the experimental results (see Appendix A3). Carry out necessary processing of the experimental data. For the experimental results, which need to be calculated numerically, the original data obtained from the measurement must be substituted into the calculation formula truthfully. For the photographs, they can be cut, and their brightness and contrast can be adjusted appropriately. They should also be numbered in turn and inserted after the corresponding text.

(6) Analysis and discussion.

The students should make a rationalanalysis and discussion of the results to understand the principles further and judge whether the experiment is credible or has some newfound.

If the measured physical quantity has a standard value (theoretical value, nominal value, recognized value, or previous measurement result), the difference shall be calculated. When there is a large discrepancy, the cause of the error should be analyzed. If there are data points or data groups with large deviations, the reason should be carefully analyzed to consider eliminating them or finding out new laws.

(7) Conclusions.

This part should provide a concise and comprehensive summary of the experiment. Describe the major results and findings, which may include a mention of the method and apparatus used.

(8) Questions and further thoughts.

Several questions have been given in the manual. They are used to guide students to expand their thinking or knowledge based on the experiment. Students should answer the questions based on their theoretical knowledge and practical experience and consult the literature by themselves.

(9) References.

The reference section is the list of all the sources cited in the experiment. These sources are related to the principles, discussion, and further thoughts in the report. The references need to be listedin common styles.

Extended readings

Baird D C. Experimentation: An Introduction to Measurement Theory and Experiment Design. 2nd Edition. Prentice-Hall, 1988.

Furr A K. CRC Handbook of Laboratory Safety. 5th Edition. CRC Press, 2000.

Hill R H, Finster D C, Laboratory Safety for Chemistry Students. 2nd Edition. John Wiley and Sons, 2016.

Mills A F, Chang B H, Error Analysis of Experiments. University of California, 2004.

Moore D, McCabe G. Introduction to the Practice of Statistics. 9th Edition. W. H. Freeman and Company, 2016.

Pugh E M, Winslow G H, The Analysis of Physical Measurements. Addison-Wesley, 1966.

Taylor J R. An Introduction to Error Analysis: The Study of Uncertainties in Physical Measurements. 2nd Edition. University Science Books, 1997.

Weissberg R. Writing Up Research: Experimental Research Report Writing for Students of English. Prentice Hall, 1990.

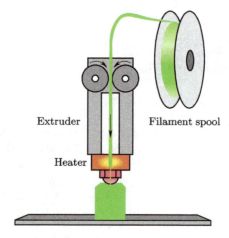

Fig. 1.2 Schematic illustration of the FDM process.

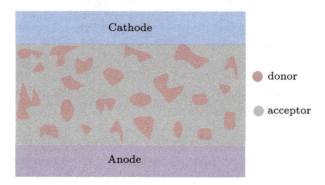

Fig. 1.7 Schematic illustration of the BHJ structure.

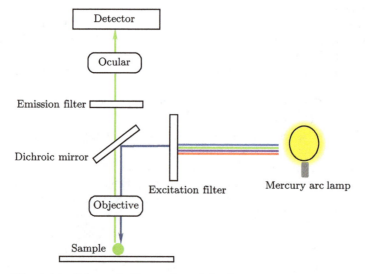

Fig. 1.14 Schematic illustration of the fluorescence microscope.

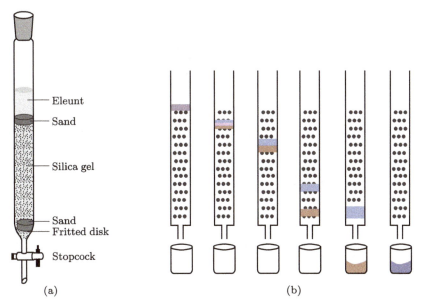

Fig. 3.16 (a) The setup of CC with silica gel as adsorbent. (b) Separation and collection of components in a mixture.

Fig. 4.8 Compare the movement of a ping-pong ball on ping-pong ball layers on the trays.

Fig. 4.9 A twin structure on tray B.

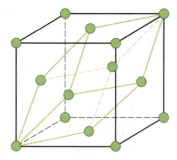

Fig. 4.10 The unit cell (outside) and primitive cell (inside) of the FCC lattice.

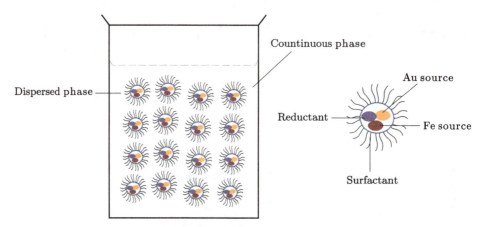

Fig. 6.4 The principle of the nanoemulsion method.

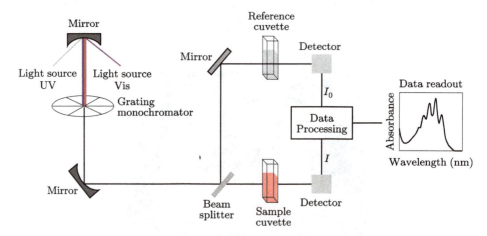

Fig. 6.5 The working principle of the UV-Vis spectrometer.

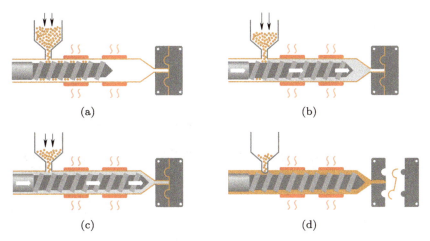

Fig. 7.16 Injection molding process with a single screw extruder: (a) add polymer material from the hopper; (b) the molten polymer fills the barrel; (c) inject the molten polymer into the mold; (d) open the mold and eject the product.

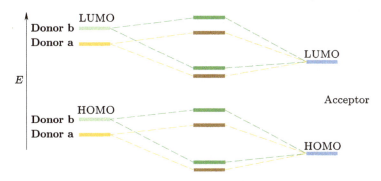

Fig. 8.8 Hybridization of the energy levels of donor and acceptor monomers to form molecular orbitals in a D-A polymer.

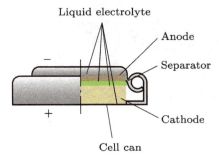

Fig. 10.2 The structure of a typical Li-ion coin-cell type battery.

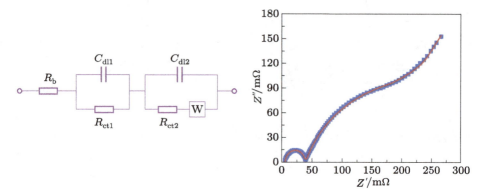

Fig. 10.6 An example of (a) simulated parallel circuit and (b) fitted Nyquist plot of a Li-ion battery.

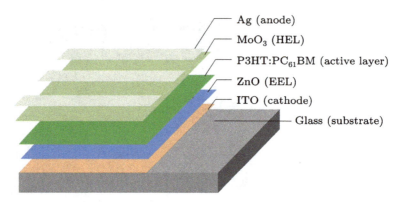

Fig. 10.11 The structure of a polymer solar cell.

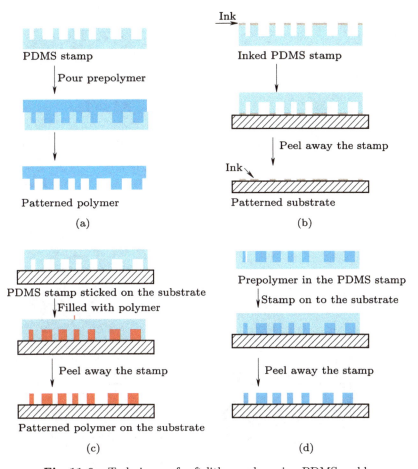

Fig. 11.6 Techniques of soft lithography using PDMS mold.

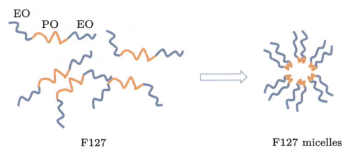

Fig. 12.9 Self-assemble of Pluronic F127 in aqueous solution.

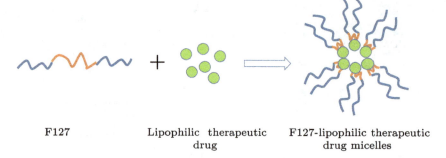

Fig. 12.10 Formation of Pluronic F127- lipophilic therapeutic drug micelles in aqueous solution.

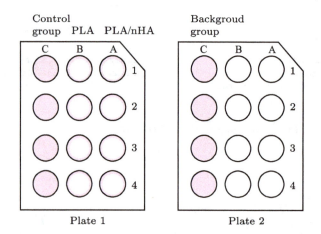

Fig. 12.17 The 12-well plates for the WST-8 assay.

郑重声明

高等教育出版社依法对本书享有专有出版权。任何未经许可的复制、销售行为均违反《中华人民共和国著作权法》，其行为人将承担相应的民事责任和行政责任；构成犯罪的，将被依法追究刑事责任。为了维护市场秩序，保护读者的合法权益，避免读者误用盗版书造成不良后果，我社将配合行政执法部门和司法机关对违法犯罪的单位和个人进行严厉打击。社会各界人士如发现上述侵权行为，希望及时举报，本社将奖励举报有功人员。

反盗版举报电话　　（010）58581999　58582371　58582488
反盗版举报传真　　（010）82086060
反盗版举报邮箱　　dd@hep.com.cn
通信地址　　　　　北京市西城区德外大街4号
　　　　　　　　　高等教育出版社法律事务与版权管理部
邮政编码　　　　　100120